SOURCE

The Prentice Hall
ENGINEERING SOURCE

A User's Guide to Engineering

James N. Jensen

University at Buffalo
Buffalo, NY

PEARSON
Prentice
Hall

Upper Saddle River, NJ 07458

Library of Congress Cataloging-in-Publication Data

Jensen, James N.
  A user's guide to engineering / James N. Jensen.
    p. cm.—(ESource—the Prentice Hall engineering source)
  ISBN 0-13-148025-1
  1. Engineering—Vocational guidance. 2. Engineering—Case studies. I. Title. II. Series.

TA157.J465 2005
620—dc22   2005048719

Editorial Director, ECS: *Marcia J. Horton*
Senior Editor: *Holly Stark*
Associate Editor: *Dee Bernhard*
Editorial Assistant: *Nicole Kunzmann*
Executive Managing Editor: *Vince O'Brien*
Managing Editor: *David A. George*
Production Editor: *Scott Disanno*
Director of Creative Services: *Paul Belfanti*
Art Director: *Jayne Conte*
Cover Designer: *Bruce Kenselaar*
Art Editor: *Greg Dulles*
Manufacturing Manager: *Alexis Heydt-Long*
Manufacturing Buyer: *Lisa McDowell*

© 2006 Pearson Education, Inc.
Upper Saddle River, New Jersey 07458

The author and publisher of this book have used their best efforts in preparing this book. These efforts include the development, research, and testing of the theories and programs to determine their effectiveness. The author and publisher make no warranty of any kind, express or implied, with regard to these programs or the documentation contained in this book. The author and publisher shall not be liable in any event for incidental or consequential damages in connection with, or arising out of, the furnishing, performance, or use of these programs.

Pearson Prentice Hall™ is a trademark of Pearson Education, Inc.

Printed in the United States of America.

ISBN 0-13-148025-1

Pearson Education Ltd., *London*
Pearson Education Australia Pty. Ltd., *Sydney*
Pearson Education Singapore, Pte. Ltd.
Pearson Education North Asia Ltd., *Hong Kong*
Pearson Education Canada, Inc., *Toronto*
Pearson Educación de Mexico, S.A. de C.V.
Pearson Education—Japan, *Tokyo*
Pearson Education Malaysia, Pte. Ltd.
Pearson Education, *Upper Saddle River, New Jersey*

About ESource

ESource—The Prentice Hall Engineering Source—
www.prenhall.com/esource

ESource—The Prentice Hall Engineering Source gives professors the power to harness the full potential of their text and their first-year engineering course. More than just a collection of books, ESource is a unique publishing system revolving around the ESource website—www.prenhall.com/esource. ESource enables you to put your stamp on your book just as you do your course. It lets you:

Control You choose exactly which chapters are in your book and in what order they appear. Of course, you can choose the entire book if you'd like and stay with the authors' original order.

Optimize Get the most from your book and your course. ESource lets you produce the optimal text for your students needs.

Customize You can add your own material anywhere in your text's presentation, and your final product will arrive at your bookstore as a professionally formatted text. Of course, all titles in this series are available as stand-alone texts, or as bundles of two or more books sold at a discount. Contact your PH sales rep for discount information.

ESource ACCESS

Professors who choose to bundle two or more texts from the ESource series for their class, or use an ESource custom book will be providing their students with an on-line library of intro engineering content—ESource Access. We've designed ESource ACCESS to provide students a flexible, searchable, on-line resource. Free access codes come in bundles and custom books are valid for one year after initial log-on. Contact your PH sales rep for more information.

ESource Content

All the content in ESource was written by educators specifically for freshman/first-year students. Authors tried to strike a balanced level of presentation, an approach that was neither formulaic nor trivial, and one that did not focus too heavily on advanced topics that most introductory students do not encounter until later classes. Because many professors do not have extensive time to cover these topics in the classroom, authors prepared each text with the idea that many students would use it for self-instruction and independent study. Students should be able to use this content to learn the software tool or subject on their own.

While authors had the freedom to write texts in a style appropriate to their particular subject, all followed certain guidelines created to promote a consistency that makes students comfortable. Namely, every chapter opens with a clear set of **Objectives**, includes **Practice Boxes** throughout the chapter, and ends with a number of **Problems**, and a list of **Key Terms**. **Applications Boxes** are spread throughout the book with the intent of giving students a real-world perspective of engineering. **Success Boxes** provide the student with advice about college study skills, and help students avoid the common pitfalls of first-year students. In addition, this series contains an entire book titled *Engineering Success* by Peter Schiavone of the University of Alberta intended to expose students quickly to what it takes to be an engineering student.

Creating Your Book

Using ESource is simple. You preview the content either on-line or through examination copies of the books you can request on-line, from your PH sales rep, or by calling 1-800-526-0485. Create an on-line outline of the content you want, in the order you want, using ESource's simple interface. Insert your own material into the text flow. If you are not ready to order, ESource will save your work. You can come back at any time and change, re-arrange, or add more material to your creation. Once you're finished you'll automatically receive an ISBN. Give it to your bookstore and your book will arrive on their shelves four to six weeks after they order. Your custom desk copies with their instructor supplements will arrive at your address at the same time.

To learn more about this new system for creating the perfect textbook, go to www.prenhall.com/esource. You can either go through the on-line walkthrough of how to create a book, or experiment yourself.

Instructor Resources

A wealth of instructor resources for the ESource books are available through the Prentice Hall online Instructor Resource Center. Resources may include text-specific products such as PowerPoint presentations, image banks, and instructor solution manuals. Visit www.prenhall.com/esource and click on Instructor Resources.

Titles in the ESource Series

Design Concepts for Engineers, 3/e
0-13-146499-X
Mark N. Horenstein

Engineering Success, 2/e
0-13-041827-7
Peter Schiavone

Engineering Design and Problem Solving, 2E
0-13-093399-6
Steven K. Howell

Exploring Engineering
0-13-093442-9
Joe King

Engineering Ethics
0-13-784224-4
Charles B. Fleddermann

Introduction to Engineering Analysis, 2/e
0-13-145332-7
Kirk D. Hagen

Introduction to Engineering Communication
0-13-146102-8
Hillary Hart

Introduction to Engineering Experimentation
0-13-032835-9
Ronald W. Larsen, John T. Sears, and Royce Wilkinson

Introduction to Mechanical Engineering
0-13-019640-1
Robert Rizza

Introduction to Electrical and Computer Engineering
0-13-033363-8
Charles B. Fleddermann and Martin Bradshaw

Introduction to MATLAB 7
0-13-147492-8
Delores Etter and David C. Kuncicky with Holly Moore

MATLAB Programming
0-13-035127-X
David C. Kuncicky

Introduction to Mathcad 2000
0-13-020007-7
Ronald W. Larsen

Introduction to Mathcad 11
0-13-008177-9
Ronald W. Larsen

Introduction to Maple 8
0-13-032844-8
David I. Schwartz

Mathematics Review
0-13-011501-0
Peter Schiavone

Power Programming with VBA/Excel
0-13-047377-4
Steven C. Chapra

Introduction to Excel 2002
0-13-008175-2
David C. Kuncicky

Introduction to Excel, 3/e
0-13-146470-1
David C. Kuncicky and Ronald W. Larsen

About the Authors

No project could ever come to pass without a group of authors who have the vision and the courage to turn a stack of blank paper into a book. The authors in this series, who worked diligently to produce their books, provide the building blocks of the series.

Martin D. Bradshaw was born in Pittsburg, KS in 1936, grew up in Kansas and the surrounding states of Arkansas and Missouri, graduating from Newton High School, Newton, KS in 1954. He received the B.S.E.E. and M.S.E.E. degrees from the University of Wichita in 1958 and 1961, respectively. A Ford Foundation fellowship at Carnegie Institute of Technology followed from 1961 to 1963 and he received the Ph.D. degree in electrical engineering in 1964. He spent his entire academic career with the Department of Electrical and Computer Engineering at the University of New Mexico (1961–1963 and 1991–1996). He served as the Assistant Dean for Special Programs with the UNM College of Engineering from 1974 to 1976 and as the Associate Chairman for the EECE Department from 1993 to 1996. During the period 1987–1991 he was a consultant with his own company, EE Problem Solvers. During 1978 he spent a sabbatical year with the State Electricity Commission of Victoria, Melbourne, Australia. From 1979 to 1981 he served an IPA assignment as a Project Officer at the U.S. Air Force Weapons Laboratory, Kirkland AFB, Albuquerque, NM. He has won numerous local, regional, and national teaching awards, including the George Westinghouse Award from the ASEE in 1973. He was awarded the IEEE Centennial Medal in 2000.

Acknowledgments: Dr. Bradshaw would like to acknowledge his late mother, who gave him a great love of reading and learning, and his father, who taught him to persist until the job is finished. The encouragement of his wife, Jo, and his six children is a never-ending inspiration.

Stephen J. Chapman received a B.S. degree in Electrical Engineering from Louisiana State University (1975), the M.S.E. degree in Electrical Engineering from the University of Central Florida (1979), and pursued further graduate studies at Rice University.

Mr. Chapman is currently Manager of Technical Systems for British Aerospace Australia, in Melbourne, Australia. In this position, he provides technical direction and design authority for the work of younger engineers within the company. He also continues to teach at local universities on a part-time basis.

Mr. Chapman is a Senior Member of the Institute of Electrical and Electronics Engineers (and several of its component societies). He is also a member of the Association for Computing Machinery and the Institution of Engineers (Australia).

Steven C. Chapra presently holds the Louis Berger Chair for Computing and Engineering in the Civil and Environmental Engineering Department at Tufts University. Dr. Chapra received engineering degrees from Manhattan College and the University of Michigan. Before joining the faculty at Tufts, he taught at Texas A&M University, the University of Colorado, and Imperial College, London. His research interests focus on surface water-quality modeling and advanced computer applications in environmental engineering. He has published over 50 refereed journal articles, 20 software packages and 6 books. He has received a number of awards including the 1987 ASEE Merriam/Wiley Distinguished Author Award, the 1993 Rudolph Hering Medal, and teaching awards from Texas A&M, the University of Colorado, and the Association of Environmental Engineering and Science Professors.

Acknowledgments: To the Berger Family for their many contributions to engineering education. I would also like to thank David Clough for his friendship and insights, John Walkenbach for his wonderful books, and my colleague Lee Minardi and my students Kenny William, Robert Viesca and Jennifer Edelmann for their suggestions.

 **Mark Dix** began working with AutoCAD in 1985 as a programmer for CAD Support Associates, Inc. He helped design a system for creating estimates and bills of material directly from AutoCAD drawing databases for use in the automated conveyor industry. This system became the basis for systems still widely in use today. In 1986 he began collaborating with Paul Riley to create AutoCAD training materials, combining Riley's background in industrial design and training with Dix's background in writing, curriculum development, and programming. Mr. Dix received the M.S. degree in education from the University of Massachusetts. He is currently the Director of Dearborn Academy High School in Arlington, Massachusetts.

 **Delores M. Etter** is a Professor of Electrical and Computer Engineering at the University of Colorado. Dr. Etter was a faculty member at the University of New Mexico and also a Visiting Professor at Stanford University. Dr. Etter was responsible for the Freshman Engineering Program at the University of New Mexico and is active in the Integrated Teaching Laboratory at the University of Colorado. She was elected a Fellow of the Institute of Electrical and Electronics Engineers for her contributions to education and for her technical leadership in digital signal processing.

 **Charles B. Fleddermann** is a professor in the Department of Electrical and Computer Engineering at the University of New Mexico in Albuquerque, New Mexico. All of his degrees are in electrical engineering: his Bachelor's degree from the University of Notre Dame, and the Master's and Ph.D. from the University of Illinois at Urbana-Champaign. Prof. Fleddermann developed an engineering ethics course for his department in response to the ABET requirement to incorporate ethics topics into the undergraduate engineering curriculum. *Engineering Ethics* was written as a vehicle for presenting ethical theory, analysis, and problem solving to engineering undergraduates in a concise and readily accessible way.

Acknowledgments: I would like to thank Profs. Charles Harris and Michael Rabins of Texas A & M University whose NSF sponsored workshops on engineering ethics got me started thinking in this field. Special thanks to my wife Liz, who proofread the manuscript for this book, provided many useful suggestions, and who helped me learn how to teach "soft" topics to engineers.

 **Kirk D. Hagen** is a professor at Weber State University in Ogden, Utah. He has taught introductory-level engineering courses and upper-division thermal science courses at WSU since 1993. He received his B.S. degree in physics from Weber State College and his M.S. degree in mechanical engineering from Utah State University, after which he worked as a thermal designer/analyst in the aerospace and electronics industries. After several years of engineering practice, he resumed his formal education, earning his Ph.D. in mechanical engineering at the University of Utah. Hagen is the author of an undergraduate heat transfer text.

 **Hillary Hart** is a Senior Lecturer with the Department of Civil Engineering (CE) at the University of Texas at Austin, where she created the CE program in Engineering Communication and developed and introduced courses at both the graduate and undergraduate levels. After the completion of the Ph.D. degree in English from Bryn Mawr College, her career has spanned work as a technical writer, editor, educator, curricula developer, and consultant. Hillary's research efforts include work in instructional-technology evaluation and in environmental and risk communication. She has conducted research on behalf of the National Science Foundation, BP Oil Co., and other major corporations and agencies. Hillary has been recognized by The University of Texas with the College of Engineering Award for Excellence in Engineering Teaching and the Faculty Excellence Award from the Graduate Engineering Council. She is an Associate Fellow of the Society for Technical Communication.

Acknowledgments: I am fortunate to work in an incredibly supportive environment and I want to thank so many of my colleagues, teaching assistants, and engineering students. Special thanks to assistants Tanya Luthi and Cynthia Duquette Smith; to engineering colleagues Desmond Lawler, Sharon Wood, Robert Gilbert, Kerry Kinney, Rebecca Richards-Kortum, Rich Corsi, Maria Juenger and John Hall; and to students Jeffrey Hunt, John Carson, Andrew Giacobe, and Beth Woodward. Big thanks to my apparently effortlessly patient husband, John.

Mark N. Horenstein is a Professor in the Department of Electrical and Computer Engineering at Boston University. He has degrees in Electrical Engineering from M.I.T. and U.C. Berkeley and has been involved in teaching engineering design for the greater part of his academic career. He devised and developed the senior design project class taken by all electrical and computer engineering students at Boston University. In this class, the students work for a virtual engineering company developing products and systems for real-world engineering and social-service clients.

Acknowledgments: Several of the ideas relating to brainstorming and teamwork were derived from a workshop on engineering design offered by Prof. Charles Lovas of Southern Methodist University. The principles of estimation were derived in part from a freshman engineering problem posed by Prof. Thomas Kincaid of Boston University.

Steven Howell is the Chairman and a Professor of Mechanical Engineering at Lawrence Technological University. Prior to joining LTU in 2001, Dr. Howell led a knowledge-based engineering project for Visteon Automotive Systems and taught computer-aided design classes for Ford Motor Company engineers. Dr. Howell also has a total of 15 years experience as an engineering faculty member at Northern Arizona University, the University of the Pacific, and the University of Zimbabwe. While at Northern Arizona University, he helped develop and implement an award-winning interdisciplinary series of design courses simulating a corporate engineering-design environment.

Douglas W. Hull is a graduate student in the Department of Mechanical Engineering at Carnegie Mellon University in Pittsburgh, Pennsylvania. He is the author of *Mastering Mechanics I Using Matlab 5*, and contributed to *Mechanics of Materials* by Bedford and Liechti. His research in the Sensor Based Planning lab involves motion planning for hyper-redundant manipulators, also known as serpentine robots.

Scott D. James is a staff lecturer at Kettering University (formerly GMI Engineering & Management Institute) in Flint, Michigan. He is currently pursuing a Ph.D. in Systems Engineering with an emphasis on software engineering and computer-integrated manufacturing. He chose teaching as a profession after several years in the computer industry. "I thought that it was really important to know what it was like outside of academia. I wanted to provide students with classes that were up to date and provide the information that is really used and needed."

Acknowledgments: Scott would like to acknowledge his family for the time to work on the text and his students and peers at Kettering who offered helpful critiques of the materials that eventually became the book.

James N. Jensen is a professor in the Department of Civil, Structural and Environmental Engineering at the University at Buffalo. Dr. Jensen received engineering degrees from Caltech and the University of North Carolina at Chapel Hill. At the University at Buffalo, he presently is Director of Undergraduate Studies and the Director of the Environmental Science Program. Dr. Jensen also has served as Director of the Center of Teaching and Learning Resources. His teaching and research interests focus on environmental engineering and science. He has published over 40 refereed journal articles and 2 books. Dr. Jensen holds two patents and has received a number of teaching and research awards, including the Chancellor's Award for Excellence in Teaching in 1995.

Joe King received the B.S. and M.S. degrees from the University of California at Davis. He is a Professor of Computer Engineering at the University of the Pacific, Stockton, CA, where he teaches courses in digital design, computer design, artificial intelligence, and computer networking. Since joining the UOP faculty, Professor King has spent yearlong sabbaticals teaching in Zimbabwe, Singapore, and Finland. A licensed engineer in the state of California, King's industrial experience includes major design projects with Lawrence Livermore National Laboratory, as well as independent consulting projects. Prof. King has had a number of books published with titles including M*atlab*, Math-CAD, Exploring Engineering, and Engineering and Society.

 **David C. Kuncicky** is a native Floridian. He earned his Baccalaureate in psychology, Master's in computer science, and Ph.D. in computer science from Florida State University. He has served as a faculty member in the Department of Electrical Engineering at the FAMU–FSU College of Engineering and the Department of Computer Science at Florida State University. He has taught computer science and computer engineering courses for over 15 years. He has published research in the areas of intelligent hybrid systems and neural networks. He is currently the Director of Engineering at Bioreason, Inc. in Sante Fe, New Mexico.

Acknowledgments: Thanks to Steffie and Helen for putting up with my late nights and long weekends at the computer. Finally, thanks to Susan Bassett for having faith in my abilities, and for providing continued tutelage and support.

 **Ron Larsen** is a Professor of Chemical Engineering at Montana State University, and received his Ph.D. from the Pennsylvania State University. He was initially attracted to engineering by the challenges the profession offers, but also appreciates that engineering is a serving profession. Some of the greatest challenges he has faced while teaching have involved non-traditional teaching methods, including evening courses for practicing engineers and teaching through an interpreter at the Mongolian National University. These experiences have provided tremendous opportunities to learn new ways to communicate technical material. Dr. Larsen views modern software as one of the new tools that will radically alter the way engineers work, and his book *Introduction to MathCAD* was written to help young engineers prepare to meet the challenges of an ever-changing workplace.

Acknowledgments: To my students at Montana State University who have endured the rough drafts and typos, and who still allow me to experiment with their classes—my sincere thanks.

 **Sanford Leestma** is a Professor of Mathematics and Computer Science at Calvin College, and received his Ph.D. from New Mexico State University. He has been the long-time co-author of successful textbooks on Fortran, Pascal, and data structures in Pascal. His current research interest are in the areas of algorithms and numerical computation.

 **Jack Leifer** is an Assistant Professor in the Department of Mechanical Engineering at the University of Kentucky Extended Campus Program in Paducah, and was previously with the Department of Mathematical Sciences and Engineering at the University of South Carolina–Aiken. He received his Ph.D. in Mechanical Engineering from the University of Texas at Austin in December 1995. His current research interests include the analysis of ultra-light and inflatable (Gossamer) space structures.

Acknowledgments: I'd like to thank my colleagues at USC–Aiken, especially Professors Mike May and Laurene Fausett, for their encouragement and feedback; and my parents, Felice and Morton Leifer, for being there and providing support (as always) as I completed this book.

 **Richard M. Lueptow** is the Charles Deering McCormick Professor of Teaching Excellence and Associate Professor of Mechanical Engineering at Northwestern University. He is a native of Wisconsin and received his doctorate from the Massachusetts Institute of Technology in 1986. He teaches design, fluid mechanics, an spectral analysis techniques. Rich has an active research program on rotating filtration, Taylor Couette flow, granular flow, fire suppression, and acoustics. He has five patents and over 40 refereed journal and proceedings papers along with many other articles, abstracts, and presentations.

Acknowledgments: Thanks to my talented and hard-working co-authors as well as the many colleagues and students who took the tutorial for a "test drive." Special thanks to Mike Minbiole for his major contributions to Graphics Concepts with SolidWorks. Thanks also to Northwestern University for the time to work on a book. Most of all, thanks to my loving wife, Maiya, and my children, Hannah and Kyle, for supporting me in this endeavor. (Photo courtesy of Evanston Photographic Studios, Inc.)

Holly Moore is a professor of engineering at Salt Lake Community College, where she teaches courses in thermal science, materials science engineering, and engineering computing. Dr. Moore received the B.S. degree in chemistry, the M.S. degree in chemical engineering from South Dakota School of Mines and Technology, and the Ph.D. degree in chemical engineering from the University of Utah. She spent 10 years working in the aerospace industry, designing and analyzing solid rocket boosters for both defense and space programs. She has also been active in the development of hands-on elementary science materials for the state of Utah.

Acknowledgments: Holly would like to recognize the tremendous influence of her father, Professor George Moore, who taught in the Department of Electrical Engineering at the South Dakota School of Mines and Technology for almost 20 years. Professor Moore earned his college education after a successful career in the United States Air Force, and was a living reminder that you are never too old to learn.

Larry Nyhoff is a Professor of Mathematics and Computer Science at Calvin College. After doing bachelor's work at Calvin, and Master's work at Michigan, he received a Ph.D. from Michigan State and also did graduate work in computer science at Western Michigan. Dr. Nyhoff has taught at Calvin for the past 34 years—mathematics at first and computer science for the past several years.

Paul Riley is an author, instructor, and designer specializing in graphics and design for multimedia. He is a founding partner of CAD Support Associates, a contract service and professional training organization for computer-aided design. His 15 years of business experience and 20 years of teaching experience are supported by degrees in education and computer science. Paul has taught AutoCAD at the University of Massachusetts at Lowell and is presently teaching AutoCAD at Mt. Ida College in Newton, Massachusetts. He has developed a program, Computer-aided Design for Professionals that is highly regarded by corporate clients and has been an ongoing success since 1982.

Robert Rizza is an Assistant Professor of Mechanical Engineering at North Dakota State University, where he teaches courses in mechanics and computer-aided design. A native of Chicago, he received the Ph.D. degree from the Illinois Institute of Technology. He is also the author of *Getting Started with Pro/ENGINEER.* Dr. Rizza has worked on a diverse range of engineering projects including projects from the railroad, bioengineering, and aerospace industries. His current research interests include the fracture of composite materials, repair of cracked aircraft components, and loosening of prostheses.

Peter Schiavone is a professor and student advisor in the Department of Mechanical Engineering at the University of Alberta, Canada. He received his Ph.D. from the University of Strathclyde, U.K. in 1988. He has authored several books in the area of student academic success as well as numerous papers in international scientific research journals. Dr. Schiavone has worked in private industry in several different areas of engineering including aerospace and systems engineering. He founded the first Mathematics Resource Center at the University of Alberta, a unit designed specifically to teach new students the necessary *survival skills* in mathematics and the physical sciences required for success in first-year engineering. This led to the Students' Union Gold Key Award for outstanding contributions to the university. Dr. Schiavone lectures regularly to freshman engineering students and to new engineering professors on engineering success, in particular about maximizing students' academic performance.

Acknowledgments: Thanks to Richard Felder for being such an inspiration; to my wife Linda for sharing my dreams and believing in me; and to Francesca and Antonio for putting up with Dad when working on the text.

David I. Schneider holds an A.B. degree from Oberlin College and a Ph.D. degree in Mathematics from MIT. He has taught for 34 years, primarily at the University of Maryland. Dr. Schneider has authored 28 books, with one-half of them computer programming books. He has developed three customized software packages that are supplied as supplements to over 55 mathematics textbooks. His involvement with computers dates back to 1962, when he programmed a special purpose computer at MIT's Lincoln Laboratory to correct errors in a communications system.

David I. Schwartz is an Assistant Professor in the Computer Science Department at Cornell University and earned his B.S., M.S., and Ph.D. degrees in Civil Engineering from State University of New York at Buffalo. Throughout his graduate studies, Schwartz combined principles of computer science to applications of civil engineering. He became interested in helping students learn how to apply software tools for solving a variety of engineering problems. He teaches his students to learn incrementally and practice frequently to gain the maturity to tackle other subjects. In his spare time, Schwartz plays drums in a variety of bands.

Acknowledgments: I dedicate my books to my family, friends, and students who all helped in so many ways.

Many thanks go to the schools of Civil Engineering and Engineering & Applied Science at State University of New York at Buffalo where I originally developed and tested my UNIX and Maple books. I greatly appreciate the opportunity to explore my goals and all the help from everyone at the Computer Science Department at Cornell.

John T. Sears received the Ph.D. degree from Princeton University. Currently, he is a Professor and the head of the Department of Chemical Engineering at Montana State University. After leaving Princeton he worked in research at Brookhaven National Laboratory and Esso Research and Engineering, until he took a position at West Virginia University. He came to MSU in 1982, where he has served as the Director of the College of Engineering Minority Program and Interim Director for BioFilm Engineering. Prof. Sears has written a book on air pollution and economic development, and over 45 articles in engineering and engineering education.

Michael T. Snyder is President of Internet startup company Appointments 123.com. He is a native of Chicago, and he received his Bachelor of Science degree in Mechanical Engineering from the University of Notre Dame. Mike also graduated with honors from Northwestern University's Kellogg Graduate School of Management in 1999 with his Masters of Management degree. Before Appointments123.com, Mike was a mechanical engineer in new product development for Motorola Cellular and Acco Office Products. He has received four patents for his mechanical design work. "Pro/ENGINEER was an invaluable design tool for me, and I am glad to help students learn the basics of Pro/ENGINEER."

Acknowledgments: Thanks to Rich Lueptow and Jim Steger for inviting me to be a part of this great project. Of course, thanks to my wife Gretchen for her support in my various projects.

Jim Steger is currently Chief Technical Officer and cofounder of an Internet applications company. He graduated with a Bachelor of Science degree in Mechanical Engineering from Northwestern University. His prior work included mechanical engineering assignments at Motorola and Acco Brands. At Motorola, Jim worked on part design for two-way radios and was one of the lead mechanical engineers on a cellular phone product line. At Acco Brands, Jim was the sole engineer on numerous office product designs. His Worx stapler has won design awards in the United States and in Europe. Jim has been a Pro/ENGINEER user for over six years.

Acknowledgments: Many thanks to my co-authors, especially Rich Lueptow for his leadership on this project. I would also like to thank my family for their continuous support.

Royce Wilkinson received his undergraduate degree in chemistry from Rose-Hulman Institute of Technology in 1991 and the Ph.D. degree in chemistry from Montana State University in 1998 with research in natural product isolation from fungi. He currently resides in Bozeman, MT and is involved in HIV drug research. His research interests center on biological molecules and their interactions in the search for pharmaceutical advances.

ESource Reviewers

We would like to thank everyone who helped us with or has reviewed texts in this series.

Christopher Rowe, *Vanderbilt University*
Steve Yurgartis, *Clarkson University*
Heidi A. Diefes-Dux, *Purdue University*
Howard Silver, *Fairleigh Dickenson University*
Jean C. Malzahn Kampe, *Virginia Polytechnic Institute and State University*
Malcolm Heimer, *Florida International University*
Stanley Reeves, *Auburn University*
John Demel, *Ohio State University*
Shahnam Navee, *Georgia Southern University*
Heshem Shaalem, *Georgia Southern University*
Terry L. Kohutek, *Texas A & M University*
Liz Rozell, *Bakersfield College*
Mary C. Lynch, *University of Florida*
Ted Pawlicki, *University of Rochester*
James N. Jensen, *SUNY at Buffalo*
Tom Horton, *University of Virginia*
Eileen Young, *Bristol Community College*
James D. Nelson, *Louisiana Tech University*
Jerry Dunn, *Texas Tech University*
Howard M. Fulmer, *Villanova University*
Naeem Abdurrahman, *University of Texas, Austin*
Stephen Allan, *Utah State University*
Anil Bajaj, *Purdue University*
Grant Baker, *University of Alaska–Anchorage*
William Beckwith, *Clemson University*
Haym Benaroya, *Rutgers University*
John Biddle, *California State Polytechnic University*
Tom Bledsaw, *ITT Technical Institute*
Fred Boadu, *Duk University*
Tom Bryson, *University of Missouri, Rolla*
Ramzi Bualuan, *University of Notre Dame*
Dan Budny, *Purdue University*
Betty Burr, *University of Houston*
Dale Calkins, *University of Washington*
Harish Cherukuri, *University of North Carolina –Charlotte*
Arthur Clausing, *University of Illinois*
Barry Crittendon, *Virginia Polytechnic and State University*
James Devine, *University of South Florida*

Ron Eaglin, *University of Central Florida*
Dale Elifrits, *University of Missouri, Rolla*
Patrick Fitzhorn, *Colorado State University*
Susan Freeman, *Northeastern University*
Frank Gerlitz, *Washtenaw College*
Frank Gerlitz, *Washtenaw Community College*
John Glover, *University of Houston*
John Graham, *University of North Carolina–Charlotte*
Ashish Gupta, *SUNY at Buffalo*
Otto Gygax, *Oregon State University*
Malcom Heimer, *Florida International University*
Donald Herling, *Oregon State University*
Thomas Hill, *SUNY at Buffalo*
A.S. Hodel, *Auburn University*
James N. Jensen, *SUNY at Buffalo*
Vern Johnson, *University of Arizona*
Autar Kaw, *University of South Florida*
Kathleen Kitto, *Western Washington University*
Kenneth Klika, *University of Akron*
Terry L. Kohutek, *Texas A&M University*
Melvin J. Maron, *University of Louisville*
Robert Montgomery, *Purdue University*
Mark Nagurka, *Marquette University*
Romarathnam Narasimhan, *University of Miami*
Soronadi Nnaji, *Florida A&M University*
Sheila O'Connor, *Wichita State University*
Michael Peshkin, *Northwestern University*
Dr. John Ray, *University of Memphis*
Larry Richards, *University of Virginia*
Marc H. Richman, *Brown University*
Randy Shih, *Oregon Institute of Technology*
Avi Singhal, *Arizona State University*
Tim Sykes, *Houston Community College*
Neil R. Thompson, *University of Waterloo*
Raman Menon Unnikrishnan, *Rochester Institute of Technology*
Michael S. Wells, *Tennessee Tech University*
Joseph Wujek, *University of California, Berkeley*
Edward Young, *University of South Carolina*
Garry Young, *Oklahoma State University*
Mandochehr Zoghi, *University of Dayton*

Contents

PART II ENGINEERING PROBLEM SOLVING   47

5
INTRODUCTION TO ENGINEERING PROBLEM SOLVING AND THE SCIENTIFIC METHOD   49

6
ENGINEERING ANALYSIS METHOD   62

7

ENGINEERING DESIGN METHOD 89

10   COMPUTING TOOLS IN ENGINEERING   147

11   FEASIBILITY AND PROJECT MANAGEMENT   162

13 WRITTEN TECHNICAL COMMUNICATIONS   199

14 ORAL TECHNICAL COMMUNICATIONS   224

http://www.prenhall.com/esource/

PART V ENGINEERING PROFESSION   239

15 INTRODUCTION TO THE ENGINEERING PROFESSION AND PROFESSIONAL REGISTRATION   241

16 ENGINEERING ETHICS   250

22 POWER TRANSMISSION CASE STUDY   296

23 WALKWAY COLLAPSE CASE STUDY   303

24 TREBUCHET CASE STUDY   311

A REVIEW OF PHYSICAL RELATIONSHIPS   318

PHOTO CREDITS

Chapter 6: Page 66, Peter James Kindersley © Dorling Kindersley; page 67, Kelley Mooney Photography, Corbis Digital Stock.

Chapter 7: Page 106, (a) University of Washington Libraries. Special Collections, UW21422; (b), © Pat O'Hara/Corbis.

Chapter 10: Page 148, Courtesy of the Library of Congress; page 157, Courtesy of Silicon Graphics, Inc.

0

Introduction

WELCOME TO ENGINEERING

Engineering is like a trek through the remote wilds of the world. It begins with a carefully constructed plan (in your case, a plan of classes to take). The plan is cherished and memorized before the adventure begins. Soon, it is amended when life's little surprises come along (like blocked registration or, heaven forbid, failing a class). The surprises are not all bad: grand opportunities will arise that you never thought possible, such as studying abroad or become editor of the engineering newspaper. You finish the adventure, full of friendships you have made, challenges you have faced successfully, and with a deep appreciation of your chosen field.

Well, that's what it says in the brochures. In reality, engineering is more like food: you love it most of the time, hate it when it is delivered too quickly to be enjoyed, and savour it more when you share it with friends.

Whatever metaphor you like, the next few years will be the foundation of what I hope is a long and rewarding career for you in engineering. This user's guide has been prepared to assist you on your journey. As with other guides, you may find it useful at various points in your career. I hope it will help you now at the beginning of your adventure. Don't forget to refer back to this guide as you continue your trek. You may find portions of it even more useful as you become acquainted with engineering practice.

As with any good adventure or any luscious food, enjoy your experiences in engineering. Keep this guide with you for advice, inspiration, or as a punching bag if the surprises get you down. Good luck!

HOW TO USE THIS BOOK

This book is organized into six parts and the parts are divided into chapters. Six devices have been used to assist in communicating the main ideas of each chapter. First, each chapter begins with an introduction and ends with a summary. The chapter introductions set the stage for each chapter and provide a guide to the material in the chapter. Chapter summaries are concise reviews of the main points in each chapter.

Second, the text contains the definitions of important terms inserted into the margin. Terms to be defined in the margin are indicated in the text by a ***boldface italic*** font. The marginal definitions provide a good study aid for reviewing the important new terms in each chapter.

Third, the text also contains key ideas inserted into the margin. The key ideas are compiled at the end of every chapter. As with the definitions, reading the key idea summaries provides a good review of the important new concepts in each chapter.

Fourth, the text is interrupted frequently by boxes containing a **Ponder This**. Please try to answer the question in the **Ponder This** before continuing to read. The questions in the **Ponder This** boxes always are answered in the following text. The **Ponder This** boxes are used to simulate the classroom interaction between instructor and student.

Fifth, almost all chapters have problems that are worked out in the text. These problems reinforce and extend the concepts in the text.

Last, most chapters of this book contain special sections called *Focus On*. The *Focus On* sections are short stories about engineers and engineering that help bring to life an important lesson in the part.

ENGINEERING CASE STUDIES

An important feature of this text is the collection of engineering case studies in Part VI. The case studies present stories of young engineers faced with challenges that can be solved by applying the fundamental ideas presented in the text. The case studies can be used to present or integrate course material. The *link icons* in the case study chapters refer you to the appropriate background material in the text.

ACKNOWLEDGMENTS

A number of people have contributed significantly to this book. I thank Linda Chattin and Michael Ryan for introducing me to the joys of educating freshman engineers. The current and past staff of the EAS 140 (Engineering Solutions) course — Erik Althoff, Stelios Andreadis, Rajan Batta, Kevin Burke, Carl Lund, James Sarjeant, Brian Zabel, and Jennifer Zirnheld — have been instrumental in refining and focussing the text.

Floreal Prieto and Pneena Sageev from the School of Engineering and Applied Sciences' Center for Technical Communications at the University at Buffalo provided extremely valuable input into Part IV. I thank my faculty and staff colleagues in the Department of Civil, Structural and Environmental Engineering for their patience and support.

A number of faculty members at the University at Buffalo contributed ideas and time toward the case studies in Part VI. Their contributions are acknowledged in Part VI.

The anonymous reviewers of the text provided extremely valuable advice. This book is much improved because of their efforts. As always, any errors that remain in the text are mine.

Finally, I thank the former students of EAS 140 for their patience and suggestions. Other freshmen engineers will benefit from your input over the years.

This text is dedicated to the memory of Dr. Thomas Hill. Tom's dedication to teaching, learning, and the welfare of his students has been an inspiration to me. His imprint is seen through this text, particularly in Chapter 6.

PART I
Discovering Engineering

en·gi·neer—a designer or builder of engines; a person who is trained in or follows as a profession a branch of engineering; a person who carries through an enterprise by skillful or artful contrivance.
Merriam-Webster

Engineer Identification Test
You walk into a room and notice that a picture is hanging crooked. You . . .

A. Straighten it.
B. Ignore it.
C. Buy a CAD (computer-aided design) system and spend the next six months designing a solar-powered, self-adjusting picture frame while often stating aloud your belief that the inventor of the nail was a total moron.

The correct answer is "C," but partial credit can be given to anybody who writes "It depends" in the margin of the text or simply blames the whole stupid thing on "Marketing."
Scott Adams

1

Introduction to Discovering Engineering

1.1 INTRODUCTION

You are beginning an exploration of engineering that will continue throughout your life. This chapter will explore the ways that this journey may begin. In Section 1.2, you will be asked to examine your own motivation for becoming an engineer. In Section 1.3, some surprising advice on how to discover engineering is shared. In Section 1.4, the clock is turned back 175 years to show how history affects your engineering education.

1.2 WELCOME TO ENGINEERING

In some ways, it is amazing that you found engineering in the first place. Most people select careers and academic programs based on their high school experiences. You probably took math and science classes in high school, perhaps even some technology classes. However, you probably did not take *engineering* classes in high school.

Some of your high school friends may be comfortable with their exposure to chemistry or French or English literature in secondary school. They may be looking forward to majoring in one of those fields in college. Although their college experiences will challenge and extend them, your friends probably have a pretty good idea what to expect in college based on their high school experiences.

You may be a little envious. After all, you took a chance on a field that is a little less familiar to you. Your motivation for doing so is unique to you.

OBJECTIVES

After reading this chapter, you will be able to:

- identify why people choose engineering as a career;
- find engineers to speak with about engineering;
- list what you should expect in your engineering education.

PONDER THIS

What is *your* motivation for becoming an engineer?

Key idea: Engineering is the right place for people who have curiosity, a strong work ethic, a desire to help other people, and a deep respect for math and science.

Many students pursue a degree in engineering because they performed well in math and science classes in high school. Some engineering students have relatives who are engineers. Some pursue engineering because the job opportunities and salaries for recent engineering graduates are pretty good. Whatever your motivation, you are taking a small leap of faith in entering a profession that may seem a mystery to you right now. Have no fear: if you have curiosity, a strong work ethic, a desire to help other people, and a deep respect for math and science, then you have found a home in engineering. For a few stories about why working engineers chose engineering, see *Focus on Choosing Engineering: So Why Did You Become an Engineer?*

FOCUS ON CHOOSING ENGINEERING: SO WHY DID YOU BECOME AN ENGINEER?

Every engineer has a unique answer to the question: Why did you choose engineering? Compare your motivation for becoming an engineer with the following stories from practicing engineers.

Helping People

My long-term goal has been to work for and with people, that's why I became an engineer. I like helping people, and that's why I want to use my talents to better other people's lives. This project is exactly the kind of thing I want to do with the rest of my life.

—*From a Northwestern University senior, commenting on her involvement in a project to build a toy car for disabled children* (http://www.asme.org/mechanicaladvantage/fall98/CHILD'SPLAYCARPAGE.HTM)

Problem Solving

When the tragic Challenger disaster occurred in 1986, I found myself not only touched by the loss, but also driven to understand why, and motivated to ensure such a tragedy did not happen again.

—*From a Lockheed Martin mechanical engineer* (http://www.lmaeronautics.com.about/eweek/why)

Math and Science

I always liked science and math anyway, so the idea of working in a profession where one can apply the laws of nature for the benefit of mankind was very inspiring.

—*From a civil engineer working at the Philadelphia District of the U.S. Army Corps of Engineers* (*District Observer ONLINE*, Jan–Feb 2000)

Curiosity

Growing up on a 200-acre cotton farm in middle Georgia, I became fascinated and elated with the mechanical equipment that was becoming available to do work on the farm.

—*From another Lockheed Martin mechanical engineer* (http://www.lmaeronautics.com/about/eweek/why)

Impact

One of the reasons why I became an engineer is because I believe engineering is a profession that allows you to predict the future. As I tell my students, " . . . (Y)ou can use (the laws of physics) to predict how something . . . should work. Then, you can . . . build that something and test it. If it works the way you thought it should, then you effectively forecast the future."

—*From Dr. James Meindl, electrical engineering professor at the Georgia Institute of Technology* (*Georgia Tech Alumni Magazine Online*, Vol. 72, No. 1, Summer 1995)

1.3   HOW TO DISCOVER ENGINEERING

Key idea: Discover engineering by talking to engineers.

So how do you learn more about the engineering profession? First, *put down this book*. No words on the page can help you realize the richness and satisfaction of your career. No book can bring alive the dramatic and compelling history of engineering, where two steps forward are invariably followed by one step backward. And no mere textbook can do justice to the triumphs, diversity, and human stories of the engineers themselves.

Textbook authors usually do not tell you to stop reading their text. But engineering is not about textbooks. It is about people: people who learn, people who translate ideas into reality, people who solve problems, people who communicate their ideas to others, and people who behave responsibly. To truly discover engineering, you must talk to engineers. But how do you find them?

PONDER THIS

How can you find engineers to speak with about engineering?

Key idea: Learn more about your future career by finding engineers at a university.

The best place to find engineers is at your university. Almost all engineering faculty are trained engineers,* and many have work experience outside of the university. Find a faculty member to help you understand the profession. Many universities have freshman mentoring programs, where freshmen are assigned faculty mentors or advisors. If your school does not have such a program, read the departmental brochures or Internet information and find a professor in an area that interests you. Call him or her for an appointment. Be persistent: the faculty are as busy as you are.

Engineering societies are another great source of information about this practice. These societies are professional organizations; some societies are general in scope, while others focus on a specific discipline. The major discipline-specific engineering societies are listed in Table 1.1. Other discipline-specific organizations are listed in Table 1.2. Table 1.3 lists the main engineering societies that are not associated with a specific engineering field. In Table 1.4, engineering societies focused on increasing the diversity of engineers are shown. For more information about any of these organizations, search on the Internet. For a story about historical diversity in engineering, see *Focus on Diversity in Engineering: The Real McCoy?*

TABLE 1.1 Major Discipline-Specific Engineering Societies

Name	Date Started	Number of Members
American Institute of Chemical Engineers (AIChE)	1908	58,000
American Society of Civil Engineers (ASCE)	1852	120,000
American Society of Mechanical Engineers (ASME)	1880	125,000
Institute of Electrical and Electronics Engineers (IEEE)[a]	1884	330,000
Institute of Industrial Engineers (IIE)	1948	24,000

[a]The American Institute of Electrical Engineers (founded in 1884) and the Institute of Radio Engineers merged in 1962 to form IEEE.

---

*You must be a registered professional engineer to use the title "professional engineer." Almost all engineering faculty have received training as engineers, but not all faculty are registered professional engineers.

TABLE 1.2 Other Discipline-Specific Engineering Societies

Name	Discipline[a]
American Academy of Environmental Engineers (AAEE)	civil
American Ceramic Society (ACerS)	several
American Institute for Medical and Biological Engineering (AIMBE)	several
American Institute of Aeronautics and Astronautics (AIAA)	mechanical
American Institute of Mining, Metallurgical, and Petroleum Engineers (AIME)	several
American Nuclear Society (ANS)	several
American Public Works Association (APWA)	civil
American Society for Quality (ASQ)	industrial
American Society of Agricultural Engineers (ASAE)	civil
American Society of Heating, Refrigerating and Air Conditioning Engineers (ASHRAE)	mechanical
American Society of Naval Engineers (ASNE)	several
American Society of Safety Engineers (ASSE)	several
Associated General Contractors of America (AGC)	civil
Association for Facilities Engineering (AFE)	industrial
Biomedical Engineering Society (BMES)	several
Human Factors and Ergonomics Society (HFES)	industrial
National Association of Power Engineers (NAPE)	several
Society of American Military Engineers (SAME)	several
Society of Automotive Engineers (SAE)	mechanical
Society of Fire Protection Engineers (SFPE)	several
Society of Manufacturing Engineers (SME)	several
Society of Petroleum Engineers (SPE)	chemical
Society of Plastics Engineers (SPE)	chemical
SPIE—The International Society for Optical Engineering[b]	several

[a]"Discipline" refers to the major engineering area(s) (chemical, civil, electrical, industrial, and mechanical engineering) targeted by the society.
[b]Originally, this was the Society of Photo-Optical Instrumentation Engineers.

TABLE 1.3 General Engineering Societies

Name	Date Founded
American Association of Engineering Societies (AAES)	1979
American Consulting Engineers Council (ACEC)	1910
American Society of Engineering Education (ASEE)	1893
Junior Engineering Technical Society (JETS)	1957[a]
National Council of Examiners for Engineering/Surveying (NCEES)	1920
National Society of Professional Engineers (NSPE)	1934
Tau Beta Pi[b] (ΤΒΠ)	1885

[a]JETS was established in 1950 and incorporated in 1957.
[b]Tau Beta Pi is the national engineering honor society. Several disciplines have their own national honor societies (e.g., Sigma Gamma Tau for aerospace engineering and Eta Kappa Nu for electrical engineering).

TABLE 1.4 Engineering Societies Focused on Diversity in Engineering

Name	Date Founded
American Indian Science and Engineering Societies (AISES)	1977
Mexican American Engineers and Scientists (MAES)	1974
National Action Council for Minorities in Engineering (NACME)	1974
National Society of Black Engineers (NSBE)	1976
National Organization of Gay and Lesbian Scientists and Technical Professionals (NOGLSTP)	1983
Society of Hispanic Professional Engineers (SHPE)	1974
Society of Woman Engineers (SWE)	1950

student chapter: a student-run organization or club associated with a national society.

The easiest way to meet engineers in professional societies is through student chapters. A **student chapter** is a student-run organization affiliated with a national society. Student chapters of engineering societies typically have a faculty advisor and a practicing engineer who serves as a liaison to the parent society. Most of the organizations listed in Tables 1.1 through 1.4 have student chapters. The student chapters may invite practicing engineers to share their experiences with students. Look for opportunities to speak with the presenters.

Another way to learn from engineers in professional societies is through participation in National Engineers Week. National Engineers Week was established by the National Society of Professional Engineers in 1951 to increase public awareness of the profession. It is held each year during the week of George Washington's birthday (February 22) to acknowledge Washington's contributions as a surveyor. The local activities during National Engineers Week will provide a great opportunity to learn from professional engineers.

Key idea: To learn more, speak with professionals at engineering firms or engineering departments.

Finally, practicing engineers are a wonderful source of information. Many offices, departments of large companies, and government departments offer office tours and internship programs. Summer jobs and co-op programs provide good opportunities to ask questions. The telephone book and Internet are useful guides to engineering practice in your area. In addition, ask the career planning staff at your university to help locate practicing engineers who have volunteered to act as student mentors.

1.4 ENGINEERING EDUCATION: WHAT YOU SHOULD EXPECT

For many engineers, the discovery of engineering begins with their college years. To see what is in store for you as you begin your engineering education, it is instructive to look back in time. Engineering education has a long history in the United States. Engineering education, as with engineering itself, began with military applications.

In the United States, the first formal training program for engineers began in 1794, when Congress added the rank of cadet to the Corps of Artillerists and Engineers. The Corps was assigned to the garrison at West Point. A four-year degree program began at West Point in 1817, under the direction of Sylvanus Thayer (1785–1872). Civilian education in engineering began in 1820 at the American Literary, Scientific, and Military Academy (now Norwich University) in Norwich, Vermont, under the guidance of Alden Partridge (1785–1854).

The work of Thayer and Partridge expanded the traditional university curriculum to educate soldiers and citizen-soldiers about the applied sciences. A different approach was developed by Amos Eaton and Stephen Van Rensselaer. In 1824, they founded the Rensselaer School (now called Rensselaer Polytechnic Institute in Troy, New York) to "teach the application of science to the common purposes of life" (Griggs, 1997).

FOCUS ON DIVERSITY IN ENGINEERING: THE REAL MCCOY?

BACKGROUND

Engineering has made great strides in becoming more diverse and increasing the participation of previously underrepresented groups. Few people realize that one of the most productive engineers of the post–Civil War era was African-American. Elijah McCoy was born in Colchester, Ontario, on May 2, 1844. McCoy's parents were former slaves who fled from Kentucky before the outbreak of the Civil War. (In fact, McCoy's wife was born in 1846 at an Underground Railway station.) After receiving training as a mechanical engineer in Scotland, McCoy moved to Detroit and obtained a job as a fireman on the Michigan Central Railroad.

THE PROBLEM AND ITS SOLUTION

In his job, McCoy became aware of a problem that plagued the railroads and other industries that relied on steam engines. Steam engines required lubrication, which, in the mid-19th century, was usually accomplished by hand. Hand lubrication meant that the machinery had to be turned off or idled to be oiled. McCoy realized that a well-designed automatic lubricator would solve the problem and allow equipment to be run continuously.

McCoy's solution was to improve the hydrostatic lubricator based on a drip cup. In a steam engine, steam from the boiler fills the cylinder and pushes the piston back. In the McCoy lubricator, a small portion of the steam was used to pressurize the lubricator body containing the oil. The pressurized oil drips continuously into the cylinder, thus lubricating the cylinder and piston. The steam condenses into water inside the lubricator body and the oil floats on top. Eventually, the water drains off and the oil is replenished.

The device was patented in 1872. With McCoy's improved lubricator, the continuous operation of steam engines was easier and the transcontinental railroad (completed in 1869) was exploited.

MCCOY'S CONTRIBUTIONS

Elijah McCoy's contributions to lubrication have been exaggerated by some historians and underemphasized by others. While McCoy did not invent the hydrostatic lubricator, he contributed significantly to its optimization and usage. In fact, McCoy's original patent was titled *"Improvement* in Lubricators for Steam-Engines" (emphasis added). Elijah McCoy was a prolific inventor. He was eventually responsible for 57 patents, most involving lubrication equipment. One of his greatest contributions to the field was the graphite lubricator. By suspending graphite in oil, McCoy developed a device to lubricate the then-emerging superheated steam engines.

As with many inventors, McCoy had to assign a number of his patents to investors in his companies. As a result, he did not reap great financial benefits from his inventions. McCoy died at age 85 and was inducted into the National Inventors Hall of Fame in 2001.

IS HE THE "REAL MCCOY"?

The story goes that McCoy's device was prized for its performance, even above the many other automatic lubrication devices that were patented later. Engineers were purported to have asked if their machinery was equipped with "the real McCoy."

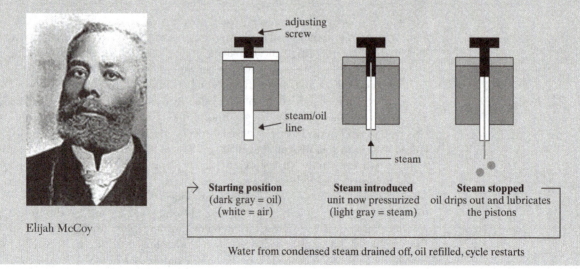

Elijah McCoy

adjusting screw

steam/oil line

steam

Starting position
(dark gray = oil)
(white = air)

Steam introduced
unit now pressurized
(light gray = steam)

Steam stopped
oil drips out and lubricates
the pistons

Water from condensed steam drained off, oil refilled, cycle restarts

Charles "Kid" McCoy

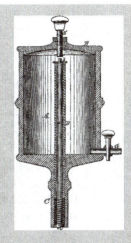

McCoy's lubricator

Did Elijah McCoy's inventions contribute to the popularization of the phrase "the real McCoy"? This question may be impossible to answer. Several people (and objects) from the mid-19th century could be the source of the phrase. Explanations range from Elijah's lubricator to boxer Norman Selby (the Light Heavyweight Champion of the World in 1904, who fought under the name "Kid McCoy") to Mssrs. Mackay's whiskey (made in Edinburgh and marketed as "the real Mackay.")

Regardless of the etymology of "the real McCoy," it is remarkable that a black railroad fireman could have a significant impact on railroad operations within a dozen years of the end of the Civil War. Elijah McCoy stands as a testament to problem solving, perseverance, and intelligence. In these characteristics, he truly was a real McCoy.

Amos Eaton based the program on five rules of education. Eaton's rules are reproduced next, with the original spelling and punctuation (Griggs, 1997). Although the engineering profession and engineering education have changed greatly in almost 200 years,* Eaton's rules are still meaningful today. In this section, Eaton's rules shall be interpreted for the 21st-century engineering curriculum.

1.4.1 Eaton's First Rule: "... make practical applications of all the sciences ..."

Key idea: Discover engineering through problem solving and hands-on work.

"Let the student make practical applications of all the sciences, with the immediate direction and shewing [showing] of the teacher, before studying any elementary rules. For example, shew him in taking the courses and distances around a field with the compass and chain, before he studies any of the rules of surveying—let him measure a pile of wood and attempt to calculate it before teaching him duo-decimals; let him use optical instruments, under the teacher's shewing, before studying optics, let him give an experimental course on chemistry, before reading any work on chemistry; excepting a textbook of experimental description while in the course of experimenting."

Engineering education must be a marriage of fundamental science and practical applications. *Let the applied problem serve as your introduction to and motivation for theory and analysis.* As Amos Eaton might have put it: engage the hands first and the mind will follow.

---

*For example, Eaton speaks of students using only the pronouns "he" and "him," since women engineers were rare in the late 19th century. In addition, the original engineering program at the Rensselaer School could be completed in one year!

1.4.2 Eaton's Second Rule: "... take the place of the teacher ... [in] exercises."

"Let the student always take the place of the teacher on his exercises. He must make every subject his own and then teach his fellow and the school-master, as though there were not a book in the world which treated on this subject, and he was the very oracle of science. Extemporaneous lectures on Tuesday, Wednesday, Thursday, and Friday, and written lectures on Mondays, is a good exercise for that student, in the acquisition of knowledge. He must not speak without a specimen in hand, or the apparatus before him."

Key idea: Learn by teaching others.

Teaching others is a valuable learning strategy. To master engineering fundamentals, practice explaining your reasoning to faculty and fellow students. Most professors feel that teaching deepens their understanding and appreciation of any material. In the last sentence of the quote, note again Eaton's emphasis on the practical problem.

1.4.3 Eaton's Third Rule: "... attend to but one branch of learning at the same time."

"Let a student attend to but one branch of learning at the same time. Personal exercise in the afternoon, at surveying, engineering, collecting plants and minerals, inspecting factories, machines, and agricultural operations, may be permitted. For, although reflection is required, such exercises call such different faculties into action, that the mind is not thereby burdened or fatigued."

Unfortunately, engineering curricula usually do not allow the luxury of immersing the students in only one subject at a time. In fact, the modern engineering curriculum ensures that the courses you will take are integrated together and build on one another. However, *you can focus on one topic at a time*. A key to success in engineering is mastering the material from one lecture or set of readings before the next lecture occurs.

Key idea: Take the time to master material before new concepts are presented.

It is absolutely critical that you give yourself time to master the material. As Eaton said, "reflection is required." In high school, you may have been able to master some material by sitting passively in class and listening to your teacher. In your engineering courses, you will need to think and talk about the material *outside* of the classroom. You must give yourself the time to think, to make mistakes, and to explore. In reading this text, use the *Ponder This* questions as a springboard for reflection.

In this rule, Eaton once again urges you to return to practice. Talking to professionals, going on field trips at every opportunity, and searching the Internet are some 21st-century versions of Eaton's "personal exercises" (see also Section 1.2).

Finally, Eaton's advice about the importance of using "different faculties" of the mind should serve as a guide to your college education. Use liberal arts and social science courses to exercise the different parts of your brain. These courses are every bit as valuable as your technical courses.

1.4.4 Eaton's Fourth Rule: "Let the amusements and recreation of students be of a scientific character."

"Let the amusements and recreation of students be of a scientific character. Collecting and preserving minerals and plants, surveying, and engineering, are good amusement."

Eaton's idea of amusement may not jibe with *your* idea of amusement, but the sentiment is important. Eaton's comment focuses on how you spend your time away from the classroom. In modern times, this rule should remind you to get involved with

Key idea: Join student clubs and enter student engineering contests.

student clubs and student engineering contests. In addition, consider getting involved in service-oriented organizations (such as Habitat for Humanity) where your engineering skills can be used to help others immediately. Also, look for engineering in your everyday life—from the automatic teller machine to the roller coaster at your local amusement park to your cell phone.

Eaton also is speaking about commitment. To be a successful engineer, it is *not* necessary to spend every waking moment with your nose in a technical journal. However, *you must make a commitment to getting your degree* or graduation day will never come. Success in engineering is all about using your time wisely to achieve your goals. Start now: make earning an engineering degree a high priority in your life.

1.4.5 Eaton's Fifth Rule: "Let every student daily criticize those whose exercise he has attended."

Key idea: Challenge your instructors and respect the value of good technical communication skills.

"Let every student daily criticize those whose exercise he has attended. Such as to point out all errors in language, gesture, position, and manner of performing experiments, etc. The teacher must always preside during the hour of criticism. No exercise sharpens the faculty of discrimination like this, while it causes each student to be perpetually on his guard."

Give constructive feedback to your instructors. Challenge them as they challenge you. Learning requires two-way communication, giving you the responsibility to interact with your instructors. Eaton's fifth rule is also a reminder of the importance in the engineering profession of both *technical communications* (i.e, avoiding "errors in language, gesture, position") and *data collection* (i.e., avoiding errors in the "manner of performing experiments").

1.5   SUMMARY

Your discovery of engineering has begun. It is sincerely hoped that your sense of discovery will be just as keen 40 years from now as it is today. Although you may have some trepidation as you begin your exploration of engineering, know that engineering is the place for you if you are curious, have a strong work ethic, wish to help other people, and respect and enjoy math and science. Discover engineering by talking with engineers, including your professors, members of professional societies (and the student chapters of professional societies), and practicing engineers. Consider the wisdom of Amos Eaton when discovering engineering in your classes.

SUMMARY OF KEY IDEAS

- Engineering is the right place for people who have curiosity, a strong work ethic, a desire to help other people, and a deep respect for math and science.
- Discover engineering by talking to engineers.
- Learn about your future career by finding engineers at a university.
- To learn more, speak with professionals at engineering firms or engineering departments.
- Discover engineering through problem solving and hands-on work.
- Learn by teaching others.
- Take the time to master material before new concepts are presented.
- Join student clubs and enter student engineering contests.
- Challenge your instructors and respect the value of good technical communication skills.

Problems

1.1. Find and record the Internet home page of each society listed in Table 1.1. Organizations often write a mission statement that succinctly states their goals and aspirations. Read the mission statement of each organization.

1.2. Using Table 1.1, pick two societies that interest you the most and explain why they interest you.

1.3. Summarize the purpose and goals of three societies listed in Tables 1.2 through 1.4.

1.4. Using the library and the Internet, write a short essay on the contributions to engineering education from Sylvanus Thayer, Alden Partridge, Amos Eaton, or any other pioneering educator in a technical field.

1.5. Devise a way to teach a high school student about a technical topic using the approach suggested in Eaton's first rule. Pick a topic that you learned about in high school or are learning about now. Possible topics might be Newton's laws of motion, Boyle's law, or Ohm's law.

1.6. State an applied problem that interests you. Looking at the curriculum for your field of study, list the courses that you think will help you solve this problem.

1.7. How can you use the ideas in Eaton's second rule to study engineering?

1.8. Write a paragraph on the opportunities at your university to teach others.

1.9. Make a list of the liberal arts and social science courses that you plan to take and explain why they interest you.

1.10. Ask an engineering professor how teaching deepens his or her understanding and appreciation of engineering.

1.11. Make a list of "amusements and recreation … of a scientific character" in your community. Pick an activity to participate in this year.

1.12. Attend a meeting of a service-oriented organization in your community. Write a short paragraph about how you might use your engineering training to help the organization.

1.13. Attend a meeting of an engineering student club. Write a short paragraph about the plans of the student club for the year.

1.14. Make a list of local engineering firms in the discipline of most interest to you. Visit a local office and report on your visit.

2

What Is Engineering?

2.1 INTRODUCTION

The question posed by the title of this chapter may seem a bit strange. After all, you do not have to ask the meaning of brain surgery, soccer, or veterinary science—and there are many more engineers in the world than brain surgeons, professional soccer players, or veterinarians. Your familiarity with engineered *systems* (highways, buildings, computers, and factories, to name a few) does not tell you much about the *process* that made those systems. The process is engineering. In this chapter, you will explore the characteristics that engineers and engineering disciplines have in common.

2.2 DEFINING ENGINEERING

What is engineering? This simple question has a very complex answer. Engineering is a diverse collection of professions, academic disciplines, and skills. You can start your exploration of engineering with the dictionary. Your ego may be boosted to learn that the word "engineering" stems from the Latin *ingenium*, meaning skill. (Other words sharing this Latin root include "ingenious" and "ingenuity.") Engineers are skilled at what they do. But what do they do? The dictionary offers you further insight. The Latin root *ingenium* comes from *in* + *gignere*, meaning to produce or beget (also the source of the words "generate" and "kin"). Thus, engineers are skilled producers or creators of things.

This exercise in word origins does not do justice to the field of engineering. Many definitions of "engineering" and "engineer" are possible. Most definitions have some elements in common.

OBJECTIVES

After reading this chapter, you will be able to:

- identify the elements that all engineering disciplines have in common;
- describe how engineers help others.

Based on your experiences, what is your definition of engineering?

Key idea: Engineers are professionals who apply science and mathematics to useful ends, solve problems creatively, optimize, and make reasoned choices.

Common elements in the definitions include the following:

- Engineers apply science and mathematics to useful ends.
- Engineers solve problems creatively.
- Engineers optimize.
- Engineers make choices.
- Engineers help others.
- Engineering is a profession.

You will examine each of these elements in more detail in this chapter.

2.3   ENGINEERING AS AN APPLIED DISCIPLINE

2.3.1   Knowledge Generation versus Knowledge Implementation

Almost everyone would agree that engineering is the application of science and mathematics to practical ends. Indeed, the emphasis on practice and application always is in the mind of the engineer. They care more about *using* basic knowledge than *generating* basic knowledge. They care more about converting basic science into technology and converting technology into useful products than in expanding basic science.

However, the emphasis on application tells only part of the story of engineering. The pure engineer may be concerned only with practice, just as the pure scientist is concerned only with generating new knowledge. In reality, both practicing scientists and engineers contribute to the complicated and rewarding process of converting ideas into reality. The pure scientist and the pure engineer are extremes of a spectrum of skills required to make new things.

2.3.2   The Role of Engineering

The role of the engineer in turning ideas into usable ideas or objects° is illustrated in Figure 2.1. Both scientists and engineers use mathematics and natural sciences as their tools. Engineers focus on answering the questions that lie on the more applied side of the spectrum. In your career as an engineer, it is likely that you will help develop and implement technology. You will likely work from the middle to the right side of the spectrum

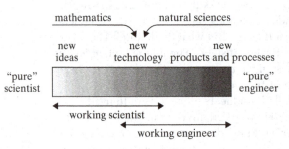

Figure 2.1.   Spectrum of Skills in Engineering and Science.

°Engineers develop both *products* (e.g., ballpoint pens, toasters, microprocessors, and satellites) and *processes* (e.g., better ways to stock inventory, new approaches to manufacturing compact discs, and innovative ways to treat wastewater).

in Figure 2.1. As an engineer, you will acquire a pool of skills required to translate new knowledge into usable ideas.

You are urged to return to Figure 2.1 throughout your career. It may help you remember that knowledge generation and product design are two ends of the same spectrum. Neither skill is useful without the other.

Figure 2.1 is a good road map for getting the most out of your courses. For example, if you are suffering from motivational problems in your science and mathematics courses, think about how the material can be applied. Speak to your science and mathematics professors about the practical use of the material. Ask engineering mentors how they use basic science and mathematics in their everyday professional lives.

EXAMPLE 2.1: APPLICATION OF SCIENTIFIC PRINCIPLES

In your high school or freshman chemistry course, you may have learned about the unit of chemical concentration called the *mole*. At first glance, the use of molar units of concentration may seem very theoretical and not applied. Give an example of how each main engineering discipline can use this concept.

SOLUTION

Lithium iodine batteries as small as 4 mm thick have been used in implantable cardiac pacemakers for over 20 years (Photo courtesy of Greatbatch, Inc.)

Engineers use different sets of units to solve different kinds of problems. Molar units express the proportions in which chemicals combine. Therefore, they are very useful for solving problems involving *combining proportions*.

Almost every engineering discipline uses molar units for some applications. For example, a civil engineer (in the environmental engineering specialty) might use molar units for determining doses of chemicals to react with pollutants and solve pollution problems. A chemical engineer may use moles to determine the ratios of chemicals used to synthesize polymers on a commercial scale. An electrical engineer would use molar units to determine the number of electrons per time (also called the current) produced by an electrochemical cell (such as a battery). Molar units could be used by an industrial engineer to design monitoring devices to make the workplace safer. Mechanical engineers use the concept of the mole to optimize material properties. All engineers may work together to use molar units to design sensors capable of detecting chemical or biological agents.

2.4   ENGINEERING AS CREATIVE PROBLEM SOLVING

2.4.1   Solving Problems

Engineers solve problems. These three simple words have far-reaching ramifications on the life of an engineer. First, since engineers solve problems, engineering work is usually motivated by a concern or roadblock. This idea may conjure up an image of lone warriors furiously performing dense calculations in a cubicle under a green-shaded desk lamp.

No image of engineering could be more incorrect. In reality, engineers often solve other people's problems. Thus, engineers must be able to *listen to a concern* and *map out a solution*. Whether the problem is to make cars that pollute less or to make oil refining more efficient or to reduce the manufacturing cost of a child's toy, engineers must be able to understand problems that their clients face. In this way, engineering is a very people-oriented profession.

2.4.2   Standard Approaches to Solving Problems

Second, engineers must be skilled in using standard approaches to solve problems. We all respect the brilliant physician who can leap to a diagnosis using intuition and experience. We marvel at the aircraft mechanic who spots the mechanical problem by "feel."

However, we also know that *every* physician must be able to follow standard diagnostic procedures and *every* mechanic must be familiar with the inspection checklist. Similarly, engineers must know the well-established protocols used in solving many problems.

2.4.3 Creative Approaches to Solving Problems

Key idea: The solution to engineering problems involves both standard and creative approaches.

Third, engineers must be creative in solving problems. Just like the physician and aircraft mechanic, engineers must supplement the standard solution methods with creativity and insight. Engineering is a highly creative profession. As Theodore von Kármán (1881–1963), a well-known Hungarian-born specialist in fluid mechanics and aerodynamics, put it, "The scientist describes what is; the engineer creates what never was" (Mackay, 1991). This quotation should not be interpreted as minimizing the creativity of scientists. Rather, it points out that engineers must have vision to create something that did not previously exist.

2.5   ENGINEERING AS CONSTRAINED OPTIMIZATION

2.5.1   Constraints

constrained optimization: determining the best solution to a problem, given limitations on the solution

Engineering, like life, is about **constrained optimization**. In high school, it was likely that you did not strive to be the *best* student you could be. Rather, you strived to be the best student you could be *given that* you had to work part-time or you had family obligations or you were active in community groups. In other words, your time available for studying was *constrained* by other activities.

Key idea: Engineering solutions are often constrained.

Similarly, engineers always face constraints in solving problems. As an example, electrical engineers rarely seek to design the fastest computer chip. To be useful, computer chips must exhibit other characteristics as well.

PONDER THIS

> **List some constraints on computer chip design.**

We could list the various constraints on computer chip design, but a better goal is just to say that we seek to develop the fastest computer chip of sufficiently small size with adequate heat dissipation characteristics that can be mass-produced at a reasonable cost.

Is it *ever* valuable to build the fastest chip? Absolutely! An electrical engineer specializing in the research side of research and development (R&D) may, in fact, seek to design the fastest computer chip. Some major breakthroughs in engineering have been generated by engineers and scientists who ignored constraints. However, most engineers seek to put ideas into practice. This means taking the real world and its constraints into account when designing engineered systems.

Key idea: Engineering solutions must take into account the probability of failure.

One aspect of the constrained nature of engineering is that engineers live in a probabilistic world. In other words, engineers must consider the *chances* of certain events occurring, including the probability of failure. A civil engineer does not design a bridge that will never fall down. Such a bridge would be infinitely expensive. Rather, the civil engineer examines the probabilities that certain loads will occur on the bridge from traffic, earthquakes, and wind. A bridge is designed to perform acceptably for a specified period of time under the anticipated loads and stresses. Similarly, an environmental engineer does not design a drinking water treatment plant to remove *all* pollutants completely. Such a plant is probably not possible (and if it was possible, the drinking water it produced would be unaffordable). Instead, engineers design treatment plants to meet water quality standards and minimize risk at a socially acceptable cost.

Due to constrained optimization in a probabilistic world, engineers must constantly ask: How strong is strong enough? How clean is clean? Have I thought of everything that could go wrong?* An example of extremely constrained optimization is given in the *Focus on Constrained Optimization: A Square Peg in a Round Hole*.

2.5.2   Feasibility

The ability of an engineering project to meet its constraints is often expressed in terms of feasibility. There are several aspects of feasibility, which will be introduced here. *Technical* (or engineering) *feasibility* measures whether or not a project meets its technical goals. It addresses several questions, such as "Does the new road handle the traffic?" and "Is the upgraded electrical transmission system more efficient?"

Key idea: To be successful, engineering projects must be technically, economically, fiscally, socially, politically, and environmentally feasible.

Most of your undergraduate course work is focused on technical feasibility. However, it is not sufficient for an engineering project to be technically feasible. Engineering projects also must be economically feasible. *Economic feasibility* addresses whether the project benefits outweigh the project costs. In the examples above, economic feasibility addresses whether the road benefits (e.g., tolls collected, elimination of slowdowns, and increased safety) are greater than the road construction and maintenance costs or whether the money saved from the more efficient transmission systems will pay for the upgrade work. Sometimes, the benefits and costs are difficult to quantify. What is the value of a five-minute reduction in commuting time or one less incidence of cancer for every one million people? To answer these questions, engineers may seek the advice of social scientists and economists.

Another factor to consider is *fiscal feasibility*. Fiscal feasibility measures whether sufficient funds can be generated to build the project. Many engineering projects would be profitable (i.e., are economically feasible), but are not built because start-up money cannot be acquired. The difference between economic and fiscal feasibility is important. For large, multimillion-dollar engineering projects, obtaining money through loans or bonds to achieve fiscal feasibility may be the critical step. Engineers who ignore fiscal feasibility will never see their ideas translated into reality.

The last type of feasibility is social, political, and environmental feasibility. Engineers cannot work in a vacuum. Engineering projects must be socially acceptable, have political backing, and result in an acceptable environmental impact. Many engineering projects remain only on paper because societal and political support was lacking. Should you, as an engineer, be upset because some projects die due to nontechnical issues? No. It should remind you that engineers are part of the fabric of society. The public cares about the impact of engineering projects. As a result, you must consider the social consequences of your proposal along with the technical details.

FOCUS ON CONSTRAINED OPTIMIZATION: A SQUARE PEG IN A ROUND HOLE

BACKGROUND

Engineering is about constrained optimization. The need to "make the best with what you have" is demanding when the constraints are the most severe. For example, a space vehicle located 200,000 nautical miles from Earth presents some of the most severe constraints that an engineer will face.

Such was the case with *Apollo 13*, launched April 11, 1970. The crew of the spacecraft—Commander James A. Lovell, Lunar Module Pilot Fred W. Haise, Jr., and Command Module Pilot John L. Swigert, Jr.— was hard at work and enjoying the ride. Suddenly, about 56 hours into the flight, the crew heard a loud noise (which is never a good sign in a spacecraft). The pressure in Cryogenic Oxygen Tank 2 had begun to rise

---

*For an insightful and entertaining discussion of this question as it pertains to civil engineering, see Petroski (1992).

Launch of Apollo 13, Saturday, April 11, 1970

very quickly. Within two minutes, the tank lost pressure. Why did this matter? Electricity on *Apollo 13* was generated by a fuel cell, where oxygen and hydrogen were combined. No oxygen meant no power—and no way to return to Earth.

PROBLEMS AND SOLUTIONS

The ground crew quickly assessed the situation. The three people in space required three things to return to Earth alive: power, water (to drink and to cool the equipment), and oxygen. With the fuel cells virtually inoperable, the only source of power was the batteries in the Lunar Module (LM, the *Aquarius*). It became clear that the LM, with its own ample oxygen supplies, would become the lifeboat for the crew. But the LM had its own problems. Its batteries would need to be recharged to provide enough power for the journey home. However, there was no direct electrical connection between the Command Service Module (CSM, the *Odyssey*) and the LM to recharge the batteries. Engineers on the ground discovered a way to leak current slowly from the CSM to the batteries. By turning off nonessential equipment,

the crew limped home with an amazing 20% of the LM power left.

The problem with water could be addressed only by drinking less. The crew cut its water ration to 200 milliliters per person per day (a little over one-half of a soft drink can). The crew lost a collective 31 pounds on the trip home, arriving in poor health and with 10% of the water supply remaining.

The LM had a sufficient oxygen supply, but the ground crew soon realized that *another* air supply problem would threaten the astronauts: the build-up of carbon dioxide (CO_2) exhaled by the crew. The CO_2 was removed by lithium hydroxide (LiOH) canisters through a chemical reaction. The LM was designed to transport two members of the crew from lunar orbit to the surface of the Moon. It had a sufficient canister capacity to remove the CO_2 produced by two people for about 30 hours, not the CO_2 exhaled by three people for the long trip back to Earth. Even by allowing the CO_2 levels to rise a little, the canisters could operate for only about 187 person-hours, when at least 288 person-hours would be needed. The solution? The CSM had its own LiOH canisters. But as luck would have it, the CSM canisters had *square* connectors that would not fit in the *round* fittings of the LM.

CONSTRAINED OPTIMIZATION

In a brilliant feat of constrained optimization, the ground crew had to develop an interface between the square CSM canisters and the round LM fittings from material available to the astronauts. (The near-impossibility of this task is shown dramatically in a famous scene from the movie *Apollo 13*.) The adapter, called the "mailbox," was designed by ground engineer Ed Smylie. (In NASA-speak, the adapter is known officially as the "supplemental carbon dioxide removal system.") It was made of two CSM canisters, a space suit exhaust hose, cardboard from instructional cue cards in the LM, plastic stowage bags from liquid-cooled undergarments, and one roll of duct tape. (Ironically, much of this material would have been otherwise unused by the astronauts. The cue cards contained instructions for lifting off from the moon and the undergarments were to be worn on moonwalks.)

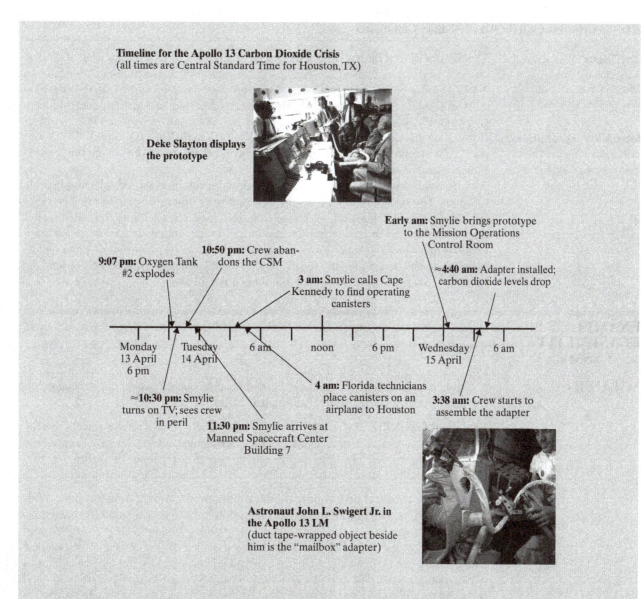

Timeline for the Apollo 13 Carbon Dioxide Crisis
(all times are Central Standard Time for Houston, TX)

Deke Slayton displays the prototype

Early am: Smylie brings prototype to the Mission Operations Control Room

9:07 pm: Oxygen Tank #2 explodes

10:50 pm: Crew abandons the CSM

3 am: Smylie calls Cape Kennedy to find operating canisters

≈4:40 am: Adapter installed; carbon dioxide levels drop

Monday 13 April 6 pm — Tuesday 14 April — 6 am — noon — 6 pm — Wednesday 15 April — 6 am

≈10:30 pm: Smylie turns on TV; sees crew in peril

11:30 pm: Smylie arrives at Manned Spacecraft Center Building 7

4 am: Florida technicians place canisters on an airplane to Houston

3:38 am: Crew starts to assemble the adapter

Astronaut John L. Swigert Jr. in the Apollo 13 LM
(duct tape-wrapped object beside him is the "mailbox" adapter)

The efforts of engineer Ed Smylie, the other Houston personnel, and the astronauts were truly unbelievable. Only about 30 hours elapsed between the time Smylie turned on his television set to learn of *Apollo 13's* problems and the time that the astronauts finished building the mailbox in space. (See the accompanying timeline.) Smylie and his assistant, Jim Correale, had no operational LiOH canisters to test the interface. Working canisters from Florida (intended for *Apollo 14* or *Apollo 15*) were airlifted to Houston to enable testing of the device. After testing on the ground, the ground crew issued about an hour's worth of instructions by radio so that the astronauts could construct the adapter in space.

The *Apollo 13* flight had a very happy ending, due to the bravery of the astronauts and the ingenuity of the engineers. When faced with an overwhelmingly constrained problem, the engineers created a solution that saved the lives of three American heroes.

2.6 ENGINEERING AS MAKING CHOICES

Key idea: Engineers make recommendations by selecting from a list of feasible alternatives.

The discussion thus far has centered on what engineering *is* rather than what engineers *do*. So what *does* an engineer do? Engineers listen carefully to the problem. (See Section 2.4.1.) Using accepted and creative methods (see Sections 2.4.2 and 2.4.3), they develop a list of feasible solutions or alternatives. Here, "feasible" means that each solution is technically, economically, fiscally, and socially/politically/environmentally feasible. (Evaluating whether a project is feasible is called a *feasibility assessment*. See Example 2.2 for an example.) Finally, engineers select an alternative from among the feasible solutions and recommend it to their client.

feasibility assessment: evaluation of the feasibility of an engineering project

In a real sense, engineering is about generating alternatives and selecting feasible solutions. The selection step sets engineers apart from other professionals (e.g., technicians and designers) who may be trained to do the calculations and run the software, but may not be trained to make recommendations. To recommend an alternative, an engineer has to balance the technical, economic, fiscal, and social/political/environmental issues. A person trained only to crunch numbers will fail in this critical decision-making task. A person trained only to crunch numbers is not an engineer.

EXAMPLE 2.2: FEASIBILITY ASSESSMENT

Conduct a feasibility assessment for buying a used car to commute to a part-time job.

SOLUTION

A feasibility assessment determines the technical, economic, fiscal, and social/political/environmental feasibility of an alternative.

Technical feasibility: Technical feasibility probes whether the alternative will solve the problem. In this case, you need to ask whether buying the car will allow you to commute to work safely and reliably. Perhaps the car you can afford will not be sufficiently reliable. Perhaps other more reliable alternatives exist, such as public transportation.

Economic feasibility: Economic feasibility questions whether the benefits of the alternative exceed its costs. The costs of car ownership include depreciation, financing, insurance, taxes and fees, fuel, maintenance, and repairs. For a used 2000 Honda Accord two-door LX coupe in Buffalo, New York, the purchase and ownership costs average about $7,570 per year for the first five years (as determined by the cost calculator at www.edmunds.com). The benefits include your ability to get to your part-time job and the freedom and convenience that car ownership engenders. Other alternatives (such as a monthly bus pass) may have lower costs, but they do not have the freedom and convenience of car ownership.

Fiscal feasibility: Fiscal feasibility probes whether you can get the start-up funds to finance the project. In this example, purchasing the car is fiscally feasible if you can qualify for a reasonable car loan.

Social/political/environmental feasibility: In this example, social/political/environmental feasibility centers on environmental impact. In spite of the large social, political, and environmental costs of reliance on the internal combustion engine, car ownership remains socially acceptable in North America. You could consider, as an alternative, a car with lower environmental impact. (Buying a new Toyota Prius has about twice the purchase + interest + depreciation costs of the Honda, but about half the fuel + maintenance + repair costs.)

2.7 ENGINEERS AS HELPING OTHERS

Professions can be characterized in many ways. Some people are attracted to the so-called caring professions.

Make a list of caring professions.

Caring professions include medicine, nursing, social work, and teaching. Did your list include engineering?

Engineering is also one of the caring professions. Why? Nearly every project that an engineer completes satisfies a need or concern of the public. For example, if you become an electrical engineer, you may develop sensors to make more powerful neonatal incubators. Perhaps as a civil engineer, you may work on earthquake-resistant buildings or develop drinking-water treatment systems for less developed countries. (Water-related diseases, the leading cause of death globally, is responsible for 14,000 deaths *per day* because more than one billion people on the planet lack access to safe drinking water.) Maybe you will become a chemical engineer and work on ways to mass-produce HIV medications, making such drugs affordable to every HIV-positive person in the world. Perhaps you will go into industrial engineering and devise systems to help nonprofit organizations better serve their clients. (Volunteers from several professional and student chapters of the Institute of Industrial Engineers recently helped make a women's shelter in Pittsburgh become more efficient to save resources.) Or maybe you will become a mechanical engineer and develop a robust heart valve for premature infants. Whatever field of engineering interests you, know that you can use your training to make the world a better place and to make people's lives healthier and more fulfilling.

2.8 ENGINEERING AS A PROFESSION

Finally, engineering is a profession. This means, of course, that engineers get paid for what they do. In addition, it means that to be called an engineer, you must meet certain requirements. Just as the public must be assured that a person called a dentist or lawyer is fully certified, so the public must know that a person using the title "engineer" has been trained properly. The process of meeting the requirements is called *registration*. A registered engineer holds the title *professional engineer*, or PE.

All professions have ethical standards. Engineers, as professionals, must meet high standards of professional ethics.

2.9 SUMMARY

The dictionary tells us that engineers give birth to things creatively. In particular, the work of engineers is characterized by six elements. First, engineers apply science and mathematics to useful ends. Second, engineers solve problems using both standard and creative approaches. Third, engineers optimize solutions subject to the constraints of the real world. The constraints are often grouped under the headings of technical feasibility (will the system perform the task for which it was designed?), economic feasibility (do benefits outweigh costs?), fiscal feasibility (are start-up funds available?), and social/political/environmental feasibility. Fourth, engineers make reasoned choices. They select and recommend feasible alternatives. Fifth, engineers help others. Without a public to serve, engineering as a profession would not exist. Finally, engineers are professionals. This means that engineers may seek professional registration and must meet ethical standards.

SUMMARY OF KEY IDEAS

- Engineers are professionals who apply science and mathematics to useful ends, solve problems creatively, optimize, and make reasoned choices.
- The solution to engineering problems involves both standard and creative approaches.
- Engineering solutions are often constrained.
- Engineering solutions must take into account the probability of failure.
- To be successful, engineering projects must be technically, economically, fiscally, socially, politically, and environmentally feasible.
- Engineers make recommendations by selecting from a list of feasible alternatives.
- Engineering is a profession and engineers have ethical responsibilities.

Problems

2.1. What are the six main elements of engineering?

2.2. Some pharmaceuticals are manufactured by genetically engineered bacteria to produce the drug. Discuss the role of the engineer (if any) in the following steps of the development of a new drug:

- Synthesis of the drug for animal tests
- Genetic engineering of the bacteria
- Mass production through bacterial synthesis
- Clinical trials
- Development of the time-release capsules and transdermal patches
- Efficiency study of the manufacturing process, and
- Design of the marketing strategy

2.3. Explain the differences in the contributions to society of scientists and engineers. Which contribution appeals more to you and why?

2.4. Give an example of constrained optimization in an engineering problem.

2.5. For Problem 2.4, what is a possible solution, given the constraints? How would the solution change if the constraints were different?

2.6. From your local newspaper, find an example of an engineering project that was not implemented because it was not economically or fiscally feasible.

2.7. Explain the difference between economic feasibility and fiscal feasibility.

2.8. Give an example of an engineering project that is economically feasible, but not fiscally feasible. Give an example of an engineering project that is fiscally feasible, but not economically feasible.

2.9. From your local newspaper, find an example of an engineering project that was not implemented because it was not socially, politically, or environmentally feasible.

2.10. For your answer in Problem 2.9, how would you change the project to make it feasible?

2.11. Talk with an engineer in government service (e.g., a town, city, or county engineer) about the difference between economic and fiscal feasibility. Illustrate the difference with an example from your community.

2.12. Two towns are separated by a river and wish to exchange goods. List several alternative solutions to this problem. Perform a feasibility assessment and rank the alternatives according to their feasibility. (Be sure to include all types of feasibility.) Recommend a solution to the problem.

2.13. Make a list of the professions that are licensed by your state. What do the professions have in common? Which licenses are in technical fields?

2.14. Which agency in your state licenses engineers? (Try searching the Internet for your state name and the phrase "professional engineer.") How many engineers are licensed in your state?

2.15. How do engineers in your area participate in public service?

3

Engineering Careers

3.1 INTRODUCTION

You have learned about the activities and approaches common to many engineers. This chapter explores the kinds of careers for which an engineering education prepares you. Engineering jobs will be discussed in general terms in this chapter. Jobs specific to each engineering discipline will be presented elsewhere.

3.2 ENGINEERING JOBS

3.2.1 Availability of Jobs

What do engineering students do when they graduate? Fortunately, they have the opportunity to work in their field if they wish. In a recent study, a sample of the 109,200 people receiving baccalaureate degrees in engineering in 1999 and 2000 were surveyed to find out what they were doing in 2001. Ninety-three percent of the recent graduates were working. Of those employed, 84% found jobs in engineering and science (NSF, 2003; see Figure 3.1). Most of the employed engineers (68%) found jobs in the same discipline as their degree. The bottom line? Trained engineers can usually find jobs as engineers.

3.2.2 Introduction to Engineering Jobs

Where are engineers employed? The range of jobs performed by engineers is truly amazing. They have optimized devices as simple as the pencil and developed systems as complex as the Space Shuttle.* For example, some engineers work on systems as small as very large scale integration (VLSI)

OBJECTIVES

After reading this chapter, you will be able to:

- list the types of jobs available to engineers;
- explain why job satisfaction is high in engineering;
- describe the future of engineering employment.

---

*The Space Shuttle *Endeavour*, built to replace the Space Shuttle *Challenger*, was first flown in May 1992. This Space Shuttle has several hundreds of thousands of parts, and it was constructed with the assistance of over 250 subcontractors at a cost of approximately $1.7 billion.

Key idea: People trained as engineers generally can find jobs as engineers.

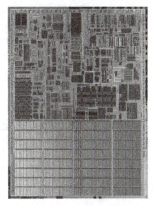

Intel Corp.'s Itanium® 2 microprocessor contains 221 million transistors in an area less than a square inch (421 mm²).

Arecibo observatory (photo courtesy of the NAIC-Arecibo Observatory, a faculty of the NSF)

Key idea: About half of all engineers work in manufacturing, producing motor vehicles, aircraft, electrical and electronic equipment, and industrial or computing equipment.

Key idea: About 28% of engineers work in the service sector, primarily as consulting engineers.

Key idea: Many engineers are small business owners.

consulting engineer: an engineer providing professional advice to clients

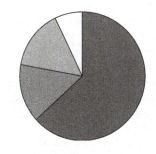

- ■ Employed in same discipline as degree
- ■ Employed in another S&E discipline
- ■ Employed outside of S&E
- ☐ Unemployed

Figure 3.1. Employment Statistics for Baccalaureate Engineers (S&E = science and engineering)

chips, with millions of transistors and other circuits on each 0.25-inch square speck of silicon. Engineers are helping to develop *nanomachines:* futuristic mechanical devices consisting of only a few thousand atoms. Others work on systems as large as the 305-meter (1,000-foot) radio telescope near Arecibo, Puerto Rico.

Engineers have done work as helpful as creating assistive devices for the disabled and inventing innovative approaches to clean polluted air. In fact, engineers work on systems as controversial as nuclear weapons and nerve gases.

With such a range of experiences, how can you make sense of engineering jobs? It is instructive to examine the distribution of engineering jobs in various economic sectors. In 2002, about 1,769,000 engineers were employed in the United States (BLS, 2004), including about 611,000 computer software engineers. About half of these jobs were in manufacturing industries. A little more than one-quarter of the jobs were in service industries. Government agencies at all levels (federal, state, and local governments) employed about 12% of all engineers in 2000.

3.2.3 Engineers in Industry

Engineers in industry are involved mainly with the manufacture of products. In 2001, there were 1,000 engineers making toys and sporting goods and 57,000 engineers involved in manufacturing electronic components and accessories. Manufacturing accounts for about half of engineering jobs. Most of these jobs are in transportation equipment (mainly motor vehicles and aircraft), electrical and electronic equipment, and industrial equipment (including computing equipment). Other engineers in industry work in nonmanufacturing areas such as construction and mining.

3.2.4 Engineers in Service

Engineers in service act as consultants, contribute to product marketing and sales, or conduct research. The service sector accounts for about 28% of engineering jobs, mostly in engineering and architecture, business services (mainly computer/data processing and personnel supply services), and research and testing services.

Most engineering companies are small. In 2001, 59% of engineering firms had fewer than 5 employees, and 35% of the firms employed 5 to 50 people (Rosenbaum, 2002). It is not uncommon to find very small "mom-and-pop" engineering consulting firms.

Consulting engineers may own their own business. About 43,000 engineers in 2000 were self-employed. Most of these engineers worked as consultants.

3.2.5 Engineers in Government

Government engineers may be employed by federal, state, or local agencies. The military also employs both civilian and active-duty engineers. Government engineers work

Key idea: About 12% of engineers work for government agencies.

for a wide range of agencies, from city engineering offices to the U.S. Army Corps of Engineers to the Peace Corps.

As stated in Section 3.2.2, about one out of every eight engineering jobs is in government. Over half of these jobs are in the federal government, primarily in such agencies as the Departments of Defense, Transportation, Agriculture, Interior, and Energy and in the National Aeronautics and Space Administration (NASA) (BLS, 2004). Engineers employed by state and local governments typically work in highway and public works departments.

3.2.6   Other Engineering Jobs

Transportation and public utilities account for another 5% of engineering jobs. Most of these jobs are with electric and telephone utilities. The remaining engineering jobs are mainly in wholesale/retail trade and construction.

3.2.7   Engineering Education as a Route to Other Fields

Key idea: An engineering education is excellent preparation for many non-engineering professions.

An engineering education provides a strong background in quantitative skills and problem solving. Such tools are highly valued in many fields. Thus, an engineering education is a good start for many nonengineering professions. Trained engineers are well suited to pursue professional degrees in other fields, including law, medicine, business, and education. For example, an engineering background is very desirable for patent and environmental law. Engineers can go on to medical school and contribute to such fields as biomechanics and neurobiology. Many business schools encourage potential students to pursue technical degrees at the undergraduate level. An engineering degree is a fine route to a teaching career in secondary or higher education.

The tools and approaches learned in engineering classes have led to successful careers in other fields. U.S. Presidents Herbert Hoover (mining) and Jimmy Carter (nuclear) were trained as engineers. Other engineers that became politicians include John Sununu (mechanical), Yasser Arafat (civil), Leonid Brezhnev (metallurgical), and Boris Yeltsin (civil).

Several famous entertainers started out as engineers. For example, both *Star Trek* creator Gene Roddenberry (aeronautical) and Academy Award-winning director Frank Capra (chemical) had engineering degrees. Other engineers who landed in entertainment-oriented careers include film directors and Roger Corman (industrial) and Alfred Hitchcock (studied at the School of Engineering and Navigation in London), jazz musician Herbie Hancock (electrical), talk show host Montel Williams (general), and television star Bill Nye, "The Science Guy" (mechanical). For another interesting story of an engineering entertainer, see the *Focus on Nonengineers*.

Engineering careers are not limited to Earth. Nearly all the early astronauts in the Mercury, Gemini, and Apollo programs were engineers. All but four of the 39 U.S. astronaut pilots in the Space Shuttle program in 2004 had engineering degrees.

Herbert Hoover

3.3   JOB SATISFACTION IN ENGINEERING

3.3.1   What Does "Job Satisfaction" Mean to You?

As discussed in Section 3.2.1, people trained as engineers can generally obtain jobs as engineers. Thus, you are not just pursuing a *degree* in engineering, but you are also traveling the road towards a *career* in engineering. One measure of a successful career is a love for the jobs you will have as you progress in your chosen field.

FOCUS ON NONENGINEERS: "IT'S NOT HEDY, IT'S HEDLEY"

Fans of Mel Brooks's notoriously crude movie *Blazing Saddles* (1974) will recognize the title of this section. The crooked attorney general Hedley Lamarr (played outrageously by Harvey Korman) repeatedly has to remind everyone how to pronounce his name: "It's not Hedy, it's Hedley."

Hedy Lamarr (photo courtesy of Anthony Loder).

So who was Hedy Lamarr, the source of the attorney general's confusion? And what does she have to do with engineering? Hedy Lamarr was born as Hedwig Eva Maria Kiesler in Vienna in 1913. She made dozens of movies, including the 1933 Austrian–Czech film *Ecstasy*, which featured one of the first nude scenes in a movie. Lamarr married arms manufacturer Fritz Mandl in 1933. To escape the Nazi regime and her domineering husband, Lamarr escaped to London in 1938. There, she met movie mogul Louis B. Mayer. Mayer gave her the stage name we know now and took her to Hollywood. She became a famous actress and pin-up girl in World War II, and she was active in the sale of millions of dollars in war bonds.

But Hedy Lamarr was not just a glamorous film star of Hollywood's Golden Age. In fact, Lamarr had the temperament, if not the education, of an engineer. At a dinner party shortly after the onset of World War II, Lamarr was talking with her friend, composer George Antheil. Using the analogy of a player piano, Lamarr and Antheil reasoned that communication between a submarine and a torpedo could be made secret if the information was scrambled by hopping it between frequencies. The pair eventually patented the idea. It became U.S. Patent Number 2,292,387, filed June 10, 1941, and issued August 11, 1942, to Hedy Kiesler Markey (her legal name at the time) and George Antheil for a "Secret Communications System."

The U.S. military decided that the idea of "frequency hopping" was impractical. In fact, given the technology of the time, frequency hopping probably was nearly impossible to implement. The idea, however, would revolutionize communication. The patent expired in 1959 and the concept was generalized as "spread spectrum technology."[*] Soon, the technology caught up with the idea. Spread spectrum technology has been used in military applications from the Cuban Missile Crisis to the 1991 Gulf War. It was released to the public domain in the 1980s and is now used in applications as diverse as pagers and traffic signals.

Spread spectrum communication is the basis for the operation of garage door openers, cell phones, and wireless Internet communication. Why? In addition to the security of the signal, spread spectrum technology has a number of advantages. By spreading the signal across a number of frequencies, this communication tool makes efficient use of the clogged radio frequency band. In addition, because the devices transmit only for a short time at any one frequency, the signals appear as background noise rather than interfering with existing transmissions at that frequency. This property allows spread spectrum devices (for example, digital spread spectrum [DSS] cordless telephones) to operate at a higher power (and hence, have a longer range) and with less interference.

It is not an exaggeration to say that Lamarr and Antheil's idea for defeating Nazism is the cornerstone of modern digital communication. Not bad for a glamorous actress and avant-garde composer. Hedy Lamarr and George Antheil were honored with an International Pioneer Award in 1997 from the Electronic Frontier Foundation. Hedy Lamarr died in 2000, remembered as an actress and inventor.

---

[*]The term "spread spectrum technology" refers to any technology where information is distributed over a wide signal bandwidth according to a pattern independent of the data transmitted. One way to accomplish this is to use the frequency-hopping approach of Hedy Lamarr and George Antheil.

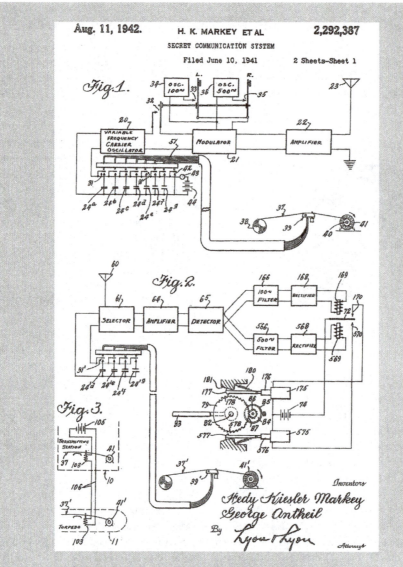

Hedy Lamarr/George Antheil patent

PONDER THIS

When you get your first engineering job, how are you going to measure your own job satisfaction?

Most people measure their job satisfaction in three categories:

- Accomplishments: what they have contributed to society
- Work environment: independence, responsibility, the degree to which they are challenged by their work, and where they work
- Monetary issues: salary, benefits, and opportunities for promotion

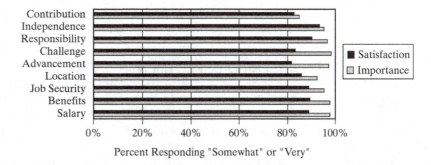

Figure 3.2. Importance in Job Selection and Level of Satisfaction of Recent Engineering Graduates

Key idea: Job satisfaction includes contributions to society, degree of independence, level of responsibility, intellectual challenge, opportunities for promotion, benefits, and salary.

Key idea: Job satisfaction among engineers is very high.

Key idea: Engineers are well compensated, with salaries varying by discipline, educational level, and experience.

Key idea: Employers expect to receive more value from you than you are paid.

It is not uncommon for people to have very high expectations of their future careers, only to have their visions squashed by reality. Entry-level engineers also have high expectations. In 2001, the National Science Foundation (NSF) surveyed 1999 and 2000 engineering graduates about their job satisfaction (NSF, 2003). The elements of job satisfaction included contribution to society, degree of independence, level of responsibility, intellectual challenge, opportunities for advancement, location, job security, benefits, and salary. As shown in the gray bars in Figure 3.2, most factors were very important to the recent engineering graduates. (Figure 3.2 shows the percentage of respondents that said the factor was "very important" or "somewhat important" to them.)

The good news is that their level of job satisfaction was high. The average job satisfaction rating for the nine factors was 88%. (For scientists in the same survey, the average job satisfaction was 82%.) The entry-level engineers were particularly satisfied with the degree of independence and level of responsibility provided in their jobs.

3.3.2 Engineering Salaries

Compared with other fields, engineers are paid well. Salaries vary by discipline, educational level, and experience. So how much do engineers make? It would be misleading to list detailed salary data here. Information about salaries can become outdated very quickly and should be interpreted with caution. It is almost impossible to know the true average salary of any group as large as the engineering community. Salary surveys invariably cover different populations and may or may not include some benefits. In general, engineers have and continue to be valued in our society. Compensation is expected to remain good for the foreseeable future.

While the higher salaries of engineers are an attraction, avoid choosing a profession (or engineering discipline) strictly on the basis of salary. Your job satisfaction will be low (and your life miserable) if you go for the bucks without listening to your heart.

One final note on salaries: The higher salaries of engineers come with some expectations.

PONDER THIS

What do employers expect of you when they offer you, say, $45,000 as a starting salary?

It means that they expect to receive more than $45,000 worth of value from you. For a company to make money, it must receive more value (in sales or client fees) from you than you are being paid. Remember this fact when you interview for jobs: salaries create an obligation to work responsibly, diligently, and ethically.

3.4 FUTURE OF ENGINEERING EMPLOYMENT

Key idea: Engineering jobs are expected to grow over 7% from 2002 to 2012, with over about 5% growth in the major engineering disciplines.

The future is good for the employment of engineers. In total, 109,000 engineering job openings are expected from 2002 to 2012. Most of the major engineering disciplines are predicted to show over about 5% growth in that period, with engineering as a whole having a 7.3% job growth. According to the best estimates of the economists, engineers trained today should have reasonable expectations of a job tomorrow.

3.5 SUMMARY

The present and future states of engineering employment are strong, and students should expect to find jobs in their fields. Engineering jobs may be found in manufacturing (e.g., mainly motor vehicles, aircraft, electrical and electronic equipment, and industrial equipment and computing equipment), services (e.g., consulting services, marketing and sales, and research), government, and other areas (e.g., transportation, public utilities, wholesale/retail trade, and construction).

Engineers are paid well, with salaries varying by discipline and increasing with educational level and experience. While engineering salaries are good, never forget that employers expect you to contribute in value to the company more than they pay you. Job availability and salaries are expected to increase in the future. Finally, an engineering education can lead to rich and rewarding careers in nonengineering fields.

SUMMARY OF KEY IDEAS

- People trained as engineers generally can find jobs as engineers.
- About half of all engineers work in manufacturing, producing motor vehicles, aircraft, electrical and electronic equipment, and industrial or computing equipment.
- About 28% of engineers work in the service sector, primarily as consulting engineers.
- Many engineers are small business owners.
- About 12% of engineers work for government agencies.
- An engineering education is excellent preparation for many nonengineering professions.
- Job satisfaction includes contributions to society, degree of independence, level of responsibility, intellectual challenge, opportunities for promotion, benefits, and salary.
- Job satisfaction among engineers is very high.
- Engineers are well compensated, with salaries varying by discipline, educational level, and experience.
- Employers expect to receive more value from you than you are paid.
- Engineering jobs are expected to grow over 7% from 2002 to 2012, with over about 5% growth in the major engineering disciplines.

Problems

3.1.   For an engineering discipline of your choice (chemical, civil, electrical, industrial, or mechanical engineering), list two jobs from each job sector in Section 3.2. Which job sector appeals to you and why?

3.2.   Interview practicing engineers from two job sectors listed in Section 3.2. Write a paragraph explaining how the engineers you interviewed got from their baccalaureate degree to their current job.

3.3.   Which of the job satisfaction criteria discussed in Section 3.3.1 is most important to you?

3.4.   Describe your ideal engineering job. As a guideline, use the job satisfaction criteria discussed in Section 3.3.1.

3.5.   Using the Internet, research jobs in the discipline of most interest to you. How closely do the jobs align with the ideal engineering job you described in Problem 3.4?

3.6.   You have received two job offers: one from Technolico, Inc., at $44,000 per year and one from Engionics at $46,000 per year. What criteria should you use in deciding which job to take?

3.7.   List four questions you would like to ask during an engineering job interview.

3.8.   A small consulting firm wishes to expand. If they hire an entry-level engineer, they can increase their revenue by $4,750 per month. The annual salary for an entry-level engineer in their community is $47,000. Benefits (e.g., health insurance and contributions to retirement accounts) cost 25% of the employee's salary. Can the firm afford to hire an entry-level engineer at the going rate?

3.9.   How much would revenues have to grow for a small consulting firm to justify hiring an entry-level engineer? Assume that the typical annual salary for an entry-level engineer is $48,500 and benefits amount to 30% of the employee's salary.

3.10.   Ask a practicing engineer whether a master's degree in engineering is desirable for engineers in your area.

3.11.   Ask a practicing engineer whether an **MBA** (master of business administration) is desirable for engineers in your area.

4

Engineering Disciplines

4.1   INTRODUCTION

Earlier, you were given some advice: to discover engineering, put down this book and talk to an engineer. Similarly, to find out more about any particular engineering discipline, speak to an engineer engaged in that discipline. Only a practitioner can give you the depth, history, and potential future of a field. Only an engineer can impart the excitement, challenges, and occasional frustrations of working in a given area. Only a working engineer can tell you what he or she does every day on the job.

So why read the rest of this chapter? The remainder of the chapter is devoted to giving you a brief taste of each of the major engineering disciplines. Its purpose is to whet your appetite rather than answer all your questions about a field of particular interest to you. This chapter is intended to motivate you to seek out engineers in the fields that pique your interest.

There are dozens of specific fields, each with dozens of types of engineers. Section 4.2 discusses how the many types of engineers can be organized into a small number of primary disciplines and emerging fields.

The other sections in the chapter are devoted to a general description of each of the principal engineering disciplines. The primary *technical areas* will also be presented to shed light on the core of each discipline. These technical areas are the tools that define what makes a chemical, civil, electrical/computer, industrial, or mechanical engineer. The technical areas are combined into *applications*, which show the fields in which each type of engineer works and illustrate the diversity of the engineering discipline. Finally, unique elements of the *curriculum* leading to an engineering degree in each discipline will be discussed. Only the typical core coursework is presented. For further information, investigate the Web page of the pertinent engineering department

OBJECTIVES

After reading this chapter, you will be able to:

- list the principal engineering disciplines;
- discuss the technical areas, applications, and curricula for chemical, civil, electrical and computer, industrial, and mechanical engineering;
- explain how new engineering disciplines emerge.

Key idea: To learn about a specific discipline, speak with an engineer working in that discipline.

at any major university. It is important that you have a general knowledge of all engineering disciplines, since engineers of different backgrounds frequently work together to solve problems.

4.2 HOW MANY ENGINEERING DISCIPLINES EXIST?

To answer this question, consider a seemingly unrelated question: how long is the coastline from Portland, Maine, to Miami, Florida? To estimate the coastline length, you might start with a globe and measure the distance with a straight ruler. This would be a crude estimate of the distance. You could refine your measurement by using a road map of the eastern United States. Now your measurement would take into account more details, such as the fishhook of Cape Cod and the coastline of the Chesapeake Bay. As a result, your measurement would likely be larger than the estimate from the globe. If you used state or local road maps, your measurement would continue to be refined and continue to grow. If, in desperation, you crawled the entire trip, measuring around each grain of sand, you would come up with a different (and larger) value.°

Key idea: The five principal engineering disciplines are chemical, civil, electrical and computer, industrial, and mechanical engineering.

Counting engineering disciplines is like measuring the distance along a coastline: as the scale narrows, the number of disciplines increases. One of the broadest views on the number of disciplines looks at the number of accredited engineering programs in the United States. The number of accredited programs is shown in Figure 4.1. In this text, the top five engineering disciplines (chemical, civil, electrical and computer, industrial, and mechanical engineering) will be called the *principal engineering disciplines*.

The engineering profession defines an engineer as someone eligible for professional registration. The Fundamentals of Engineering Examination (FE Exam) is required for professional registration as a licensed professional engineer. The specialty portion of the FE Exam is offered in the five principal engineering disciplines plus environmental engineering.

Using professional registration as a guide, your list of engineering fields can be expanded by examining the areas in which it is possible to take the Principles and Practice

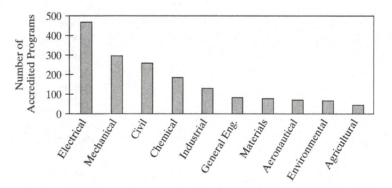

Figure 4.1. Number of Accredited Engineering Programs in the United States by Discipline (*Electrical* includes electrical and computer engineering; other disciplines include related fields. Data obtained from www.abet.org.)

°An object that looks the same at every degree of magnification (or, formally, exhibits self-similarity across scales) is said to be *fractal*. Fractal analysis is used in signal processing to analyze anything from Internet traffic data to biomedical data.

Examination (PP Exam). The PP Exam is the last step in the professional registration process. The 17 different PP Exams are listed in Table 4.1.

Examining engineering disciplines on a smaller scale, you could make a list of relatively common engineering fields. A good list is shown in Table 4.2, with most fields listed under their closest principal engineering disciplines.

TABLE 4.1 Several Lists of Major Engineering Disciplines

Principal Disciplines	Fundamentals of Engineering Exam Areas	Principles and Practices Exam Areas
		Agricultural
		Architectural
Chemical	Chemical	Chemical
Civil	Civil	Civil
		Control Systems
Electrical/Computer	Electrical	Electrical and Computer
	Environmental	Environmental
		Fire Protection
Industrial	Industrial	Industrial
Mechanical	Mechanical	Mechanical
		Metallurgical
		Mining and Mineral
		Naval Architecture and Marine
		Nuclear
		Petroleum
		Structural (I and II)

Note: The manufacturing engineering FE Exam was discontinued after October 2003.

TABLE 4.2 Common Engineering Fields by Principal Discipline

Chemical	Civil	Electrical and Computer
Biological	Architectural	Control Systems
Biomedical	Construction	Electronics
Ceramic	Environmental	Signal Processing
Control Systems	Geotechnical	
Petroleum	Sanitary	
Plastics	Structural	
Polymer	Transportation	

Industrial	Mechanical	Other
Human Factors	Aeronautical	Agricultural
Operations Research	Aerospace	Fire
Production Systems	Automotive	Military
	Biomechanical	Mining
	Heating	Naval
	Manufacturing	Nuclear
	Materials	Ocean
	Metallurgical	Plant
	Robotics	Safety

Note: The "other" category contains unique or multidisciplinary fields.

A more complete list of engineering jobs has been developed by the Bureau of Labor Statistics (BLS). They list an unbelievable 431 engineering jobs, from absorption engineering to mud engineering to zoning engineering.[*]

So how can you make sense of all these disciplines, fields, subfields, and subsubfields? In this text, the five principal engineering disciplines will be discussed in Sections 4.3 through 4.7. The major subdisciplines will be explored in Section 4.8. The process of how new disciplines evolve will be discussed in Section 4.9.

4.3   CHEMICAL ENGINEERING

4.3.1   Technical Areas

Key idea: Chemical engineers work with the transformation of chemicals to form useful products or processes.

Chemical engineers work with the transformation of chemicals to form useful products or processes. The main technical areas are catalysis and reaction engineering, heat transfer and energy conversion, and separations. As you can see by these areas (and the name of the discipline), chemical engineers tend to have a strong interest in chemistry.

Reaction engineering and catalysis refers to the design and construction of engineered systems to bring about chemical change. In catalytic processes, materials are added to make the chemistry proceed more quickly. This can make the difference between an economically feasible and an economically infeasible process.

Chemical changes are driven by energy. Some chemical engineers specialize in *heat transfer*. Other chemical engineers focus on the conversion of energy from one form to another (especially in the presence of chemical transformations).

In the field of *separations*, chemical engineers assist in product purification. Examples of products benefiting from separation systems designed by chemical engineers range from perfume to gasoline to beer.

4.3.2   Applications

chemical process industry: any industry where chemicals are extracted, isolated, or combined (the primary application of chemical engineering)

Chemical engineers apply their technical skills to almost every industry. The primary application is in the **chemical process industries**—that is, any industry where chemicals are extracted, isolated, or combined. The main industries employing chemical engineers manufacture agricultural chemicals; food; industrial gases; petrochemicals and petroleum products; pharmaceuticals and personal care products; polymers (including plastics and rubber); pulp and paper; soaps and other fats; and synthetic fibers.

Other areas where chemical engineers work include biotechnology, environmental engineering, nuclear engineering, and advanced materials. Biotechnology is the synthesis of new products by using living organisms. Chemical engineers have teamed with scientists to develop new approaches to product development through biotechnology. From the development of artificial skin and blood to the biosynthesis of pharmaceuticals, chemical engineers are at the forefront of biotechnology. In fact, many chemical engineering departments have changed names from "chemical engineering" to "chemical and biological engineering."

pollution prevention: a contribution of chemical engineers, where industrial processes are modified to minimize pollution

Chemical engineers specializing in the environmental area often work with industry to make industrial processes cleaner (called **pollution prevention**). Chemical engineers also have contributed to nuclear engineering in its various forms (e.g., power generation, propulsion, and commercial uses). Chemical engineers have led the development of new materials with highly specialized properties for the aerospace, automotive, and photographic industries, as well as many others.

---

[*]To be fair, most of the BLS jobs are not engineering positions because they employ people who are not eligible for professional registration.

4.3.3 Curriculum

Chemical engineering students usually take engineering courses in thermodynamics; conservation and transport of mass, momentum, and energy; unit operations (i.e., industrial processes such as separations); and design. Science courses in organic chemistry and biochemistry frequently are required.

4.4 CIVIL ENGINEERING

4.4.1 Technical Areas

Key idea: Civil engineers are involved in the analysis, design, and construction of public works.

Civil engineering takes its name from Latin *civis*, meaning citizen. As with many technological fields, engineering once had a mainly military focus. As engineering applications broadened, engineering was divided into two areas: military engineering and civil (i.e., nonmilitary) engineering. The other four principal engineering disciplines eventually split off, leaving what we now know as civil engineering. As a result, civil engineering is one of the broadest engineering disciplines. In a nutshell, civil engineers are involved in the analysis, design, and construction of public works.

solid mechanics: the behavior of solids at rest and in motion

Although diverse, civil engineering shares the core technical area of engineering mechanics. This core is divided into three areas: solid mechanics, fluid mechanics, and soil mechanics. **Solid mechanics** refers to the behavior of solids at rest and in motion. It is the primary tool used to analyze the structural integrity of structures both large (such as buildings, roads, and bridges) and small (such as printed circuit boards). *Fluid mechanics* is used to understand the behavior of water and air in both natural systems (such as rivers and the atmosphere) and engineered systems (such as pipes and blowers). *Soil mechanics* studies the behavior of soils in response to stress. It is an important tool in the design of foundations and earthen structures (such as landfills).

4.4.2 Applications

The main applications of civil engineering are construction, environmental, geotechnical, structural, transportation, and water resources engineering. In the *construction engineering* area, civil engineers optimize the use of materials, money, and people in construction projects. *Environmental engineers* develop and design treatment processes for a variety of pollutants and follow the fate of pollutants in the environment (see Section 4.8.4). The **geotechnical** specialty in civil engineering is concerned with the design of foundations, embankments, retaining walls, and landfills. *Structural engineering* is one of the largest specialties in civil engineering. Structural engineers focus on the design of buildings, bridges, dams, and other systems. *Transportation engineers* study both traffic patterns and construction materials to maintain and improve transportation and other delivery systems. In the *water resources* area, civil engineers plan, manage, and design systems for the use and management of lakes, rivers, groundwater, storm water, and reservoirs. Examples include irrigation systems, dams, and storm water retention ponds.

geotechnical engineering: a specialty of civil engineering concerned with the design of foundations, embankments, retaining walls, and landfills

4.4.3 Curriculum

The civil engineering curriculum reflects both the core area of mechanics and the wide variety of applications. Civil engineering students generally take engineering courses in statics, mechanics (usually separate courses in solid mechanics, fluid mechanics, and soil mechanics), materials, structures, transportation, and environmental engineering. Often, courses in project management and foundation engineering are required.

4.5   ELECTRICAL AND COMPUTER ENGINEERING

4.5.1   Technical Areas

Key idea: Electrical engineers focus on the transmission and use of electrons and photons.

Electrical engineers focus on the transmission and use of *electrons* and *photons*. This simple statement does not do justice to the richness of electrical engineering. We use electric power through many different types of communication, computational, industrial, and consumer devices. Electrical engineering often includes the related field of *computer engineering*. In addition, electric energy is often transformed prior to use. Examples include alternating and direct current conversions, analog and digital transformations, and the interconversion of electricity and magnetism, sound, and light.

physical electronics: the study of solid-state electronic devices, such as transistors

electromagnetics: the study of the complex relationships between electricity and magnetism

The core technical areas of electrical engineering include circuits, physical electronics, signal processing, and electromagnetics. *Analog and digital circuits* are fundamental tools of the electrical engineer. The circuits area involves the analysis of networks, power delivery, specialized circuits (alternating current, radio frequency, and microwave circuits), and circuit design. **Physical electronics** includes the understanding of solid-state electronic devices, such as transistors. *Signal processing* (information engineering) involves the interpretation of time-dependent voltages and currents. **Electromagnetics**, the complex relationship between electricity and magnetism, is the theory that underlies much of modern-day electrical engineering.

4.5.2   Applications

To appreciate the far-reaching applications of electrical engineering, you only have to look around your home, office, or school. Electronic devices surround you. The diversity in the field is shown by the fact that the main professional society in electrical engineering (the Institute of Electrical and Electronics Engineers or IEEE) contains 38 technical societies and councils. To simplify this broad branch of engineering, consider four basic applications: communications, digital electronics, microelectronics and photonics, and power systems.

The *communications* area includes applications as varied as communications theory, telecommunications, and optical communications systems. Electrical engineers specializing in *digital electronics* focus on digital circuit design and instrumentation, control systems, image processing, and computationally efficient architectures. In the *microelectronics and photonics* area, electrical engineers develop microprocessor architecture, work with machine- and assembly-language programming, and develop systems with optical fibers using lasers and other light sources. The field of *power systems* includes the generation, transmission, and distribution of electrical power.

4.5.3   Curriculum

Electrical engineering students usually take courses in analog and digital circuit analysis and design, microprocessors, signal processing, electromagnetics, and programming. Common electives include power systems and communications.

4.6   INDUSTRIAL ENGINEERING

4.6.1   Technical Areas

Industrial engineers seek to analyze, design, model, and optimize complex systems. The traditional role of industrial engineers has been in the manufacturing sector. However, industrial engineering also is applied to areas as diverse as transportation, service industries (e.g., hospitals and banks), and supply and distribution problems.

The core technical areas in industrial engineering are simulation, statistics, and engineering economics. *Simulation* is important to the industrial engineer, because many improvements in manufacturing or other systems must be simulated (usually through

Key idea: Industrial engineers seek to analyze, design, model, and optimize complex systems.

computer models) before they are accepted and implemented. *Statistics* plays a large role because of the random nature of errors and disruptions in the work environment. More than the other engineering branches, industrial engineers address economic feasibility in their work with manufacturing and service industries.

4.6.2 Applications

ergonomics: relationships between people and the jobs they perform (now commonly called human factors engineering)

operations research: optimization of complex systems to meet one or more goals

The tools of industrial engineering are combined and applied in three basic areas: human factors engineering, operations research, and production systems engineering. *Human factors engineering* (or **ergonomics**, literally the management of work) concerns the relationship between people and the things they use or the jobs they perform. **Operations research** deals with the optimization of complex systems to meet one or more goals (called *objective functions*). In *production systems engineering*, the industrial engineer focuses on optimizing both the physical facilities (e.g., plant layout and materials handling) and production scheduling.

4.6.3 Curriculum

Industrial engineers typically take courses in probability and statistics; engineering economics; human factors; production systems and facilities planning; operations research; and simulation. Electives in economics, computer science, and psychology are common. Since knowledge of the manufacturing site is at the heart of industrial engineering, many industrial engineering programs offer cooperative experiences (co-ops) or internships in industry.

4.7   MECHANICAL ENGINEERING

4.7.1   Technical Areas

Mechanical engineering is synonymous with machinery. In fact, the words "mechanical" and "machine" both come from the Greek *mechos*, meaning expedient (because mechanized devices are often the most efficient way to carry out a task). Mechanical engineers develop, design, and manufacture machines.

Key idea: Mechanical engineers develop, design, and manufacture machines.

The primary technical areas in mechanical engineering are mechanics and thermodynamics. In mechanical engineering, the fields of *solid mechanics* and *fluid mechanics* are applied to machines rather than to structures (as they are in civil engineering). Since machines use, generate, or transmit power, mechanical engineers are interested in the use of energy and its loss as heat. Thus, *thermodynamics* is a primary tool of the mechanical engineer.

4.7.2   Applications

Mechanical engineers generally specialize in one or more of the following areas: applied mechanics, bioengineering, fluids engineering, heat transfer, tribology, and aeronautics. Using *applied mechanics*, engineers analyze and design machines. In *bioengineering*, mechanical engineers apply the principles of mechanics to solve problems with human anatomy (e.g., prosthetic and assistive devices) and physiology (e.g., devices to improve heart and lung function). *Fluids engineering* refers to the flow of fluids (e.g., air, ink, or blood) in mechanical systems. Mechanical engineers specializing in fluids engineering may perform complex calculations to predict fluid flow, a subspecialty called *computational fluid dynamics*. *Heat transfer* involves the study of heat transport by conduction, convection, and radiation. The study of heat transfer is critical, since all processes lose energy as

tribology: the study of friction, wear, and lubrication

heat. **Tribology** (from the Greek *tribein*, to rub) concerns friction, wear, and lubrication. Aeronautics and astronautics are major subdisciplines of mechanical engineering. (See Section 4.8.3.)

Robotic arm on the Mars rover *Spirit* reaches out to a rock called Adirondack. Placing a rover on Mars required experts in nearly all fields of mechanical engineering (as well as other types of engineers). (Image credit: NASA/JPL.)

4.7.3 Curriculum

In accordance with the technical fields and primary applications, mechanical engineers usually take courses in machines, fluid mechanics, thermodynamics, instrumentation, materials, systems engineering, and design. Common technical electives include applied mathematics, aeronautics, and biomechanics.

4.8 MAJOR ENGINEERING SUBDISCIPLINES

Key idea: The major engineering subdisciplines are materials, aeronautical/aerospace, environmental, agricultural, and biomedical engineering.

4.8.1 Introduction

As shown in Figure 4.1, there are a number of accredited engineering programs in the United States outside of the principal engineering disciplines. The major subdisciplines—materials, aeronautical/aerospace, environmental, agricultural, and biomedical engineering—may exist as separate departments in universities, but more frequently are found within departments representing the five principal disciplines. Each major subdiscipline will be discussed in more detail in this section.

4.8.2 Materials Engineering

Materials engineers contribute to all aspects of materials used for engineering purposes, from the origin of a material (extraction or synthesis), through its transformations (processing, design, and manufacture), to its applications. Often called *material science and engineering*, the field emphasizes the science behind the properties of materials.

Materials engineering programs often are housed in mechanical engineering departments. However, the home department sometimes depends on the materials. For example, the materials engineering aspects of metals and their alloys often are taught in mechanical engineering departments (sometimes under the heading *metallurgical engineering*). *Polymers, ceramics,* and *biomaterials* may be the purview of chemical engineers. Civil engineers have been making more use of *composite materials*, while electrical engineers study and use *semiconductors*.

4.8.3 Aeronautical, Astronautical, and Aerospace Engineering

The fields of aeronautical and astronautical engineering are related subdisciplines of mechanical engineering. Aeronautics (from the Greek *aero*, air, + *nautes*, sailor) is the study of mechanized flight. Astronautical engineering, on the other hand, deals with mechanized flight beyond the Earth's atmosphere. (The term *aerospace*, first used in 1958, refers to both the atmosphere and outer space.)

One way to look at the diversity in aerospace engineering is to examine its largest professional society, the American Institute of Aeronautics and Astronautics (AIAA). The AIAA has over 60 technical committees, covering issues from lighter-than-air systems to modeling and simulation to space colonization.

4.8.4 Environmental Engineering

As stated in Section 4.4.2, environmental engineers both design treatment processes and model the fate of pollutants in the environment. Environmental engineering (previously called sanitary engineering) is traditionally part of civil engineering. However, some areas are associated with other principal engineering disciplines (e.g., pollution prevention with chemical engineering).

In the water treatment area, environmental engineers develop and design treatment facilities for drinking water, wastewater, and industrial wastes. Environmental engineers also work with treatment systems for air and contaminated soil. In the modeling area, environmental engineers track everything from global warming (caused, in part, by carbon dioxide emissions to the atmosphere) to leaking underground storage tanks at your neighborhood gas station.

4.8.5 Agricultural Engineering

Agricultural engineering (sometimes called food, biological, bioresource, or biosystems engineering) involves the solution of problems in the production and processing of food, fiber, timber, and renewable energy sources. Agricultural engineers seek to produce agricultural products from natural resources efficiently, while minimizing environmental impact. The major areas of agricultural engineering are food engineering, power systems and machinery design, and forest engineering.

4.8.6 Biomedical Engineering

Biomedical engineers use engineering and scientific principles to solve problems in medicine and biology. Two of the largest areas in this diverse field are biomaterials engineering and biomechanical engineering. In the biomaterials area, engineers (typically with training in chemical engineering) develop implants from both living tissue and artificial materials. Biomaterials engineers have contributed significantly to the development of artificial skin and artificial blood. In the biomechanical engineering area, engineers study fluid flow and materials properties in the human body. Biomechanical engineers have contributed greatly to the development of the artificial heart and artificial joint replacements. Electrical and computer engineers specializing in the biomedical engineering area have contributed to new imaging and surgical technologies, such as CAT (computer-aided tomography) scans and remote surgery.

4.9 HOW DO EMERGING ENGINEERING DISCIPLINES EVOLVE?

4.9.1 Introduction

Key idea: New engineering fields can be formed by the budding off of existing disciplines or by the creation of interdisciplinary fields.

A cursory glance at engineering curricula in the United States reveals that new types of engineering programs are constantly being developed. *Engineering disciplines evolve over time.* Why? If problems emerge that require more specialization, then new engineering fields will evolve to address the challenges. But how do new engineering disciplines form?

Engineering fields form in two ways. First, new engineering disciplines are created by *splitting off* from existing disciplines. Second, new engineering disciplines form when two or more fields *combine*.

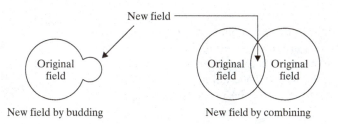

New field by budding   New field by combining

4.9.2   Creation of New Field by Budding

In some engineering fields, a subfield will mature and split off to form its own discipline like an amoeba reproducing. In fact, each of the five principal engineering disciplines formed in this way. The first split occurred when "engineering" (once used almost exclusively for warfare) divided into military engineering and "civilian" (civil) engineering. Subsequently, the other disciplines we know today budded off from civil engineering.

More recently, a number of engineering disciplines have been created by budding off of existing fields. From mechanical engineering, the field of robotic engineering has developed some independence. Photonic engineering is in the process of separating itself from electrical engineering. Photonic engineers use photons for useful purposes just as electrical engineers use electrons for useful purposes. Civil engineering has produced the developing discipline of renewal engineering. Renewal engineering is the study of the aging of engineered systems and the specification of repairs. Only time will tell if robotic, photonic, and renewal engineering will blossom into separate engineering disciplines or whether they will remain subfields of existing disciplines.

4.9.3   Creation of New Fields by Merging

Other new engineering disciplines are created as *interdisciplinary* fields. This occurs when skills from many different fields must be applied to a problem. In fact, *a great deal of the progress in science and engineering occurs where fields intersect*. Examples of emerging interdisciplinary engineering fields are bioengineering and nanoengineering. Nanoengineering is discussed in more detail in the *Focus on Emerging Disciplines: So You Want to Be a Nanoengineer?*

FOCUS ON EMERGING DISCIPLINES: SO YOU WANT TO BE A NANOENGINEER?

As stated in this chapter, engineering is a constantly evolving field. Advances in technology have opened up opportunities that would have been impossible to pursue even a few years ago. One of the most recent emerging fields is nanoengineering. The concept of nanoengineering was first described by Nobel laureate physicist Richard Feynmann in 1959. The prefix "nano-" (from the Greek *nanos*, dwarf) often refers to 10^{-9}. Thus, a nanometer is 10^{-9} m. How big is the nanometer scale? The hydrogen–oxygen bond length in water is about 0.1 nm. Thus, nanoengineering refers to building objects from the atomic level up.

To get an idea of the length scales possible with nanotechnology, the picture that follows shows the URL of the National Nanotechnology Initiative in letters written with a carbon nanotube tip. The letters are about 7 to 8 nm thick and 20 nm tall. The diameter of a carbon atom is about 0.15 nm, so each letter is about 50 carbon atoms thick and 130 carbon atoms tall.

Nanolithography: writing on the nanometer scale

Working on the atomic or molecular scale changes the nature of engineering. For example, a biomedical engineer may look at the transport of red blood cells (about 2.5 μm or 2,500 nm in size), while a nanoengineer might consider the mechanical manipulation of DNA (about 2.5 nm in size). In addition, nano-sized systems often exhibit unique characteristics.

To illustrate the potential of nanoengineering, three existing product classes will be discussed: nanoparticles, nanostructured polymer films, and quantum dots. Nanoparticles are particles with diameters on the nanometer scale. Such particles have a great deal of surface area per gram of material. As a result, nanoparticles make very efficient catalysts. They are now being used in oil refining and to remove oxides of nitrogen in automobile exhaust.

Nanostructured polymer films are being used to make displays for a wide array of digital consumer devices (e.g., cell phones and digital cameras). The polymers are formed into organic light-emitting diodes (OLEDs) to produce brighter, lighter displays.

To illustrate the potential impact of nanoengineering, consider quantum dots. Quantum dots are nano-sized semiconductors that can trap a small number of electrons. These structures may someday be used to compute on a single-electron scale (called *quantum computing*). Quantum dots in the form of nanocrystals can be used to image biological processes. They can be designed to emit certain wavelengths based on their size. (For example, a 3-nm cadmium selenide particle

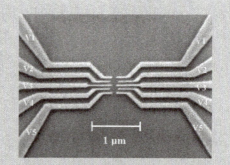

Quantum Dots. The small wires in this picture are 50 nm wide. They form two quantum dots (center of picture) that contain about 20–40 electrons. In the future, isolated electrons in the dots may be used for computers based on the behavior of single electrons. (Photo courtesy of Dave Umberger.)

emits at 520 nm—green light—while a 5.5-nm cadmium selenide particle emits at 630 nm—red light.) By using a variety of particle sizes, an array of biological processes can be imaged.

Engineers of your generation will certainly contribute to harvesting the potential of nanoengineering. Perhaps you will major in nanoengineering. (This is possible: the first undergraduate nanoengineering program was started by the University of Toronto in the fall of 2001.) Nanoengineering is an emerging, interdisciplinary engineering field with the potential to revolutionize the way we think about the world.

4.10 SUMMARY

This chapter focused on the five principal engineering disciplines: chemical, civil, electrical and computer, industrial, and mechanical engineering. For each branch, a general description of the discipline was presented, along with the major technical areas, applications, and curricula. Several major subdisciplines (materials, aeronautical/aerospace, environmental, agricultural, and biomedical engineering) were also explored.

Engineering is not static. New engineering fields evolve by splitting off of the major disciplines or by bringing together a variety of skills to address emerging challenges or to explore new opportunities.

This chapter ends with the same advice given in its beginning: to find out more about any particular engineering discipline, speak to an engineer engaged in that discipline.

SUMMARY OF
KEY IDEAS

- To learn about a specific discipline, speak with an engineer working in that discipline.
- The five principal engineering disciplines are chemical, civil, electrical and computer, industrial, and mechanical engineering.
- Chemical engineers work with the transformation of chemicals to form useful products or processes.
- Civil engineers are involved in the analysis, design, and construction of public works.
- Electrical engineers focus on the transmission and use of electrons and photons.
- Industrial engineers seek to analyze, design, model, and optimize complex systems.
- Mechanical engineers develop, design, and manufacture machines.
- The major engineering subdisciplines are materials, aeronautical/aerospace, environmental, agricultural, and biomedical engineering.
- New engineering fields can be formed by the budding off of existing disciplines or by the creation of interdisciplinary fields.

Problems

4.1. Describe your interest in two technical areas from two different engineering disciplines. Are there common elements in the two technical areas that interest you?

4.2. Discuss how *each* of the five principal engineering disciplines can contribute to the following applications:

   a. Development of a manned space station
   b. Design and production of an electric car
   c. Design of a new artificial heart valve
   d. Development of a motorized scooter aimed at the 16–25-year-old market

4.3. A company wishes to develop and sell a new ultracaffeinated beverage. Which principal engineering discipline(s) would be involved in the design of each of the following?

   a. Syrup formulation
   b. Syrup–water mixing apparatus
   c. Carbonation system
   d. Bottling line
   e. Bottling plant

4.4. Using the Internet, find two jobs performed by chemical engineers. Explain how the jobs use the technical areas listed in Section 4.3.

4.5. Using the Internet, find two jobs performed by civil engineers. Explain how the jobs use the technical areas listed in Section 4.4.

4.6. Using the Internet, find two jobs performed by electrical and computer engineers. Explain how the jobs use the technical areas listed in Section 4.5.

4.7. Using the Internet, find two jobs performed by industrial engineers. Explain how the jobs use the technical areas listed in Section 4.6.

4.8. Using the Internet, find two jobs performed by mechanical engineers. Explain how the jobs use the technical areas listed in Section 4.7.

4.9. Explain how new engineering disciplines form. Interview a professor to find how new fields are formed in his or her field of research. Summarize your findings in a short paragraph.

4.10. Summarize the state of the field of robotic, photonic, or renewal engineering.

PART II
Engineering Problem Solving

Believe nothing merely because you have been told it. Do not believe what your teacher tells you merely out of respect for the teacher. But whatsoever, after due examination and analysis, you find to be kind, conducive to the good, the benefit, the welfare of all beings—that doctrine believe and cling to, and take it as your guide.
Buddha (attributed)

Always design a thing by considering it in its next larger context—a chair in a room, a room in a house, a house in an environment, an environment in a city plan.
Eliel Saarinen

5

Introduction to Engineering Problem Solving and the Scientific Method

5.1 INTRODUCTION

5.1.1 Engineering Problems

Engineers are problem solvers. So how do engineers solve problems? In some ways, you know the answer to that question, since you have been solving problems all your life. In recent years, it is likely that you toiled long hours to solve problems given by your teachers. You probably approached those problems with several assumptions. First, you likely assumed that the problem *could be solved* and probably had *one correct answer*. Second, you may have assumed that you were given all the information you needed to solve the problem. Third, you probably started working on the problem in a way similar to that used by your teacher in class.

Solving engineering problems is a bit different: at least one of the three assumptions usually is not true. In some cases, engineers do not know if a suitable solution to an engineering problem exists. In many other cases, multiple solutions are available. If multiple solutions are possible, then your job as an engineer becomes to develop alternatives and recommend one answer.

Engineers almost never encounter problems for which they have all the information needed for a solution. In fact, a major element in engineering problem solving is collecting the required information.

Finally, engineers may face problems that no one has attempted to answer before. What do you do if faced with a unique problem? You can make good progress by becoming familiar with solution strategies for related problems. However, engineering problems often are unique enough that you cannot use a canned approach to solve them.

OBJECTIVES

After reading this chapter, you will be able to:

- compare the three main approaches to engineering problem solving;
- list the steps in the scientific method;
- solve engineering problems using the scientific method.

Key idea: Engineering problems may have one, many, or no solutions and often require data gathering to obtain the information needed to solve the problem.

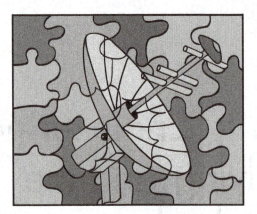

Does a unique solution exist?

Do you have all the information you need?

5.1.2 The Art and Science of Engineering Problem Solving

Key idea: Engineering problems are solved by a combination of science (knowledge of the principles of mathematics, chemistry, physics, mechanics, and other technical subjects) and art (creativity, judgment, experience, and common sense).

Good problem-solving skills are essential characteristics for a successful engineer. Early in your engineering education, you must develop the ability to solve and present the solutions of both simple and complex problems in an orderly, logical, and systematic way. Solving engineering problems involves a combination of *science* and *art*. *Science* refers to a knowledge of the principles of mathematics, chemistry, physics, mechanics, and other technical subjects you will learn during the first two or three years of your engineering studies. These principles become some of the tools of your profession.

However, science is not enough. Engineering problem solving is also an art. *Art* means the creativity, judgment, experience, and common sense to use the scientific tools to solve real-life problems effectively. You must learn how to translate a real-world problem to a form that can be solved using scientific tools. Successful problem solving also requires good judgment; you must know whether the result is a reasonable solution to the original problem. The art of problem solving depends on experience and common sense. However, the art is more effective if approached in a logical and organized manner.

Key idea: Engineering problems should be approached in a logical and organized manner.

5.1.3 Engineering Solution Methods

Key idea: Engineers use the scientific method, engineering analysis method, and engineering design method to solve problems.

No problem-solving strategy fits all engineering problems, so engineers use several types of solution methods. To meet the wide variety of challenges in engineering, three common problem-solving techniques are used: the scientific method, the engineering analysis method, and the engineering design method. While the procedures for solving different types of problems vary, all three problem-solving techniques are step-by-step processes. In this chapter, differences between the scientific, analysis, and design methods are explored. The scientific method is then discussed in more detail.

5.2 APPROACHES TO ENGINEERING PROBLEM SOLVING

5.2.1 Introduction

To illustrate the different methods of engineering problem solving, consider the following example. Suppose you are a chemical engineer at a large specialty chemical company. The management staff wishes to increase the production of an enzyme used in the synthesis of an anti-cancer agent. You are charged with answering the following three questions:

- Will an increase in operating pressure increase enzyme production significantly?
- What are the optimal operating conditions for the enzyme production line?
- Are any alternative production processes superior to the existing process?

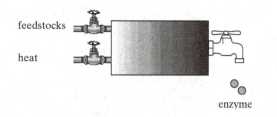

feedstocks

heat

enzyme

Would you use the same problem-solving approach to answer each of the three questions?

It is likely that a practicing engineer would use several *different* approaches to solve these three questions. Each question will be discussed in more detail later in this section.

5.2.2 Scientific Method

Key idea: The scientific method involves generating and testing hypotheses.

The first question (whether an increase in operating pressure would increase production) can be translated into a hypothesis. A hypothesis (discussed further in Section 5.3) is a trial statement about the behavior of a system. Hypotheses can be tested. The formulation, testing, and acceptance or rejection of hypotheses are at the heart of the scientific method.

How could you test the hypothesis that an increase in operating pressure will increase enzyme production significantly? There are many approaches. The simplest test would be to increase the operating pressure and observe the production rate. But this approach might be difficult in practice: the equipment may not tolerate the higher pressure or, more likely, the company may not be able to accept the downtime required to perform the tests.

numerical experiments: experiments performed with mathematical models rather than laboratory equipment

Alternatively, you might create a *mathematical model* from the known chemistry of the system. By including the effects of pressure on the chemical reactions, you could test the hypothesis with computer-based experiments (sometimes called ***numerical experiments***).

Finally, you might build a scale model (called a *physical model*) of the critical chemical reactor and test it separately to elucidate the effects of operating pressure. The use of models to test hypotheses is discussed in another chapter.

5.2.3 Engineering Analysis Method

What about the question regarding the optimal operating conditions of the enzyme production line? This question is much more complicated, since now you are being asked to examine *all* aspects of the production line (e.g., pressure, temperature, chemical feed rates, energy consumption, and heat transfer efficiency). The question requires the analysis method. With this method, the problem is defined, data gathered, analysis tools selected, and a solution calculated.

In the example at hand, you might use a combination of mathematical models and physical models to probe the effects of operating conditions on production efficiency. Before starting the analysis, you must develop a quantitative indicator of success. Are you trying to produce as much enzyme as possible, regardless of cost? Are you trying to produce the enzyme at the lowest cost per kilogram? Is the demand for the enzyme constant or do you favor a production line that runs intermittently (say, two shifts per day)? These questions need to be translated into mathematical statements called *objective functions* (mathematical statements of the state of the project). You may be able to develop mathematical models that allow you to calculate the pressure, temperature, and other operating conditions satisfying the objective functions.

5.2.4 Engineering Design Method

Key idea: Design
problems require the
generation of alternatives.

A different approach would be needed to develop and evaluate alternative production lines (as in the third question in Section 5.2.1). This is a design problem, requiring the engineering design method to develop a recommendation. In the engineering design method, you define the problem carefully, collect data, generate alternatives, analyze and select a solution, implement the solution, and evaluate the solution.

A key to the solution of design problems is defining the problem. Are you free to evaluate *any* alternative production lines? Are you limited by the availability of certain chemicals, certain equipment, or the reaction conditions?

After collecting the pertinent data, you must generate ideas for alternative production lines. How do engineers create new ideas? Perhaps you would select people with the required expertise and hold a brainstorming session.

After generating alternative production lines, you might repeat your analysis from Section 5.2.3 with each alternative. This would allow you to select the most *feasible* solution.

Even if the recommended solution is deemed to be acceptable, your job is not over. You must now *implement* the solution; in other words, you must build and operate the new production line. One of the most enjoyable aspects of engineering is watching your ideas take physical form. After the new line is built, you need to evaluate its performance to ensure that the anticipated benefits are being realized.

5.2.5 Need for Innovation

On some occasions, the traditional analysis and design methods fail to take into account the realities of the marketplace. Innovative design approaches are needed to ensure that product manufacturing, marketing, and disposal are considered in the design process. In addition, engineers sometimes get stuck using analysis and design approaches that are no longer valid for the situation at hand. In these events, innovative methods are required.

5.3 INTRODUCTION TO THE SCIENTIFIC METHOD

5.3.1 Introduction

Key idea: The scientific method is used to evaluate explanations of observed phenomena.

Of the methods of inquiry used by engineers, the scientific method is probably the most familiar to you. This method is used to evaluate explanations or trial statements of observed phenomena. The explanations are called *hypotheses*.

Key idea: Solve problems by using the scientific method: define the problem, formulate a hypothesis, and reject or conditionally accept the hypothesis.

Engineers employ the scientific method in several settings. Research engineers use it to extend our knowledge of engineered systems. For example, they may use the scientific method to test the idea that a new semiconductor will outperform existing materials. Practicing engineers use the scientific method to confirm the underlying causes of observed behavior. For example, a chemical engineer may operate a lab-scale distillation facility at several temperatures to test the idea that certain operating temperatures will improve separation of the marketable product from unwanted reaction byproducts.

5.3.2 Scientific Problem-Solving Process

The following four-step process can be used to apply the scientific method:

1. Define the problem
2. Formulate a hypothesis

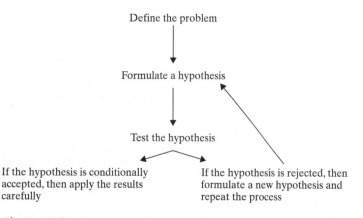

Figure 5.1.   Steps in the Scientific Method.

3. Test the hypothesis
4. Reject or conditionally accept the hypothesis

If the hypothesis is rejected, then a new hypothesis might be developed. Using the scientific method in this iterative mode is shown in Figure 5.1. Each of the steps in the scientific problem-solving method will be described in more detail.

5.4   PROBLEM DEFINITION

5.4.1   Introduction

Problem definition is the first step in all problem-solving methods. Defining the problem, often in the form of a question, is a key step in the scientific method. However, there is a tendency to think of the problem definition phase as trivial and unimportant. But if the problem is not well defined, then you are doomed to expending considerable effort to eliminate extraneous factors and focus on the root problem. If the problem is not well defined, then you may spend significant time and effort solving the wrong problem!

Key idea: Problem definitions must be specific.

How can you decide if your problem definition is useful? Problem definitions must be *specific*. A specific problem definition is one that includes all relevant solutions, but excludes irrelevant solutions.

5.4.2   Inclusive and Exclusive Definitions

As stated in Section 5.4.1, problem definitions should be *inclusive* (to include all relevant solutions or hypotheses) and *exclusive* (to exclude irrelevant solutions or hypotheses). Suppose you are conducting a study of severe accidents caused by double-wide tractor trailers.

PONDER THIS

Evaluate the following problem definition: "Do trucks cause most of the severe highway accidents?"

This problem definition is not inclusive enough, since *all* severe traffic accidents (not just highway accidents) are of interest. It is also not exclusive enough, since the study is focused only on double-wide tractor trailers (and not *all* trucks). A better problem definition would be: "Do double-wide tractor trailers cause most of the severe traffic accidents?"

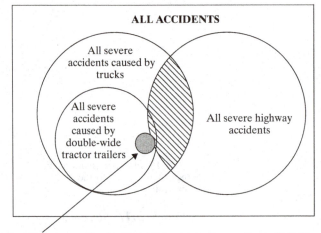

All accidents caused by double-wide tractor trailers on Route 34 in Thayer
County between January 1, 2002 and December 31, 2004

Figure 5.2. Problem Definitions for the Accident Example.

Key idea: Problem definitions should include the constraints of the problem.

To be even more specific, the problem definition should include the constraints of the problem. For example, it is likely that the truck problem has both temporal (time) and spatial (geographic) constraints. Thus, an even better problem definition would be "Did double-wide tractor trailers cause most of the severe traffic accidents on Route 34 in Thayer County between January 1, 2002, and December 31, 2004?"

The increasing specificity of the problem definition is illustrated in the Venn diagram in Figure 5.2. The problem definition in the *Ponder This* ("Do trucks cause most of the severe highway accidents?") is shown by the shaded area. The gray circle represents the most constrained problem definition.

5.4.3 Disadvantages of Definitions that Are Not Specific

If the problem definition is too inclusive, then the answer might be incorrect. As an example, suppose you wish to improve the efficiency of combustion in the internal combustion engine of an automobile. If you define the problem too broadly, then you may fail to take into account the differences in engine type. For example, conventional engines ignite the air–fuel mixture with a spark plug, while diesel engines ignite fuel with hot, compressed air. Also, the "cylinders" differ in shape (rectangular cross section for conventional, domed for the HEMI design, and oval for rotary engines). As this example shows, the answer you get may depend on the problem definition.

Key idea: Problem definitions should not limit the solutions.

Excessively exclusive problem definitions also can be misleading. One common way in which design problems are defined inappropriately is when the *needs* to be satisfied are confused with possible *solutions*. Needs should be broadly defined. When the needs are expressed in terms of solutions, the problem definition may inappropriately constrain the solution.

For example, a common problem in hotel management is the limitation on the transport of guests to the upper floors. In the past, this problem was defined as the need to add an additional elevator shaft and elevator.

PONDER THIS

How could the hotel problem be redefined to open up additional solutions?

When the problem was redefined as a requirement for additional transport capacity, an elegant solution—the outside elevator—became possible and has become quite popular. Previously, the *need* (transport capacity) was confused with a *solution* (elevator shaft and elevator), thus confining the solution to indoor elevators.

Key idea: Problem definitions should quantify the objectives.

Problem definitions should also quantify the objectives. In the transport capacity example, it is not sufficient to seek to "increase the transport capacity for guests." It is unlikely that the project would be successful if it allows the transport of only one additional guest per week to the penthouse suite. A more complete problem definition would be: "Transport at least 200 additional guests per day from the lobby to the top five floors with a project that must be completed in 18 months with a budget of $1 million." Another example of a problem definition is shown in Example 5.1.

EXAMPLE 5.1: PROBLEM DEFINITION

You have been charged with designing a new type of family automobile. The new car should combine the passenger capacity of a minivan with the sportiness of a sport utility vehicle (SUV). Evaluate the following design problem definitions:

1. Design an automobile with a length halfway between a minivan and SUV.
2. Redesign a minivan with a more attractive front grille.

SOLUTION

The first definition is **too constrained**. It assumes that the length of the new car will be the average of the lengths of the minivan and SUV. The second definition confuses the problem definition with a solution. **It assumes that the front grille limits the sportiness of the minivan**.

A better problem definition is: **design a family automobile with the passenger capacity of a minivan and the sportiness of an SUV**.

5.5 FORMULATE A HYPOTHESIS

5.5.1 Introduction

A hypothesis (from the Greek *hypo-* + *tithenai*, to put under or suppose) is an educated guess of the reason for observed phenomena. While problem definitions are usually questions, hypotheses are usually statements. Thus, a problem statement might be: "Why did the Space Shuttle *Discovery* disintegrate upon reentry on February 1, 2003?" A hypothesis concerning this topic would be a statement of an educated guess: "The Space Shuttle *Discovery* disintegrated upon reentry because falling foam hit the leading edge of the wing and damaged the heat-resistant tile."

Hypotheses generally come from the science behind the engineering. In other words, science tells you how the system should behave if you have identified all pertinent features of the system.

5.5.2 Hypotheses as Testable Statements

The most important characteristic of a hypothesis is that it is testable. For example, the hypothesis "Removal of control rods will result in the core meltdown of a nuclear power facility" is not testable under any reasonable system of ethics (except through mathematical models). The hypothesis "Cam shaft misalignment causes vibrations at vehicle speeds above 50 miles per hour" is testable. If you cannot test it, then your statement is not a hypothesis.

Key idea: Hypotheses must be capable of being tested.

5.6   TEST THE HYPOTHESIS

5.6.1   Testing a Hypothesis by Experiment

Key idea: Hypotheses can be tested by experiment or analysis.

An important step in the scientific method is testing the hypothesis. This step can be achieved in several ways. Most commonly, hypotheses are tested by conducting experiments. In this context, an *experiment* is the probing of an engineered system. For example, to test whether a new inventory system improves throughput, you might isolate a production line and try out the new system. Experiments often are performed with smaller scale models of a system or with mathematical models.

5.6.2   Hypothesis Testing by Analysis

On other occasions, a hypothesis might be tested by using the analysis techniques of the engineering analysis method. Suppose you wish to design a disinfection device for the treatment of water after a natural disaster (such as a tsunami or an earthquake). Water flows continuously through the device and is well mixed, with bacteria killed by a disinfectant inside. You hypothesize that the fraction of surviving bacteria decreases proportionally with the residence time in the device. (The residence time is the average time spent by the bacteria in the device.) In other words, if you double the residence time, then the fraction of surviving bacteria decreases by one-half.

You could, of course, test this hypothesis experimentally. You might also test it by analysis. Performing a balance on the number of bacteria and assuming that no bacteria stay in the water purifier, you know that

rate of bacteria going into the device = rate of surviving bacteria exiting the device

+ rate of kill of bacteria   (5.1)

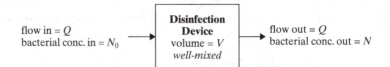

The rate of bacteria going into the device (in number per second) is equal to the flow Q (in L/s) multiplied by the number concentration of bacteria going into the device, N_0 (in number/L). Similarly, the rate of bacteria exiting the device is equal to QN, where N is the number concentration of surviving bacteria exiting the device. The rate of kill commonly is modeled as kNV, where k is a disinfection rate constant and V is the volume of the device. Thus, the number balance in Eq. (5.1) becomes

$$QN_0 = QN - kNV$$

Rearranging terms, we find that the fraction of surviving bacteria ($=N/N_0$) is given by

$$\frac{N}{N_0} = \frac{1}{1 + k\theta}$$

where $\theta = V/Q$ = residence time in the device.

This analysis tells you that the hypothesis is not completely correct: if you double the residence time, the fraction of surviving organisms will decrease by a factor of $(1 + k\theta)/(1 + 2k\theta)$, not by a factor of one-half. For example, if k was 0.1 per minute, then increasing the residence time from two to four minutes would decrease the fraction surviving from $1/[1 + (0.1 \text{ min}^{-1})(2 \text{ min})] = 0.83$ to $1/[1 + (0.1 \text{ min}^{-1})(4 \text{ min})] = 0.71$.° Thus, analysis can be used to test hypotheses.

° Note that the fraction of surviving organisms will decrease by a factor approaching $\frac{1}{2}$, as hypothesized, only when $k\theta \gg 1$.

5.7   DRAWING CONCLUSIONS FROM HYPOTHESIS TESTING

5.7.1   Rejecting a Hypothesis

Based on the results of the tests, a hypothesis may be rejected or conditionally accepted. If the data collected through experiments do not support the hypothesis, then the hypothesis is rejected. For example, suppose you are trying to improve engine performance by increasing the engine cooling capacity. You hypothesize that a new antifreeze formulation will increase the engine cooling capacity. The experiments you conduct in a test facility reveal that the new antifreeze *decreases* fuel mileage by 10–25% over a large number of test conditions. You may reasonably reject the hypothesis that the new formulation improves fuel efficiency.

5.7.2   Conditionally Accepting a Hypothesis

If tests conducted under two conditions show that the new antifreeze *increases* fuel efficiency, you would not reject the hypothesis. You might conclude that the new antifreeze formulation shows promise and should be evaluated in greater detail.

Key idea: Accept hypotheses only for the conditions under which they were tested.

Be very careful about *accepting* a hypothesis. In all cases, accept a hypothesis only for the conditions under which the hypothesis was tested. Usually, we say that we "conditionally accept" a hypothesis; that is, we accept the hypothesis under the conditions tested until more data are collected.

A classic example of the conditional acceptance of a hypothesis was developed by the British philosopher Karl Popper (1902–1994). He asked his readers to consider the hypothesis: "All swans are white." How many white swans would you have to see to accept this hypothesis? 10? 100? 1,000? Your answer may depend on your patience with counting swans. No matter how many white swans you count, you can only conditionally accept the hypothesis: seeing *one* black swan disproves the hypothesis.[*]

Similarly, in your engineering career, you may find yourself conditionally accepting hypotheses until they are proven false. For example, you may conclude that temperature does not affect the gear assembly in a snowmobile. One day during testing, the temperature may drop to a certain point where the mechanical properties of the metal change precipitously and the gears seize. You now reject a hypothesis that you previously accepted conditionally (namely, you reject the hypothesis that temperature does not affect the gear assembly).

5.8   EXAMPLES OF THE USE OF THE SCIENTIFIC METHOD

Two examples will illustrate the use of the scientific method. Suppose you wish to explore whether human activities have led to global warming. The issues involved in understanding global warming are very broad, but a problem can be defined for a very focused question based on science.

Suppose you formulate the following problem definition: "Does the Earth's temperature depend on the composition of the atmosphere?" A corresponding hypothesis might be "The average global temperature of the Earth depends on the composition of the atmosphere."

---

[*]European scientists were convinced that all swans were white, until black swans were observed in Australia by Dutch explorers in 1697. Black swans live in Australia, New Zealand, and Tasmania.

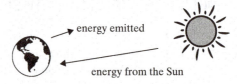

This hypothesis can be tested by analysis. The key scientific concept to test this hypothesis is the energy balance: the energy input from the Sun to the Earth must equal the energy emitted from the Earth (called *back-radiation*).

The energy input from the Sun is given by

energy input from the Sun = (cross-sectional area of the Earth)
$$\times \text{ (solar output)}(1 - \text{reflectivity of the Earth})$$

or

$$\text{energy input from the Sun} = (\pi R^2)S(1 - A) \qquad (5.2)$$

where R = Earth's radius, S = solar output (also called the solar *constant*) = 1,367 W/m^2 (watts per square meter), and A = reflectivity (also called the albedo) of the Earth = 0.31. The energy emitted from the Earth is given by

energy emitted from the Earth = (constant)(surface area of the Earth)
$$\times \text{ (fraction of back-radiation that escapes}$$
$$\text{the atmosphere of the Earth)}$$
$$\times \text{ (temperature of the Earth)}^4$$

or

$$\text{energy emitted from the Earth} = \sigma(4\pi R^2)eT^4 \qquad (5.3)$$

where σ = Stefan–Boltzmann constant = 5.5597×10^{-8} W/m^2-K^4 (watts per square meter per Kelvin to the fourth power), e = fraction of back-radiation that escapes the atmosphere of the Earth (also called the emissivity), and T = temperature of the Earth in K. (Recall that K = °C + 273.16.) Equation (5.3) is called the *Stefan–Boltzmann equation*. To perform the energy balance, set Eq. (5.2) equal to Eq. (5.3):

$$(\pi R^2)S(1 - A) = \sigma(4\pi R^2)eT^4$$

Solving for temperature yields

$$T = \left[\frac{S(1 - A)}{4e\sigma}\right]^{1/4}.$$

Substituting in the values given for S, A, and σ, and assuming that $e = 1$ (all energy escapes the atmosphere) gives $T = 255$ K or -18°C. Clearly, the average temperature of the Earth is not -18°C. (If it were this cold, then all fresh water would freeze.)

In fact, the atmosphere *retains* some of the back-radiated energy. This process is called the *greenhouse effect* and is due mainly to water vapor and carbon dioxide in the atmosphere. From the natural levels of water vapor and carbon dioxide in the atmosphere (i.e., before human activity), the value of e is about 0.615 (i.e., about 38.5% of the energy is trapped and 61.5% of the energy is transmitted through the atmosphere). Recalculating with $e = 0.615$ yields $T = 288$°K or $+15$°C.

This model predicts the actual long-term average global temperature of the Earth fairly well. The long-term average temperature is 13.9°C. The average global temperature in 2003 was 0.56°C above the long-term average. This represents about a 1%

decrease in e. This analysis shows that the average global temperature depends on e and e depends on the atmospheric composition. Thus, you would conditionally accept the hypothesis that "The average global temperature of the Earth depends on the composition of the atmosphere."

Another example of the use of the scientific method is given in Example 5.2.

EXAMPLE 5.2: APPLICATION OF THE SCIENTIFIC METHOD

Using the scientific method, analyze why some photographs exhibit "red eye" (i.e., red-colored pupils in the subjects).

SOLUTION

First, define the question. In this case, the question is obvious from the problem statement: why do some photographs exhibit red eye? Second, formulate a hypothesis. (You may wish to formulate your own hypothesis at this point.) One hypothesis states that red eye is caused by reddish color in the camera's flash. Third, test the hypothesis. The preceding hypothesis could be tested by using different filters on the flash and observing whether red eye is related to the wavelengths of light reaching the subject.

Note: Experiments would show that the stated hypothesis is false and should be rejected. An alternative hypothesis for red eye is discussed in Problem 5.2.

5.9   SUMMARY

Engineers solve problems. Engineering problems are solved by a combination of science (knowledge of the principles of mathematics, chemistry, physics, mechanics, and other technical subjects) and art (creativity, judgment, experience, and common sense). Engineering problems involving the testing of hypotheses are solved by the scientific method. In the scientific method, you define the problem, pose a hypothesis, test the hypothesis, and reject or conditionally accept the hypothesis.

An important step in all of the problem-solving methods is defining the problem. Problem definitions must be specific, should include the constraints of the problem, and should quantify the objectives. In addition, problem definitions should not limit the solutions. In other words, the problem definition should not overly constrain possible solutions.

Once the problem is defined, hypotheses can be generated. Hypotheses must be capable of being tested, usually by experiment or analysis. In interpreting the results of hypothesis testing, be sure to accept hypotheses only for the conditions under which they were tested.

SUMMARY OF KEY IDEAS

- Engineering problems may have one, many, or no solutions and often require data gathering to obtain the information needed to solve the problem.
- Engineering problems are solved by a combination of science (knowledge of the principles of mathematics, chemistry, physics, mechanics, and other technical subjects) and art (creativity, judgment, experience, and common sense).
- Engineering problems should be approached in a logical and organized manner.
- Engineers use the scientific method, engineering analysis method, and engineering design method to solve problems.
- The scientific method involves generating and testing hypotheses.

- Design problems require the generation of alternatives.
- The scientific method is used to evaluate explanations of observed phenomena.
- Solve problems by using the scientific method: define the problem, pose a hypothesis, test the hypothesis, and reject or conditionally accept the hypothesis.
- Problem definitions must be specific.
- Problem definitions should include the constraints of the problem.
- Problem definitions should not limit the solutions.
- Problem definitions should quantify the objectives.
- Hypotheses must be capable of being tested.
- Hypotheses can be tested by experiment or analysis.
- Accept hypotheses only for the conditions under which they were tested.

Problems

5.1. State whether you would use the scientific method, engineering analysis method, or engineering design method to solve each of the following engineering problems. Explain your reasoning in each case.

  a. Find the best cylinder arrangement to maximize the output of a racing car engine.
  b. Decide whether temperature influences the performance of an electromagnet.
  c. Determine the response of a building to an earthquake.

5.2. An alternative hypothesis for red eye (see Example 5.2) is that red eye is caused by the reflection of light from the flash off of the blood vessels at the back of the retina. If the flash is rapid, then the pupil remains dilated and the surface area for reflection is relatively large. If the angle formed by the flash, subject's eye, and camera lens (see accompanying figure) is too small, then the light is reflected back to the camera. Restate this information as a hypothesis and discuss how you would test it. Research red-eye reduction techniques on the Internet to verify your hypothesis-testing ideas.

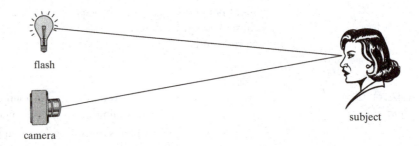

flash

camera

subject

5.3. A traffic signal was designed to turn red for cars every time the "Walk" button is pressed. The traffic engineers are considering a change in the traffic signal operation. In the new system, the traffic signal will turn red for cars every minute. Develop a problem statement and hypothesis regarding the effects of the traffic signal operation on the average wait time of a car.

5.4.   Develop a problem statement and hypothesis regarding the effects of current on the power output of a resistive heater. Assume that the resistive heater has constant resistance. Test the hypothesis using the following equations: power = current × voltage and voltage = current × resistance.

5.5.   List the characteristics of a good problem definition. Illustrate with a proper problem definition and problem definitions that are deficient.

5.6.   The television show *MythBusters* tests urban myths. Watch an episode or visit their Web site (http://dsc.discovery.com/fansites/mythbusters/mythbusters.html) and select three myths they have tested. Evaluate their problem statements, hypotheses, and testing methodologies. Did they have sufficient evidence to accept or reject their hypotheses?

5.7.   Does tapping on the side of a soda can prevent it from foaming over when you open it? Develop a problem statement, hypothesis, and testing methodology. Test your hypothesis and explain why you reject or conditionally accept the hypothesis.

5.8.   The problem described in Problem 5.7 was evaluated at the Web site http://www.snopes.com/science/sodacan.htm. Evaluate their problem statement, hypothesis, and testing methodology. Did they have sufficient evidence to accept or reject their hypothesis?

5.9.   Do toilets flush clockwise in the northern hemisphere and counterclockwise in the southern hemisphere? Develop a problem statement, hypothesis, and testing methodology. Test your hypothesis and explain why you reject or conditionally accept the hypothesis.

5.10.   Do eelskin wallets made from electric eels demagnetize credit cards? Develop a problem statement, hypothesis, and testing methodology. Test your hypothesis and explain why you reject or conditionally accept the hypothesis.

6

Engineering Analysis Method

6.1 INTRODUCTION

6.1.1 Introduction to Engineering Analysis

Analysis is the application of mathematical and scientific principles to solve a technical problem. The word analysis comes from the Greek *ana- + lyein*, literally to loosen or break up. You analyze problems by breaking them up into parts.

In an ***analysis problem***, the system has been defined and your job is to determine specific characteristics of the system. Just about every homework problem you have done in the past is an analysis problem. A typical analysis problem might read: "Given a, b, and c, determine x, y, and z."

Analysis problems have two important characteristics. First, the system and problem context are usually well defined. Second, analysis problems usually have only one solution. Think back to the homework assignments in your high school physics class. You might have been given an object's acceleration and been asked to determine its position at some time. The system is well defined (i.e., you know exactly what to calculate) and has only one solution. (In this case, the position is given by $x = \frac{1}{2}at^2$, where $a =$ acceleration and $t =$ time. For each value of t, this is only one possible value of x.) Example engineering analysis problems include determining the stresses in a truss for a proposed bridge design, calculating the thrust needed to propel a satellite into orbit, or estimating the reliability of a computer network.

You have so much experience with analysis problems that you may wonder whether *all* engineering problems are well defined with one possible answer. In fact, a large number of engineering problems (called design problems) have *many* solutions. Consider two problems: (1) Determine the force on a pulley, and (2) devise wheelchair controls to be operated by a quadriplegic person. Clearly, the pulley problem

OBJECTIVES

After reading this chapter, you will be able to:

- list the steps in the engineering analysis method;
- identify the kinds of engineering problems for which the engineering analysis method is appropriate;
- solve engineering problems using the engineering analysis method.

analysis: application of mathematical and scientific principles to solve technical problems

analysis problem: a problem where the system is well defined and system characteristics must be determined

Key idea: Analysis problems are usually well defined and have only one solution.

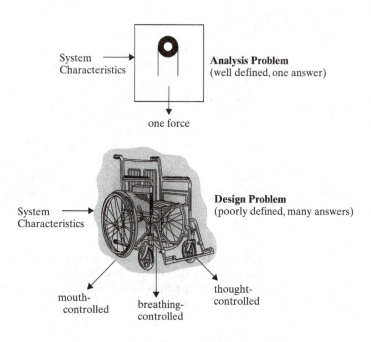

System Characteristics → **Analysis Problem** (well defined, one answer)

one force

System Characteristics → **Design Problem** (poorly defined, many answers)

mouth-controlled

breathing-controlled

thought-controlled

Figure 6.1. Illustration of Analysis and Design Problems

has one answer (one force), but many wheelchair-control designs are possible (see Figure 6.1). Design problems are addressed by the engineering design method.

6.1.2 Solving Analysis Problems

The process used to solve analysis problems is called the *engineering analysis method*. The engineering analysis method uses established mathematical and scientific principles. The analysis method is more science than art. The following six-step process should be used for solving analysis problems:

Key idea: Solve analysis problems by defining the problem, gathering data and verifying data accuracy, selecting the analysis methods, estimating the solution, solving the mathematical expressions, and checking the results.

1. Define the problem
2. Gather data and verify the accuracy of the data
3. Select the analysis method(s)
4. Estimate the solution
5. Solve the problem
6. Check the results

The last five steps in the engineering analysis method will be described in more detail next.

6.2 GATHERING DATA

6.2.1 Introduction

Before you continue in the solution process, you must collect and substantiate all data pertinent to the problem. These quantities may include physical data (e.g., dimensions, voltages, currents, temperatures, or velocities) or data from interviews. Some problems require that problem definition and data gathering steps be done separately. In other problems, defining the problem may automatically provide some or all of the data.

6.2.2 Data Collection

Engineers gather data in two ways. First, *laboratory-scale* (also called *bench-scale*) or *pilot-scale* experiments* can be performed to mimic full-scale systems. System variables can be varied and the output recorded. Experiments conducted at a scale smaller than full scale represent physical models of all or part of the system of interest.

Second, measurements can be taken in the field. Engineering measurements are as diverse as engineering itself. Measured values may range from stresses in beams to stray currents to worker's opinions of a new touch-screen layout.

Key idea: Data are gathered through experiments and field measurements.

In all cases, it is important to verify the reasonableness of the data. For measured values, look to see if the values seem reasonable and remeasure if you have questions. For data taken from other sources, look closely at the sources of data. For example, it is not always clear whether data available on the Internet come from a disinterested party or from a company wishing to sell a particular product or service. (Data sources that have not been verified independently sometimes are called *gray literature*.) Similarly, when interviewing people about a problem, be sure to note whether their statements are backed up by data or whether they have made assumptions about the on-site conditions. It is useful to ask three questions: What do you know? What do you *not* know? and What do you *think* you know?

Key idea: Test all data for reasonableness.

6.3 SELECTING THE ANALYSIS METHOD

6.3.1 Introduction

Key idea: To select an analysis method, first select the physical laws and then translate the physical laws into mathematical equations.

The selection of an analysis method is a two-step process. First, you must select the fundamental laws or principles that apply to the system. The laws or principles typically come from mathematics and science. For example, Newton's second law of motion (force = mass $\times$ acceleration or $F = ma$) can be combined with a force balance to solve many problems in statics, where system components are not moving. Equations generated from the fundamental laws are used to solve the problem.

Key idea: Engineers need quantifiable relationships between variables.

The second step is to translate the physical laws into mathematical statements (also called *mathematical models*). Engineering is a quantitative field. As an engineer, you will frequently develop mathematical models of the physical world. You may need to know how voltage depends on current, or the relationship between heat capacity and temperature, or the effects of a delay in a concrete delivery on a construction schedule. *Simply knowing that one variable depends on another is not enough.* To analyze systems that interact with the physical world, it is necessary to have *quantifiable* relationships between variables. As an engineer, you will develop and use mathematical models for relationships between variables.

6.3.2 Selection of Physical Laws

physical law: a description of nature assumed to be true

As stated in Section 6.3.1, you generally begin with a **physical law**. In engineering and science, laws are propositions about nature that are assumed to be true.

PONDER THIS

How do you *know* the laws are true?

In fact, you do not know if the laws are true! Laws are unprovable because they relate the values of real, physical properties. Due to experimental imprecision, no matter

---

*For etymology fans, the word "experiment" comes from the Latin *experiri* (to try). Small-scale experiments (smaller than full scale, but larger than laboratory scale) sometimes are called *pilot-scale* experiments. Here, the word "pilot" is used in the sense "to guide." Pilot-scale experiments guide engineers in the design and operation of full-scale engineered systems just as a pilot guides a ship or airplane.

how carefully you measure the physical properties, you cannot confirm that the values are *exactly* as stated in the law. It may seem strange that so much engineering work is based on some unproven assumptions, but these assumptions have withstood the test of time in engineered systems.

The vast majority of engineering calculations start with one of a small handful of physical laws. Three kinds of physical laws are important in engineering: laws of conservation, laws of motion, and constitutive laws. You will learn about these laws over the next few semesters. The laws will be introduced briefly here to demonstrate how physical laws are used in the engineering analysis method.

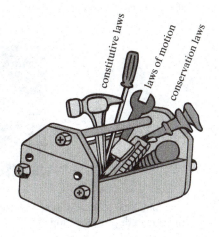

Conservation laws, laws of motion, and constitutive laws will be some of the most important tools in your engineering toolbox.

laws of conservation:
the statements that mass, momentum, energy, and charge (among other properties) are unchanged for different states of a system

The first class of important laws consists of the ***laws of conservation***. It often is assumed that several key physical parameters remain unchanged in engineered systems. Specifically, engineers assume that mass, momentum (mass × velocity), angular momentum, energy (the entity that allows the system to do work), and charge are conserved. The statement of the conservation of energy is sometimes called the *First Law of Thermodynamics*. Conserved properties will *balance* (i.e., be unchanged for different states of the engineered system).

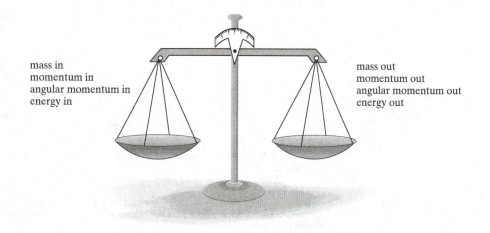

mass in
momentum in
angular momentum in
energy in

mass out
momentum out
angular momentum out
energy out

The conservation laws are behind many engineering calculations.[*] A chemical engineer might use an energy balance to design a refrigeration system. A mechanical engineer uses a momentum balance to calculate ink flow in an ink-jet printer. An industrial engineer's complex parts routing system ultimately must be based on a mass balance of the parts stored.

The conservation laws are summarized in Table 6.1. As noted in Table 6.1, some other statements called "laws" stem from the conservation laws.

Other common laws serving as starting places for engineering calculations are the ***laws of motion***. Isaac Newton (1643–1727) proposed three laws of motion. The First Law of Motion (also called the *law of inertia*) states that objects at rest will remain at rest unless acted upon by an unbalanced force. (It also states that objects in motion will remain in a straight-line motion at constant speed unless acted upon by an unbalanced force.)

laws of motion:
statements about the motion of objects in response to applied forces

Newton's First Law of Motion (velocity is constant if the sum of the "forward" forces is equal to the sum of the "reverse" forces)

TABLE 6.1 Common Laws of Conservation

Property Conserved	Example of the Conservation Law	Other Laws Based on Conservation of the Property
Mass (m)	Determining the chemical doses required for an industrial process	*Continuity equation*: for a flow of constant density, the sum of the flows in and out of a node is zero
Momentum (mv)	Determining the initial velocity of a baseball after being hit by a bat	
Angular momentum (mvr)	Determining the rotational velocity of a fan	*Kepler's Second Law*: a line between the Sun and a planet sweeps out equal areas in equal time intervals
Energy (E) (First Law of Thermodynamics)	Determining the amount of lift generated by an airplane wing	*Bernoulli Equation*: the relationship between pressure, velocity, and elevation of a fluid
		Kirchhoff's Voltage Law: the net voltage drop around any closed path must be zero
Charge (z)	Determining the pH of acid rain	*Kirchhoff's Law*: the sum of the currents in and out of a node is zero

v = velocity, r = radial distance

Namesakes of the laws and equations: Daniel Bernoulli (1700–1782), Johannes Kepler (1571–1630), and Gustav Robert Kirchhoff (1824–1887).

---

[*]You may know that Albert Einstein (1879–1955) showed that mass (m) and energy (E) are interchangeable according to $E = mc^2$, where c = speed of light. For a vast majority of engineered systems, it is still useful to write separate conservation statements for mass and energy.

Newton's second law of motion states that force (F) equals mass (m) multiplied by acceleration (a), or $F = ma$. Finally, the Third Law of Motion (also called the *law of action and reaction*) states that for every force, there is an equal and opposite force (in other words, the forces that two objects exert on each other are equal but in opposite directions).

$F = ma$

Newton's Third Law of Motion

Newton's Second Law of Motion

An electrical engineer would use Newton's first law of motion to describe the motion of electrons in a magnetic field. An aerospace engineer starts with Newton's second law of motion when designing a novel propulsion system. A civil engineer uses Newton's third law of motion to understand that a concrete floor pushes against the force exerted by a person standing on it. The laws of motion are summarized in Table 6.2.

A third class of laws consists of the relationships between the measurable properties of a system and is called **constitutive laws**. The constitutive laws are empirical; that is, they are based on observations rather than theory. Three important constitutive laws are

constitutive laws:

relationships (usually empirical) between the measurable properties of a system

Hooke's law (after Robert Hooke, 1635–1703): the force exerted by a spring (F) is proportional to the displacement of the spring (x), or $F = kx$, where k is the spring stiffness

Ohm's law (after Georg Simon Ohm, 1789–1854): the voltage (V) equals current (I) × resistance (R), or $V = IR$

Ideal gas law:* pressure × volume = number of moles × ideal gas constant × temperature, or $PV = nRT$

TABLE 6.2 Newton's Laws of Motion

Law of Motion	Statement of Law
First Law (Law of inertia)	An object at rest will remain at rest and an object in motion will remain in motion at constant speed, unless either object is acted upon by an unbalanced force
Second Law	Force (F) equals mass (m) × acceleration (a): $F = ma$
Third Law (Law of action and reaction)	For every force, there is an equal and opposite force

*The ideal gas law is consistent with Boyle's Law (pressure is proportional to 1/volume at constant temperature and amount of material, after Robert Boyle, 1627–1691) and Charles's Law (volume is proportional to temperature at constant pressure and amount of material, after Jacques Charles, 1746–1823).

A civil engineer might use a version of Hooke's Law to relate the stress in a beam (stress = force/area) to the strain (displacement/length). An electrical engineer might use Ohm's Law to determine voltage drops in electrical transmission systems. A mechanical engineer would start with the ideal gas law to determine the required thickness of a new material for storing liquid oxygen.

How do you decide which physical laws to select? In general, *seek a relationship between the components of the system that you know and the system components you are trying to calculate*. If you want to know the velocity of a crash dummy's head after a collision with a wall, you might pick the conservation of momentum statement. It will contain elements that you know (masses and initial velocities) and elements that you wanted to determine (namely, the velocity of the dummy's head). Other examples of how to select physical laws are given in Example 6.1.

EXAMPLE 6.1 SELECTING PHYSICAL LAWS

Select the physical law from Tables 6.1 and 6.2 to model the following phenomena: the pressure exerted by a waterbed mattress on the floor, the escape velocity of a rocket, and the operation of an ionization smoke detector. (You may need to use the Internet to learn how a smoke detector works.)

SOLUTION

Pressure is force divided by area. Thus, a pertinent analysis method for the waterbed problem is **Newton's Second Law of Motion**. Newton's Second Law can be used to relate what you want (force/area) with what you know (system properties of mass, acceleration, and waterbed area).

The escape velocity of a rocket is described by **conservation of energy**. The kinetic energy of the rocket is balanced by the gravitational energy pulling it back to Earth.

The ionization smoke detector works by **conservation of charge**. The detector contains about 200 μg of the radioactive element americium 241. The alpha particles released by the americium form ions from the components of air, which are trapped on charged plates to produce a background current. (The smoke detector battery charges the plates.) The current (charges per unit time) is reduced when ions adsorb to smoke particles.

Note: The waterbed and rocket problems are evaluated further in Example 6.2.

6.3.3 Translation into Mathematical Expressions

Physical laws do not help in a calculation unless you can translate the physical law into a mathematical expression involving the parameters you wish to determine. The second step in performing an engineering calculation is the *translation of the physical law into a mathematical statement*. For example, suppose that you are analyzing a household electric circuit, where the electric supply is split into two subcircuits at a node. If you know the supply current and the current of one subcircuit, how do you find the current in the other subcircuit? You need an expression that relates the currents in a circuit. Kirchhoff's Current Law will do (see Table 6.1). It states that the sum of the currents flowing into a circuit node must be equal to the sum of the currents leaving the node. In this case:

supply current = current in subcircuit #1 + current in subcircuit #2

You now have a mathematical expression involving the known and the unknown quantities. You can solve for the parameter of interest. An example of translating physical laws into mathematical expressions is given in Example 6.2.

EXAMPLE 6.2 TRANSLATING PHYSICAL LAWS INTO MATHEMATICAL EXPRESSIONS

From Example 6.1, calculate the escape velocity of a rocket. Assume that the rocket stops moving when it is infinitely far from Earth. Also, from Example 6.1, determine whether your waterbed mattress violates the building code loading of 100 pounds of force per square foot (4,780 Pa, where 1 Pa = 1 N/m^2 and 1 N = 1 newton = 1 kg-m/s^2; so the building code is 4,780 N/m^2 = 4780 kg/m-s^2). The dimensions of the mattress are 5.0 ft × 7.0 ft × 0.75 ft (1.5 m × 2.1 m × 0.23 m).

SOLUTION

Rocket problem:

At take-off, the rocket's net energy is the kinetic energy of the rocket minus the gravitational energy pulling it back to Earth. When the rocket stops infinitely far from the Earth, its total energy is zero. An energy balance reveals several factors:

$$\text{energy at take-off} = \text{energy infinitely far away from Earth}$$

$$\text{kinetic energy} - \text{gravitational energy} = 0$$

So

$$\text{kinetic energy} = \text{gravitational energy}$$

The kinetic energy at take-off is $\frac{1}{2}mv^2$, where m = rocket mass and v = escape velocity. The gravitational energy is mGM/R, where G = universal gravitational constant = 6.672×10^{-11} N-m^2/kg^2 = 6.672×10^{-11} m^3/kg-s^2, M = Earth's mass = 5.98×10^{24} kg, and R = Earth's radius = 6.37×10^6 m. The energy balance becomes

$$\tfrac{1}{2}mv^2 = mGM/R \quad \text{or} \quad v = (2GM/R)^{1/2}$$

Substituting in the values, we find that

$$v = \text{escape velocity} = \sqrt{\frac{GM}{R}}$$

$$= \sqrt{\frac{2\left(6.67 \times 10^{-11}\,\dfrac{\text{m}^3}{\text{kg} - \text{s}^2}\right)(6.98 \times 10^{24}\,\text{kg})}{6.37 \times 10^6\,\text{m}}}$$

$$= \textbf{11,200 m/s} \ (\text{about 24,000 mph})$$

Waterbed problem:

For the waterbed, the force from Newton's Second Law of Motion is $F = ma = mg$ (g = acceleration due to the Earth's gravity = 9.8 m/s^2). The pressure P is $F/A = mg/A$, where A is the area of the mattress in contact with the floor. The mass of the waterbed is mainly the mass of the water: mass = (density)(volume) = ρV. The density of water is about 1,000 kg/m^3. Therefore,

$$P = mg/A = \rho Vg/A = (1{,}000\ \text{kg/m}^3)(1.5 \times 2.1 \times 0.23\ \text{m}^3)(9.8\ \text{m/s}^2)/(1.5 \times 2.1\ \text{m}^2)$$

$$= 2{,}300\ \text{kg/m-s}^2$$

The pressure exerted by the waterbed is 2,300 kg/m-s^2 (about 47 pounds of force per square foot). Therefore, **the mattress does not violate the code of 4,780 kg/m-s^2** (100 pounds of force per square foot).

6.4 ESTIMATING THE SOLUTION

6.4.1 Introduction

An important, but often ignored, step in solving an analysis problem is developing a "ballpark" estimate of the solution. In many cases, mistakes in analysis can be caught by making an order-of-magnitude guess of the solution. Estimation serves as a check on the calculations and also helps in the development of engineering intuition. Of course, the result of the calculation may be surprising; in fact, it may be significantly different from your estimate. Still, estimation is a good "reality check" on engineering calculations.

Even in this age of speedy personal computers, some engineers still lament the passing of the **slide rule**. (Do you know what a slide rule is? Search on the Internet to learn more.) Slide rules are fairly primitive calculators, but they have one interesting feature: they do not indicate the decimal place of the answer. If you multiply 71.2 by 46.3 on a slide rule, you get the answer: 330. Is this 3.30? 33.0? 330? 3,300? With a slide rule, you have to estimate the location of the decimal place.

slide rule: a ruler scaled logarithmically with a movable centerpiece typically used for performing multiplication and division (phrase was coined in 1663)

PONDER THIS

Take a moment to estimate the product of 71.2 × 46.3. Is it closest to 3.30, 33.0, 330, or 3,300?

You might guess that 71.2 × 46.3 ≈ 70 × 50 = 3,500, so the answer from the slide rule is 3,300. Slide rules help to develop intuition in engineering calculations. Even with electronic calculators and spreadsheets, it is useful to estimate the location of the decimal place as a way to check the calculation. This is called an *order-of-magnitude estimation*.

6.4.2 Example

terminal velocity: a constant velocity sometimes achieved by falling objects (hence, the acceleration is zero)

As an example of estimation, suppose you are planning to take your first skydiving lesson and you wish to know how fast you will be falling. You remember that your diving instructor talked about reaching **terminal velocity**.

PONDER THIS

Estimate your terminal velocity during skydiving. Is it 1 mile per hour (mph)? 10 mph? 100 mph? 1,000 mph?

Your intuition probably tells you the terminal velocity will be greater than 10 mph but smaller than 1,000 mph. What is your estimate? Something in the 50 to 150 mph range sounds about right.

You can calculate the terminal velocity more accurately. Recall that the terminal velocity is a constant velocity. Thus, the acceleration (change in velocity with respect to time) is zero at the terminal velocity.

PONDER THIS

If terminal velocity means that you are not accelerating, what net force are you experiencing? Why?

If your acceleration is zero, the net force on you will also be zero (since, from Newton's Second Law of Motion, force = mass × acceleration: $F = ma$). Two forces act on you: a force from gravitational acceleration ($F_g = mg$) and a drag force ($F_d = $

$\frac{1}{2} C_d A \rho v_t^2$). The drag force depends on the drag coefficient C_d (approximately 0.22 in this example); the cross-sectional area of your falling body, A (≈ 0.8 m^2 if you curl up in a ball); the density of air, ρ; and the terminal velocity v_t.

If the net force is zero, then

$$0 = F_g - F_d = mg - \frac{1}{2} C_d A \rho v_t^2$$

Solving for the terminal velocity yields

$$v_t = \sqrt{\frac{2mg}{C_D A \rho}}$$

If you weigh 60 kg, then your terminal velocity is

$$\sqrt{\frac{2(60 \text{ kg})\left(9.8\frac{\text{m}}{\text{s}^2}\right)}{(0.22)(0.8 \text{ m}^2)\left(1{,}000\frac{\text{kg}}{\text{m}^3}\right)}} = 2.6 \text{ m/s} = 5.8 \text{ mph}$$

Did the answer 5.8 mph match your estimate? No, this answer seems *much* too small. Can you find the error in the equation? The density of water (1 kg/L = 1,000 kg/m^3) was used inadvertently instead of the density of air (1.225 kg/m^3 at 15°C). Substituting the correct density, the terminal velocity turns out to be 74 m/s or 165 miles per hour. This probably fits your preconceived notion more closely.* Another estimation example is shown in Example 6.3.

**EXAMPLE 6.3
ESTIMATING
THE SOLUTION**

In the waterbed scenario from Examples 6.1 and 6.2, you can estimate the pressure from the "rule of thumb" that a 2.3-foot column of water exerts a pressure of 1 pound of force per square inch (1 psi). (This relationship is derived easily from the density of water.)

SOLUTION

The 0.75-foot "column" of water in the waterbed is about one-third of 2.3 feet; thus, it would exert about $\frac{1}{3}$ psi or about ($\frac{1}{3}$ psi) (144 square inches/square foot) $\approx$ 50 pounds of force per square foot. With this estimate, you might conclude that the calculated value of 47 pounds of force per square foot seems reasonable.

6.5   SOLVING THE PROBLEM

6.5.1   Solving Mathematical Expressions by Isolating the Unknown

Key idea: Manipulate expressions (i.e., rearrange terms) to solve for the variable of interest.

If the previous steps result in a mathematical expression, it is normally solved by application of mathematical theory. The key to solving mathematical expressions is to isolate the unknown on one side of the expression. Always be aware of which variables you seek to calculate: "Keep your eyes on the prize."

expression: any relationship between variables (if the relationship is an equality, the expression is called an **equation**)

By convention, you usually try to isolate the parameter or variable of interest on the left-hand side of the expression. The term **expression** refers to any relationship between variables. An expression can be an inequality (e.g., $y > x^2$) or an equality (e.g., $y = mx + b$). Equality expressions are called **equations**. A few simple rules will come to your aid in manipulating expressions.

---

*Does this analysis mean that your terminal velocity falling in water is 2.6 m/s? No—estimation should tell you that 2.6 m/s seems too fast. In water, a third force is important: buoyancy. Your buoyancy in water makes your settling velocity in water less than 2.6 m/s.

6.5.2 "Golden Rule" of Expression Manipulation

Key idea: The "Golden Rule" of expression manipulation is as follows: "Do to one side of the expression that which you did to the other side."

First, obey the ***"Golden Rule" of expression manipulation***: Do to one side of the expression that which you did to the other side. Suppose you wish to know how the acetone concentration in the feedstock of an industrial process changes with time. Suppose your mathematical model yields the following relationship between the feedstock concentration C and time t:

$$C = C_0 e^{-kt} \tag{6.1}$$

where C_0 is the initial feedstock concentration and k is a rate constant (C_0 and k are always positive here). Suppose you want to find the time (called the *half life*, $t_{1/2}$) at which the feedstock concentration falls to one-half its initial value; in other words, find $t = t_{1/2}$ when $C = \frac{1}{2}C_0$. You would proceed by dividing *both sides* of Eq. (6.1) by C_0. Since $C = \frac{1}{2}C_0$, it follows that

$$C/C_0 = \frac{1}{2} = e^{-kt}$$

Using the "Golden Rule" and taking the natural log *of both sides* yields

$$\ln(\tfrac{1}{2}) = \ln(e^{-kt}) = -kt \qquad [\text{recall that } \ln(e^x) = x]$$

Dividing *both sides* by k gives

$$t_{1/2} = -\ln(\tfrac{1}{2})/k \tag{6.2}$$

Thus, employing the "Golden Rule" and performing the same operations on both sides of the equation tells you that $t_{1/2} = -\ln(\frac{1}{2})/k$.

6.5.3 Manipulating Inequalities

Key idea: Remember to reverse the inequality sign when multiplying or dividing both sides by a negative number.

The second rule of manipulating equations states that inequality signs are reversed when multiplying or dividing both sides by a negative number. Suppose you are interested in reaction times greater than the half life: $t > t_{1/2}$. Is C greater than or less than $\frac{1}{2}C_0$ at $t > t_{1/2}$? Equation (6.2) becomes

$$t > t_{1/2} = -\ln(\tfrac{1}{2})/k$$

or

$$t > -\ln(\tfrac{1}{2})/k$$

Multiplying by $-k$ (a negative quantity) yields

$$-kt < \ln(\tfrac{1}{2})$$

Notice that the "greater than" sign became a "less than" sign when you multiply the inequality by the negative value $-k$. Exponentiating *both sides* gives the following: $e^{-kt} < \frac{1}{2}$. From Eq. (6.1), $e^{-kt} = C/C_0$, so $e^{-kt} < \frac{1}{2}$ means that $C/C_0 < \frac{1}{2}$. Multiplying both sides by C_0 (a positive quantity) gives

$$C < \tfrac{1}{2}C_0$$

(Note that you do not change the inequality sign, since C_0 is positive.) Thus, if $t > t_{1/2}$, then $C < \frac{1}{2}C_0$. In other words, at times greater than the half life, the feedstock concentration is less than half the initial feedstock concentration.

6.5.4 Hints for Manipulating Equations

A common problem encountered when manipulating equations is the tendency to substitute numbers for symbols too early in the process. *Always substitute numbers for symbols only in the last step of the calculation.* Inserting values into an expression is commonly called "plugging and chugging." However, it is important to "chug" (i.e., manipulate the expression) *before* you "plug" (i.e., substitute values for symbols).

Say, for example, you want to find the current (in amperes or A) required to achieve a heating power of 1,000 watts (1,000 W) in a resistive heating device (for example, a space heater) if the resistance is 10 ohms ($= 10\ \Omega$). The relationship between power (P), current (I), and resistance (R) is

$$P = I^2 R$$

(Recall that $1\ \text{A}^2 \times 1\ \Omega = 1\ \text{W}$.) Many people make the mistake of substituting in the known values first and then solving for the unknown:

$$(1{,}000\ \text{W}) = I^2 (10\ \Omega),$$

or

$$I^2 = (1{,}000\ \text{W})/(10\ \Omega) \text{ and } I = (1{,}000\ \text{W}/10\ \Omega)^{1/2} = 10\ \text{A}$$

What is the problem with this approach? If you want to calculate the current at *another* resistance (say, 20 Ω), then you must start the calculation over again.

It is much more efficient to *solve for the unknown with symbols and then substitute in numbers during the last step in the calculation*. Thus, from $P = I^2 R$, you can derive

$$I = \sqrt{\frac{P}{R}}$$

This equation can be used to calculate the current for *any* power and resistance. The bottom line: manipulate symbols to solve for the unknown, and *then* substitute in the values of the known quantities. Another example of manipulating equations is shown in Example 6.4.

EXAMPLE 6.4: MANIPULATING EXPRESSIONS

Two resistors in series (with resistances R_1 and R_2) give an overall resistance R, with $1/R$ equal to $1/R_1 + 1/R_2$. Find the overall resistance for 1-Ω and 2-Ω resistors in series and for 2-Ω and 10-Ω resistors in series.

SOLUTION

Manipulate symbols *first*, and then plug in numbers. Thus, we begin with

$$\frac{1}{R} = \frac{1}{R_1} + \frac{1}{R_2}$$

Inverting *both* sides gives

$$R = \frac{1}{\dfrac{1}{R_1} + \dfrac{1}{R_2}} = \frac{R_1 R_2}{R_1 + R_2}$$

Plugging in values then produces the following result: **The overall resistance is 0.67 Ω for 1-Ω and 2-Ω resistors in series and 1.7 Ω for 2-Ω and 10-Ω resistors in series.**

6.6 CHECKING THE RESULTS

Key idea: Answers to engineering calculations almost always have physical meaning.

Key idea: Check engineering calculations by logic, estimation, and checking units.

6.6.1 Introduction

It is tempting to stop after computing an answer and assume that the solution is correct. However, determining *a* solution and determining the *correct* solution are two very different things: obtaining a solution does not mean that the solution is correct or even realistic.

In almost all engineering calculations, the result is a physical quantity. You are computing *something* and not just calculating for the sake of calculation. This fact puts constraints on the answers you obtain as an engineer: answers represent tangible quantities in the real world. In engineering calculations, *numbers almost always have physical meaning*. As a result, there are three tools you can use to check engineering calculations: logic, estimation, and checking units.

6.6.2 Use Logic to Avoid Aphysical Answers

aphysical: not physically possible

Key idea: Use logic to eliminate aphysical answers: always ask if your answers make sense.

Since answers in engineering calculations represent physical quantities, you must employ logic to eliminate nonsensical results. Such results are sometimes call **aphysical**, because they are physically impossible answers. One of the best ways to avoid reporting errors is to ask yourself a simple question: *does the answer make sense?* Never leave an engineering calculation without thinking about whether the answer is reasonable. In many ways, this is the most important step in the engineering calculation process.

As an example of using logic to weed out aphysical answers, suppose you and your studymates ponder the possibility of bungee jumping off the top of Ye Olde Administration Building. In a moment of clarity, you decide it would be prudent to determine if you would hit the ground at the bottom of the first bounce.

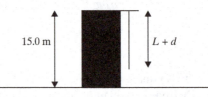

You want to know if the distance you will fall is greater than the height of the building.* At the bottom of your jump, you will have fallen a distance equal to the length of the unstretched bungee cord (L, say, 9.0 m) plus the length that the bungee cord has stretched (d). Thus, you want to know if $L + d$ is greater than the building height (say, 15.0 m). Since you know $L = 9.0$ m, you really just need to know if $d > 15.0$ m − 9.0 m = 6.0 m. So the question to be answered is simple: is d > 6.0 m?

PONDER THIS

Without performing a potentially tragic experiment on an unsuspecting classmate, how would you decide whether you would hit the ground?

potential energy: energy associated with the position in a field (for a gravitational field, potential energy = mass × gravitational acceleration × height)

Conservation of energy reveals that the loss of **potential energy** at the bottom of the jump should equal the energy stored in the bungee cord. Your loss in potential energy is $mg(L + d)$, where m = your mass (assume 70 kg), g = gravitational acceleration = 9.8 m/s^2, and $L + d$ = the distance you have fallen. The bungee cord is like a spring. Thus, the energy stored in the bungee cord is equal to $\frac{1}{2}kd^2$, where k = spring stiffness (say, 150 N/m = 150 kg/s^2). Thus,

$$\text{potential energy lost} = \text{energy stored in the bungee cord}$$
$$mg(L + d) = \frac{1}{2}kd^2$$

---

*Assume you are standing in a loop at the end of the bungee cord, so that your height does not affect the calculation.

Rearranging terms yields

$$\tfrac{1}{2}kd^2 - mgd - mgL = 0 \qquad (6.3)$$

You and your studymates realize immediately that you can solve Eq. (6.3) for d by using the **quadratic equation** (see definition). In Eq. (6.3), the symbols in the quadratic equation are

$$a = \tfrac{1}{2}k,\, b = -mg,\, \text{and } c = -mgL$$

So

$$d = \frac{mg \pm \sqrt{m^2g^2 - 2kmgL}}{k} \qquad (6.4)$$

The total distance you would fall is

$$L + d = L + \frac{mg}{k} \pm \frac{\sqrt{m^2g^2 - 2kmgL}}{k} \qquad (6.5)$$

Substituting the preceding values of m, g, L, and k into Eq. (6.5) gives

$$L + d = 9.0 \text{ m} + 4.6 \text{ m} \pm 10.2 \text{ m} = 3.4 \text{ m or } 23.7 \text{ m}$$

There are two solutions. If you fall 3.4 m, you *may* be safe. However, a fall of 23.7 m from a height of 15.0 m would be disastrous!

How do you decide which is the right answer? Remember: *The answer must have physical meaning.* From Eq. (6.4),

$$d = 4.6 \text{ m} \pm 10.2 \text{ m} = -5.6 \text{ m or } 14.7 \text{ m}$$

Clearly, d must be 14.7 m: it does not make much sense for you to "fall" 5.6 m *up* if you jump off a building! Thus, d must be 14.7 m and $L + d$ is 23.7 m. The moral of the story: *use logic to eliminate aphysical answers* (and don't bungee-jump off buildings on campus).

6.6.3 Using Logic to Check Expression Manipulation

Logic also can help you check your manipulation of mathematical expressions. One way to do this is to check whether one variable changes as expected when other variables change. In other words, you can check the predicted trends of your expressions. For example, consider a common property of matter: the ratio of the surface area of an object to its volume (called the surface-to-volume ratio or S/V). The S/V is an important quantity in engineering. It controls the functioning of objects from ball bearings to industrial catalysts to the human lung. For simple shapes, S/V is easy to calculate. For a sphere of radius r,

$$V = 4\pi r^3/3 \text{ and } S = 4\pi r^{2\circ}$$

Thus,

$$S/V = 3/r \qquad (6.6)$$

Does this equation make sense? One way to see if it makes sense is to look at the trend in S/V with the independent variable (r).

quadratic equation:
if $ax^2 + bx + c = 0$, then

$$x = \frac{b \pm \sqrt{b^2 - 4ac}}{2a}$$

Key idea: Use logic to check whether one variable changes as expected with changes in the other variables.

---

*If you have knowledge of calculus, you can show that, for a sphere, $S = dV/dr = d(4\pi r^3/3)/dr = 4\pi r^2$.

PONDER THIS

How should the surface-to-volume ratio change with the particle size?

To see how the surface-to-volume ratio changes with the particle size, compare a 3 cm by 3 cm by 3 cm cube with twenty-seven 1 cm by 1 cm by 1 cm cubes.

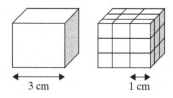

3 cm   1 cm

The large cube has the same volume as the small cubes together. However, the surface area of the large cube (1 cube × 6 faces/cube × 3 cm × 3 cm = 54 cm^2) is three times less than the surface area of the small cubes together (27 cubes × 6 faces/cube × 1 cm × 1 cm = 162 cm^2, after the small cubes are separated). The S/V of the large cube is 2 cm^{-1}, and the S/V of the small cubes is 6 cm^{-1}. Thus, you might expect the surface-to-volume ratio to *increase* with *decreasing* particle size. Equation (6.6) is consistent with this notion: as r decreases, $S/V(= 3/r)$ increases. The use of logic does not *prove* that you manipulated the equations correctly. However, it is a useful check on manipulations that will catch *some* errors.

Key idea: Use logic to check whether one variable is predicted correctly for extreme values of the other variables.

Another way to use logic to check mathematical expressions is to make sure the variable on the left side of the equation is predicted correctly for *extreme values* of the variables on the right side of the equation. As an example, consider the flight of a champagne cork. If the champagne bottle is at an angle θ with the ground and the cork has an initial velocity out of the bottle of v_0, then (ignoring air resistance) the cork will travel a distance equal to

$$\text{distance} = (2v_0^2/g)\sin(\theta)\cos(\theta) \qquad (6.7)$$

Does the dependency of the distance shot on the angle in Eq. (6.7) make sense? Try looking at extreme values of the angle. For $\theta = 0°$ (when the bottle is lying on the ground), the cork will hit the ground immediately and the distance traveled should be zero. For $\theta = 90°$ (when the bottle is vertical), the cork will go straight up and down and the horizontal distance traveled should also be zero.

Equation (6.7) predicts that the distance is equal to zero for $\theta = 0°$ and $\theta = 90°$. Thus, Eq. (6.7) matches your thoughts about the values of the distance traveled at extreme values of the angle of the bottle to the ground.* An example of using logic to check expressions is given in Example 6.5.

---

*Remember from trigonometry that $\sin(\theta)\cos(\theta) = \frac{1}{2}\sin(2\theta)$. Thus, the distance traveled is $(2v_0^2/g)\sin(\theta)\cos(\theta) = (v_0^2/g)\sin(2\theta)$. The distance is maximized when $\sin(2\theta)$ is maximized; that is, at $\theta = 45°$. To shoot a champagne cork the farthest, hold it 45° to the ground.

EXAMPLE 6.5: USING LOGIC TO CHECK EXPRESSIONS

You are helping a friend move. A box of books is sliding down the ramp of the moving van. You push against the box to hold it in place, and you remember that the force needed to stop an object from sliding down an inclined plane is

$$F = W[\sin(\alpha) - \mu\cos(\alpha)]$$

where W = weight of the object, α = angle of the inclined plane, and μ = coefficient of friction = $\tan(\varphi)$, where φ is the angle of the inclined plane at which the box starts to slide back by itself. Evaluate this formula using logic.

SOLUTION

The required force is $W[\sin(\alpha) - \tan(\varphi)\cos(\alpha)]$. Checking extremes, it is logical that the required force is zero when $\alpha = \varphi$ (the ramp is at the angle where the box starts to slide). The equation implies that the required force at $\alpha = \varphi$ is $W[\sin(\varphi) - \tan(\varphi)\cos(\varphi)] = W[\sin(\varphi) - \sin(\varphi)] = 0$, which is consistent with the logic [recall that $\tan(\varphi)\cos(\varphi) = \sin(\varphi)$].

It is also logical that the required force is W when $\alpha = 90°$ (when the ramp is straight up), since this ramp angle is equivalent to holding the box in your arms. The equation implies that the required force at $\alpha = 90°$ is $W[\sin(90°) - \tan(\varphi)\cos(90°)] = W[1 - 0] = W$, which is again consistent with the logic. **Thus, the equation checks at the extremes**.

Checking the dependency of the required force on the weight of the box, ramp angle, and coefficient of friction, it is logical that the required force should increase as the weight of the box (W) increases and as the angle increases. It is also logical that the required force should decrease as the coefficient of friction decreases. **The equation is consistent with these predictions**.

6.6.4   Using Estimation to Check Solutions

Key idea: Use estimation to reveal errors in the mathematical model.

The use of estimation was discussed in Section 6.4. Estimations can be used to check the solutions. Estimations can also reveal errors in the mathematical model. As an example, say you are a chemical engineer growing bacteria to synthesize a new pharmaceutical product. You know the bacterial population doubles in 20 minutes. If you start with one bacterium, what will be the mass of the organisms after three days of growth?

PONDER THIS

> **Based on your everyday experiences (e.g., that funny growth in your refrigerator), estimate the bacterial mass after three days. A few grams? A few kilograms? A few metric tons?**

doublings: here, the number of times a population doubles (a population doubles three times when growing from 1 to 8: 1 to 2, 2 to 4, and 4 to 8)

You need to determine how many times the bacterial population will double. The number of **doublings** is equal to the time divided by the doubling time. Now, three days is (3 days) (24 hours/day) (60 minutes/hour) = 4,320 minutes or (4,320 minutes)/(20 minutes per doubling time) = 216 doublings. A moment of reflection will confirm that the number of bacteria after n doublings is 2^n. Thus, you will have $2^{216} = 1.05 \times 10^{65}$ bacteria after three days.

You need to convert the number of bacteria into the mass of bacteria:

$$\text{total mass} = (\text{number of bacteria})(\text{mass per bacterium})$$

You can estimate the mass of a bacterium by its volume multiplied by its density (mass = volume $\times$ density). The volume of a bacterium can be found by approximating the organism as a sphere with diameter = $5\ \mu$m = 5×10^{-6} m (or a radius, r = diameter/2 = 2.5×10^{-6} m). Thus, the volume of one bacterium is about

$$(4/3)\pi r^3 = (4/3)\pi(2.5 \times 10^{-6}\ \text{m})^3 = 6.54 \times 10^{-17}\ \text{m}^3$$

A bacterium has a density near water ($1{,}000\ \text{kg/m}^3$). Thus, after three days, the bacterial mass is

$$(1.05 \times 10^{65}\ \text{bacteria})(6.54 \times 10^{-17}\ \text{m}^3/\text{bacterium}) \times (1{,}000\ \text{kg/m}^3) = 6.9 \times 10^{51}\ \text{kg}$$

This value of almost 7×10^{51} kg is probably far greater than your estimate. In fact, *this mass is about 10^{27} times the mass of the Earth*! Something must be amiss. A check of the calculations reveals no errors. In this case, the mathematical model is incorrect: bacteria do **not** double at a constant time interval at their maximum growth rate for such a large number of doublings. (Doubling at a fixed time interval is called *exponential growth*.) In reality, bacteria run out of resources (i.e., food, water, or space) or are killed off by a build-up of their waste products and natural decay processes. In this calculation, a "reality check" spurred by estimation uncovered an error in the model. For other examples of estimation, see John Harte's delightful books on problem solving (Harte, 1988 and 2001).

6.6.5   Using Units to Check Solutions

Checking units is an important tool in evaluating the solutions to mathematical expressions. Units are discussed in Section 6.7.

6.7   UNITS

6.7.1   Introduction

Most of the numbers you will deal with as an engineer have *units*. A voltmeter does not measure 6.2; it measure 6.2 *millivolts*. An old personal computer does not operate at 900; it operates at 900 *megahertz*. A successful engineering calculation results in not only the right value, but also the right *units* for that value. An example of a very expensive unit error is shown in *Focus on Units: The Multimillion-Dollar Units Mistake*.

SI units: (from Système Internationale d'Unités) a standardized set of units (see Table 6.3)

The units for many physical properties have been standardized. The standardized system of units is called the *Système Internationale d'Unités* or *SI units*. A list of SI units appears in Table 6.3.

6.7.2 Dimensional Analysis

dimensional analysis: the manipulation of units to check if the units "balance"

One tool for checking the units of an expression is *dimensional analysis*. Dimensional analysis refers to the manipulation of units without numbers. This technique can be used to determine the units of a result of an engineering calculation. In the skydiving example shown in Section 6.4.2, the terminal velocity was given by $v_t = \sqrt{\dfrac{2mg}{C_D A \rho}}$. To

Key idea: Use dimensional analysis to check engineering calculations.

check the validity of this equation, you can perform a dimensional analysis. Substitute the units of each term and check to see that the terminal velocity has units of velocity (i.e., length/time). If you denote the units of X by $\{X\}$, then

$$\{v_t\} = \sqrt{\frac{\{2\}\{m\}\{g\}}{\{C_D\}\{A\}\{\rho\}}} = \sqrt{\frac{()(kg)\left(\dfrac{m}{s^2}\right)}{()(m^2)\left(\dfrac{kg}{m^3}\right)}} = \sqrt{\frac{m^2}{s^2}} = \frac{m}{s}$$

dimensionless: having no units

(Note that the number 2 and the drag coefficient C_D are *dimensionless*; that is, they have no units.) According to the equation, the units of the terminal velocity are meters per second. These are proper units for a velocity.

Please note that there are limitations to dimensional analysis. Just because the units check, does that mean the equation is valid? No. Proper units are *necessary*, but not *sufficient*, for a valid equation. For example, the units check for $v_t = \pi\sqrt{\dfrac{2mg}{C_D A \rho}}$, but this equation is **not** correct.

Key idea: Use dimensional analysis to determine the units of an unknown quantity.

Dimensional analysis also can be used to determine the units of an unknown variable. Do you sometimes forget the units of force? Since $F = ma$, it follows that

$$\{F\} = \{m\}\{a\} = (kg)(m/s^2) = kg\text{-}m/s^2$$

(Remember that 1 kg-m/s^2 is called 1 newton = 1 N; see Table 6.3.)

As a more complex example, determine the units of viscosity. Viscosity refers to the property of a fluid offering resistance to flow. It is formally defined as the ratio of the shearing stress to the shear.[*] The shearing stress is a force per unit area and the shear is the change in the velocity with respect to distance. Thus,

---

[*]If the viscosity is independent of the shearing stress, then the fluid is called a *Newtonian fluid*. Other fluids are called *non-Newtonian fluids*. For example, certain paints are made to spread more easily when you apply more pressure to the brush. These paints are non-Newtonian fluids. Fluids like these paints, where the viscosity decreases as the shearing stress increases, are called *thixotropic* fluids.

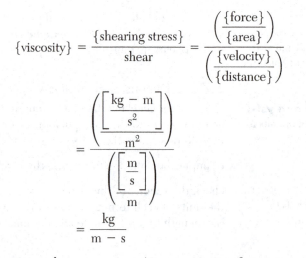

$$\{viscosity\} = \frac{\{shearing\ stress\}}{shear} = \frac{\left(\dfrac{\{force\}}{\{area\}}\right)}{\left(\dfrac{\{velocity\}}{\{distance\}}\right)}$$

$$= \frac{\left(\dfrac{\left[\dfrac{kg - m}{s^2}\right]}{m^2}\right)}{\left(\dfrac{\left[\dfrac{m}{s}\right]}{m}\right)}$$

$$= \frac{kg}{m - s}$$

The units of viscosity are kg per m per s. (Common units of viscosity are centipoise = cp = 0.01 g/cm-s. The viscosities of water, chocolate syrup, and peanut butter at room temperature are about 1 cp, 6×10^4 cp, and 2×10^5 cp.)

TABLE 6.3 SI Units (adapted from Wright, 1993)

Quantity	Unit(s)	Symbol and/or Formula
Base Units		
Length	meter	m
Mass	kilogram	kg
Time	second	s
Electric current	ampere	A
Temperature	kelvin	K
Amount of substance	mole	mol
Luminous intensity	candela	cd
Supplementary Units		
Plane angle	radian	rad
Solid angle	steradian	sr
Common Derived Units with Special Names		
Frequency	hertz	Hz (s^{-1})
Force	newton	N (kg-m/s^2)
Pressure or stress	pascal	Pa (N/m^2)
Energy or work	joule	J (N-m)
Power	watt	W (J/s)
Quantity of electricity	coulomb	C (A-s)
Electric potential	volt	V (W/A)
Capacitance	farad	F (C/V)
Electric resistance	ohm	Ω (V/A)
Conductance	siemens	S (A/V)

TABLE 6.3 (Continued)

Quantity	Unit(s)	Symbol and/or Formula
Magnetic flux	weber	Wb (V-s)
Magnetic flux density	tesla	T (Wb/m^2)
Inductance	henry	H (Wb/A)
Luminous flux	lumen	lm (cd-sr)
Illuminance	lux	lx (lm/m^2)
Common Derived Units without Special Names		
Area	m^2	
Volume	m^3	
Velocity	m/s	
Acceleration	m/s^2	
Density	kg/m^3	
Specific volume	m^3/kg	
Entropy	J/K	
Radiant intensity	W/sr	
Bending moment (or torque)	N-m	
Heat capacity	J/kg-K	

FOCUS ON UNITS: THE MULTIMILLION-DOLLAR UNITS MISTAKE

BACKGROUND

In 1993, the National Aeronautics and Space Administration (NASA) initiated the Mars Surveyor Program. The Mars Surveyor Program consisted of three missions: the *Mars Global Surveyor* (MGS, launched in 1997), the *Mars Climate Orbiter* (MCO, launched in 1998), and the *Mars Polar Lander* (MPL, launched in 1999). The Mars Surveyor Program, together with the Discovery Program's *Mars Pathfinder* lander, constituted NASA's efforts for a robotic exploration of Mars.

(Image credit: NASA/JPL.)

MCO launched from Cape Canaveral Air Force Station Space Launch Complex 17 on December 11, 1998, on a Boeing Delta II 7425 rocket. The 338-kg spacecraft consisted of a main bus (2.1 m tall by 1.6 m wide by 2 m deep) and a solar array (with a wingspan of 5.5 m). MCO's mission was threefold: provide support for the MPL, carry out scientific studies of the Martian atmosphere and climate, and serve as a relay for communication with future Mars landers.

A PRIMER ON SPACECRAFT CONTROL

Guiding a spacecraft across interplanetary space on a $9\frac{1}{2}$-month journey to Mars is no small task. To keep on course, the thrusters must be fired periodically. Data from the spacecraft were sent to the ground. The "impulse bit" (force × time) was calculated from the firing time and used to determine the change in spacecraft velocity (Δv). About four months into the flight, it was clear that the ground-based software was not modeling Δv correctly. The extent of the errors was unknown, because the thrust events occurred mainly perpendicular to the line-of-sight from the Earth. (Imagine the difficulty in determining the acceleration of a distant car moving away from you as it rounds a curve.)

THE PROBLEM

As MCO approached Mars, three steps were planned to reach the final circular orbit. First, a final trajectory correction burn would prepare the spacecraft for orbit. Second, the main Mars orbital burn (Mars orbit insertion, or MOI) would establish a highly elliptical orbit with a point of closest approach to Mars of 226 km. Finally, friction with the upper Martian atmosphere would ease MCO into a circular orbit in a process called *aerobraking*. After the final trajectory correction burn (but before the planned MOI), NASA navigators realized that the point of closest approach was going to be much smaller than anticipated—perhaps 150–170 km. As MOI grew nearer, the Martian gravitational effects increased and were included in the calculations. About one hour before MOI, navigators determined that the point of closest approach would be about 110 km, dangerously close to the minimum survivable distance of 80 km. About four minutes after the MOI burn began, MCO entered the Martian shadow (49 seconds earlier than anticipated) and communication with the spacecraft ceased. The signal was never reacquired.

THE UNITS PROBLEM

What happened? Subsequent calculations showed that the actual point of closest approach to Mars was a mere 57 km. Six days after the MOI burn, the MCO team realized that the calculated Δv values were low by a factor of 4.45. Can you account for a factor of 4.45 mistake in calculating velocity from force $\times$ time? It turns out that the computer software had been programmed to calculate forces (used to determine Δv) in units of *pounds of force*, rather than the expected unit of *newtons*. (You may wish to verify that 1 lb of force = 4.45 N.)

An error in units resulted in the loss of hundreds of millions of dollars worth of orbiter, spacecraft development, launch, and mission operations costs. Never forget: **units are important!**

6.7.3   Units and Functions

Dimensional analysis leads to four rules when you perform an operation on a number. First, you can add or subtract numbers only when they have the same units. You can use this fact to check equations: additive terms *must* have the same units.

Second, when you multiply, divide, or exponentiate (i.e., raise a number to an exponent), you must perform the same operation on the units of the number. An example was shown in Section 6.7.2 for determining the units of viscosity.

Third, some mathematical functions can operate only on dimensionless parameters. Examples of such functions include the exponential, logarithmic, and trigonometric functions. For the functions e^{ax}, $\log(y/b)$, and $\cos(2\pi\omega)$, you know that the terms ax, y/b, and $2\pi\omega$ must be dimensionless.

Fourth (a corollary to the third rule), exponential, logarithmic, and trigonometric functions also *produce* dimensionless results. In other words, the values produced by the functions e^{ax}, $\log(y/b)$, and $\cos(2\pi\omega)$ are dimensionless.

As before, dimensional analysis can be used to check units or determine units, since the collection of terms operated on and produced by exponential, logarithmic, and trigonometric functions must be dimensionless. For example, in first-order chemical reactions, concentrations decrease over time by $e^{-k_1 t}$, where k_1 is called the first-order rate constant and t is time.

Key idea: Add or subtract terms only if they have the same units.

Key idea: If you multiply, divide, or exponentiate terms, apply the same operation to their units.

Key idea: Certain functions (e.g., exponential, logarithmic, and trigonometric functions) only operate on and only produce dimensionless values.

PONDER THIS

In the expression $e^{-k_1 t}$, what is the unit of k_1?

The exponent, $-k_1t$, **must** be dimensionless. If the unit of time is the second, then the first-order rate constant must have the unit of inverse seconds (denoted 1/s or s^{-1}). In other words, first-order rate constants always have the unit of 1/time.

6.7.4   Units Conversion

English System:

a system of units in common use in the United States

You are entering the engineering profession at a unique time. In spite of numerous efforts to adopt the metric system of units (now codified as SI units), the so-called **English System** of units is still in common use in the United States. Engineers in the United States must be familiar with *both* sets of units and must be able to interconvert units with ease. You will find yourself in your career stating masses in kilograms and pounds in the same sentence. You may need to speak to a client about the power requirements of a turbine in horsepower and then call the vendor to specify the turbine power in kilowatts.

Values can be converted from one set of units to another by *conversion factors*. For example, to convert a length from inches to centimeters (cm), multiply the length in inches by the number of cm per inch:

$$\text{length in cm} = (\text{length in inches})(2.54 \text{ cm/in})$$

Note that the conversion factor here has units of cm/in and that the units in the equation check out.

Here is a more complicated example: If a spring water bottling plant produces 27,000 gallons of bottled water per day, what is the production rate in the SI base units of cubic meters per second (m^3/s)? As with many unit conversion problems, this one is best handled in steps. You can convert gallons to cubic feet and then cubic feet to cubic meters. Simultaneously, you can easily convert days to seconds. The important conversion factors for this problem are as follows: one cubic foot is equivalent to 7.48 gallons (or 1 gallon = 1/7.48 ft^3; in other words, 1/7.48 ft^3/gallon) and one day is (1 day)(24 hours/day)(60 minutes/hour)(60 seconds/minute) = 86,400 seconds (or 1/86,400 day/second). Also, one foot is equal to 0.3048 m (or 0.3048 m/ft), so one cubic foot is equal to $(0.3048 \text{ m/ft})^3 = 0.0283 \text{ m}^3$ (or 0.0283 m^3/ft^3). Thus,

$$\text{flow in ft}^3/\text{s} = (\text{flow in gallons/day})(1/7.48 \text{ ft}^3/\text{gal}) \times (1/86,400 \text{ day/s})$$

and

$$\text{flow in m}^3/\text{s} = (\text{flow in ft}^3/\text{s})(0.0283 \text{ m}^3/\text{ft}^3),$$

so,

$$\text{flow in m}^3/\text{s} = (\text{flow in gallons/day})(1/7.48 \text{ ft}^3/\text{gal})$$

$$\times (1/86,400 \text{ day/s})(0.0283 \text{ m}^3/\text{ft}^3)$$

$$= 4.38 \times 10^{-8}(\text{flow in gallons/day})$$

You have calculated the conversion factor: to convert flows from units of gallons per day to units of cubic meters per second, you simply multiply by 4.38×10^{-8}. Note that this conversion factor is **not** dimensionless: it has units of (m^3/s)/(gallons/day) or (m^3-day)/(gallons-s). Thus,

$$27,000 \text{ gallons per day} = (27,000 \text{ gallons/day})$$

$$\times [4.38 \times 10^{-8} \text{ (m}^3/\text{s})/(\text{gallons/day})]$$

$$= 1.18 \times 10^{-3} \text{ m}^3/\text{s}$$

Another example of using units to check and solve problems is given in Example 6.6.

EXAMPLE 6.6
USING UNITS
TO CHECK
SOLUTIONS

[Caution: Problem statement may contain errors!]

A great deal of aluminum has been saved by reducing the thickness of aluminum beverage cans. The current thickness of a can side is about 0.08 mm. The thickness is limited by the ability of the can to withstand internal pressures. (For example, beer often is pasteurized in the can.) What is the minimum thickness of an aluminum can? The can thickness is given by $T = \dfrac{DP}{2S}$, where D = the inside diameter of the can, P = the allowable internal pressure, and S = the tensile strength of aluminum = 3.2×10^9 g/cm-s.

SOLUTION

A look at the units reveals an error in the problem statement. Since T and D both have units of length, P and S must have the same units. Pressure is force per unit area or mass × acceleration/area or $(g)(cm/s^2)/(cm^2)$ or g/cm-s^2. Rechecking the references, the tensile strength of aluminum = 3.2×10^9 g/cm-s^2 (**not** 3.2×10^9 g/cm-s). With $D = 6.6$ cm and $P = 90$ psi $= 6.2 \times 10^6$ g/cm-s^2, the minimum thickness is about 0.06 mm. Thus, the current thickness is near the minimum thickness required to withstand an internal pressure of 90 psi.

6.8   AN EXAMPLE OF THE ENGINEERING ANALYSIS METHOD

The steps in the engineering analysis method will be applied to the following problem. You bought a used bed that is missing its legs. You plan to build four new legs out of a long pine dowel. How much will the legs compress if the total mass of the bed plus you is 100 kg? You may wish to do the analysis before reading further.

Before you begin, you should ask yourself: is this an analysis problem? Yes, the problem is amenable to the engineering analysis method, since it is reasonably well defined and one answer is sought. Using the following steps in the analysis process:

1. Define the problem.
    The problem is well defined.
2. Gather data and verify data accuracy.
    You need the dimensions of the legs. Suppose the dowel measures 1.2 m long and 2.5 cm in diameter [so each leg will be (1.2 m)/4 = 0.3 m long].
3. Select the analysis method(s).
    You need a relationship between the data (legs and bed mass) and the desired outcome (compression length). The appropriate relationship is given by the stress–strain relationship (a constitutive law similar to Hooke's Law). A common stress–strain relationship is that stress (i.e., the force applied per unit area) is proportional to strain (i.e., the length of compression divided by the length of the column). The proportionality constant is called *Young's modulus*. In other words,

$$\text{stress} = \text{force per area}$$
$$= F/A$$
$$= E(\text{strain})$$

where E is Young's modulus and strain is the compression length (δ) divided by the length of the leg (L). Rearranging terms yields

$$\delta = \frac{L\,F}{E\,A} \tag{6.8}$$

The selection of the analysis method leads to more data collection: you need Young's modulus (E) for pine and the applied force. An accepted value for Young's modulus for pine is 12,200 MPa (1 MPa $= 10^6$ Pa). The net force of the bed plus you can be calculated from the total mass of you and the bed.

4. Estimate the solution.

    Only a very small compression of the legs is expected.

5. Solve the problem.

    The force is $F = mg = (100\text{ kg})(9.8\text{ m/s}^2) = 980$ N. The force per area per leg is $(980\text{ N})/[(4\text{ legs})(\pi r^2)]$, where r = radius of the leg $= 0.0125$ m. So the pressure per leg is 5.0×10^5 N/m^2 or 0.5 MPa. From Eq. (6.8),

    $$\delta = (L/E)(F/A)$$
    $$= [(0.3\text{ m})/(12{,}200\text{ MPa})](0.5\text{ MPa})$$
    $$= 1.2 \times 10^{-5}\text{ m}$$

6. Check the results.

    This analysis is simple and no mathematical errors are apparent.

    Therefore, each leg will be compressed about 1.2×10^{-5} m or 0.012 mm. This is as small as expected in the estimation step.

6.9   SUMMARY

The analysis method is useful when the system is well defined and system characteristics can be determined through the application of mathematical and scientific principles. Analysis problems typically have one solution. They are solved by defining the problem, gathering data and verifying data accuracy, selecting analysis methods, estimating the solution, solving the mathematical expressions, and checking the results.

Engineering data can come from many sources, including measurements, interviews, and the Internet. Always remember to verify the reasonableness of the data.

Selecting an analysis method generally means selecting the physical laws that describe the system of interest. Three kinds of physical laws are important in engineering: laws of conservation, laws of motion, and constitutive laws.

It is also important to estimate the solution. Many mistakes in analysis can be caught by making even a crude guess of the solution. Estimation can help you check the calculations and the analysis method.

This chapter also provided an introduction to solving mathematical expressions. Recall that expressions can be manipulated to solve for the unknown of interest. Try to isolate the unknown by manipulating symbols, not numbers (i.e., substitute numbers at the last step of the calculation). As you explore engineering further, remember that the numbers you calculate have physical meaning. You can use logic and estimation to avoid aphysical or unreasonable solutions. Keep in mind that most numbers in engineering have units. Use dimensional analysis (i.e., manipulating units without numbers) to check equations and determine the units of unknown quantities. Remember that many functions operate on and produce only dimensionless values.

Never leave an engineering calculation without checking the results. Engineering results can be validated by logic, estimation, and checking units. Use logic to eliminate aphysical answers, test whether your answer makes sense, and check whether the variables change as expected with changes in the other variables. Use estimation to reveal errors in the mathematical model. Remember that answers must have the correct units.

SUMMARY OF KEY IDEAS

- Analysis problems are usually well defined and have only one solution.
- Solve analysis problems by defining the problem, gathering data and verifying data accuracy, selecting the analysis methods, estimating the solution, solving the mathematical expressions, and checking the results.
- Data are gathered through experiments and field measurements.
- Test all data for reasonableness.
- To select an analysis method, first select the physical laws and then translate the physical laws into mathematical equations.
- Engineers need quantifiable relationships between variables.
- Manipulate expressions (i.e., rearrange terms) to solve for the variable of interest.
- The "Golden Rule" of expression manipulation is as follows: "Do to one side of the expression that which you did to the other side."
- Remember to reverse the inequality sign when multiplying or dividing both sides by a negative number.
- Solve for the unknowns with symbols (by isolating the unknowns on one side of the expression) and *then* substitute in numbers ("chug" before you "plug").
- Answers to engineering calculations almost always have physical meaning.
- Check engineering calculations by logic, estimation, and checking units.
- Use logic to eliminate aphysical answers: always ask if your answer makes sense.
- Use logic to check whether one variable changes as expected with changes in the other variables.
- Use logic to check whether one variable is predicted correctly for extreme values of the other variables.
- Use estimation to reveal errors in the mathematical model.
- Use dimensional analysis to check engineering calculations.
- Use dimensional analysis to determine the units of an unknown quantity.
- Add or subtract terms only if they have the same units.
- If you multiply, divide, or exponentiate terms, apply the same operation to their units.
- Certain functions (e.g., exponential, logarithmic, and trigonometric functions) only operate on and only produce dimensionless values.

Problems

6.1. Using each step of the engineering analysis process, analyze the following problems.

a. What is the maximum numbers of hours you should work at a job this semester? [*Hint*: Use the 60-hour rule: for each week, the sum of the hours in class, hours studying (2 × hours in class), and hours working should be less than or equal to 60 hours.]

b. How much money do you have for entertainment? (*Hint*: Calculate an annual budget.)

    c. What is the optimal number of sides for a pencil? (*Hint:* Consider comfort and the need to prevent a pencil from sliding down a slanted drafting table.)

6.2. List at least four ways to catch errors in engineering calculations.

6.3. How can logic be used to screen for calculation errors?

6.4. The following questions concern the speed of world-class sprinters.

    a. Without any data, estimate the speed of world-class sprinters in miles per hour (mph). Is it 1 mph? 10 mph? 100 mph? Higher?

    b. A young engineer reads that the world record for the 200 m is 19.32 s. The engineer calculates the velocity as follows:

    (*Warning: The following derivation may contain one or more errors!*)

$$\text{velocity} = \text{distance/time}$$

$$\text{distance} = 200 \text{ m} = (200 \text{ m})(6.21 \times 10^{-4} \text{ miles/m}) = 0.124 \text{ mile}$$

$$\text{time} = 19.32 \text{ s} = (19.32 \text{ s})(2.78 \times 10^{-5} \text{ hours/s}) = 5.37 \times 10^{-4} \text{ hours}$$

$$\text{velocity} = (0.124 \text{ mile})/(5.37 \times 10^{-4} \text{ hours}) = 231 \text{ mph}$$

    Does this answer fit your estimate in part (a)? Identify the errors (if any) in the approach calculation.

6.5. What is the volume of your favorite 12-oz beverage in mL? (Recall that 1 gallon = 128 oz = 3.78 L.)

6.6. Show that kinetic energy ($\frac{1}{2}mv^2$) has the same unit as potential energy (mgh). In this problem, m = mass, v = velocity, g = gravitational acceleration, and h = height.

6.7. You are designing a new desktop computer monitor. To satisfy customer needs, you want the footprint of the monitor to be 400 cm^2 and the length to be 6 cm less than the width. (The footprint is the area of the desktop occupied by the monitor.) Size the unit (i.e., determine its width and height). Indicate where you are performing the same operations on both sides of the pertinent equations.

6.8. The following questions concern how high you can jump from a running start.

    a. Estimate how high you can jump from a running start. Is it 0.1 m? 1 m? 10 m?

    b. An engineering professor preferred to calculate the jump height rather than measure it. The analysis is shown below:

    (*Warning: The following derivation may contain one or more errors!*)

    Physical law: energy balance

    kinetic energy (**KE**) converted to potential energy (**PE**)

$PE = mgh = KE = mv^2/2$, where h = change in height of center of mass (CM) so

$$h = v^2/(2g)$$

$$g = 9.8 \text{ m/s}^2$$

assume

$$v = 7 \text{ m/s}$$

Thus,

$$h = 2.5 \text{ m}$$

If the CM starts at 1.0 m, then you could jump a height of

$$1.0 + 2.5 = 3.5 \text{ m}$$

Does this answer fit your estimate in part (a)? (The world record for the high jump is 2.45 m.) Identify the errors (if any) in the approach and/or calculation.

6.9.   The metric system is praised in part for the simple relationship between the volume of water and the mass of water: $1 \text{ m}^3 \approx 1 \text{ kg}$. Show that $1 \text{ ft}^3 \approx 1,000 \text{ oz}$.

6.10.   Gas mileage in the United States is often expressed in miles per gallon (mpg). What are the equivalent units using only SI base units (see Table 6.3)? If you get 30 mpg, what is your gas mileage in inverse acres (acre^{-1})?

6.11.   The farthest recorded distance for shooting a champagne cork is about 178 ft. Using the information in Section 6.6.3, what initial cork velocity (in SI base units) would be required to reach 178 ft, assuming little air resistance and the optimum launch angle?

6.12.   How long would it take for the bacteria described in Section 6.6.4 to grow to 1 gram? Estimate your answer before solving.

6.13.   For the aluminum can example in Example 6.6,

a.   Show that the conversion of psi to g/cm-s^2 is correct.

b.   Discuss whether the effects of increasing the can diameter, internal pressure, and tensile strength on the can thickness make sense.

6.14.   The crew of a blimp must know the blimp mass accurately to ensure that the landing velocity is not too large. The blimp mass is determined by allowing the blimp to "settle" and measuring its terminal settling velocity. For Goodyear's *Spirit of America* blimp, a settling velocity of 100 ft/min corresponds to a mass of about 100 pounds greater than the mass of displaced air. Find the drag coefficient for the *Spirit of America* (volume = 202,700 ft^3 and cross-sectional area = 7,540 ft^2). (*Hint*: Modify the force balance in the skydiving example of Section 6.4.2 to include a buoyancy force. The buoyancy force is equal to g times the mass of displaced air. The blimp mass is the mass of displaced air plus 100 lb.)

7

Engineering Design Method

7.1 INTRODUCTION

7.1.1 Introduction to Engineering Design

In engineering, a **design** is a description of a new or improved device or system. **Design problems** are distinguished from analysis problems by the nature of both the problems and the solutions. Design problems usually are more vaguely defined (as contrasted with the more well-defined analysis problems). While analysis problems usually have one solution, there is often no single "correct" solution to a design problem. In fact, design problems *require* you to develop several solutions or alternatives and then use a set of criteria to compare and evaluate the alternatives.

Analysis problem solving is generally more science than art, while design problem solving involves more art. Why? In meeting design challenges, it is necessary to generate new, creative ideas. Evaluating the ideas means repeating the analysis process for each alternative.

Engineering design is as varied as the engineering profession. Engineering design problems have become more complex and challenging, requiring greater specialization and teamwork. It is not uncommon for large engineering projects to be carried out by dozens or even hundreds of engineering specialists. Design problems often are attacked using an **integrated project team** (IPT). In the IPT approach, teams of engineers, scientists, and other professionals are grouped by task, not discipline. Each team, consisting of representatives from various engineering, science, and other disciplines, has the responsibility of executing a given task.

OBJECTIVES

After reading this chapter, you will be able to:

- list the steps in the engineering design method;
- identify the kinds of engineering problems for which the engineering design method is appropriate;
- solve engineering problems using the engineering design method.

design: a description of a new or improved device or system

design problem: a problem where the system is often not well defined and more than one (or sometimes no) solution is possible

integrated project team: a management strategy where personnel are organized by task rather than by discipline

7.1.2 Solving Design Problems

The solution to a design problem does not suddenly appear out of nothing. Most good design solutions are the result of a methodical process. The process may vary from engineer to engineer and problem to problem. The design method described here is quite general and can be adapted to a variety of problems. The six common steps in the engineering design process are

1. Defining the problem.
2. Gathering information.
3. Generating multiple solutions.
4. Analyzing and selecting a solution.
5. Implementing the solution.
6. Evaluating the solution.

Note that just as design is not the same as analysis, the engineering design process differs from the engineering analysis process.

PONDER THIS

> **What are the differences between the engineering design method and the engineering analysis method?**

Key idea: Solve engineering design problems by defining the problem, gathering information, generating multiple solutions, selecting a solution, implementing the solution, and evaluating the solution.

There are three main differences between the analysis and design. In analysis, you seek the *one* solution. In design, you seek to generate *multiple* solutions. In analysis, you *calculate* a solution. In design, you *select* a solution based on some evaluation criteria. Finally, in design, you *implement* the solution.

The overall engineering design method is shown schematically in Figure 7.1. In this chapter, we discuss the last four steps.

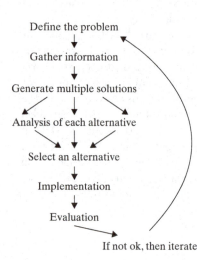

Figure 7.1. Steps in the Engineering Design Method

7.2   GENERATING MULTIPLE SOLUTIONS

7.2.1   Introduction

Once you have defined the problem and gathered information about it, you are ready to begin developing solutions or alternatives. This is when you must think creatively.[*] Techniques for generating multiple solutions are discussed in this section.

7.2.2   Brainstorming

brainstorming: a technique for generating possible solutions in a group by recording all spontaneous ideas

Several techniques can be used by groups or individuals to help generate ideas that may lead to solutions. One of the most well-known and effective (but often poorly implemented) techniques for group problem solving is **brainstorming**. The generation of ideas by brainstorming is a freewheeling process. It can be intimidating the first few times you attempt it. Successful brainstorming requires practice. It also requires a commitment to the project that is stronger than the individual egos. The following are some guidelines for a good brainstorming session.

Composition:

1. Small team
   A brainstorming team should have five to ten members to ensure enough new ideas without allowing individuals to hide in the crowd.
2. Diverse team
   Select participants from a diversity of backgrounds, including people with little direct experience with the problem.

Logistics:

1. Short meetings
   Keep brainstorming sessions shorter than one hour.
2. Record meetings
   Record the ideas for evaluation at a future session. Identify one person to record the ideas for evaluation at a future session, using a blackboard, easel, or whiteboard so that all participants may see the ideas generated.

Meeting Protocol:

1. No hierarchy
   Group members must be considered equals.
2. Nonjudgmental
   Accept all ideas without judgment or evaluation, avoiding negative comments (e.g., "That won't work," "That's stupid," "Nobody does it that way," or "We can't solve this problem").
3. Quantity over quality
   Stress the importance of the *quantity* of ideas: *generate as many ideas as possible.*
4. Build ideas
   Create new ideas by combining and building on other ideas.

---

[*]The process of looking at challenges in a new way sometimes is called *lateral thinking.*

Key idea: Generate multiple solutions through brainstorming, making checklists, listing attributes, and employing forced random relationships.

checklist: a tabulation of ways that an objective can be achieved

attribute listing: a technique for generating possible solutions where many attributes of a system are listed

Key idea: Attribute values can be combined in different ways to generate new ideas.

morphological analysis: a technique for generating alternatives where all possible combinations of solutions for all attributes are combined

random forced relationship: a technique for generating possible solutions where attributes from a random word are related to an existing problem

7.2.3 Methods for Generating New Ideas

Brainstorming sessions will stall without new ideas. How do you prevent your brainstorming session from running out of steam? Three popular techniques for generating ideas during a brainstorming session are checklists, attribute listing, and random forced relationships. (Two other methods are presented in Problems 7.5 and 7.6.) These techniques also can be used by an individual.

Checklists are a great way to encourage ideas in a brainstorming session. A **checklist** is a tabulation of ways that an objective can be achieved. For example, suppose you are asked to improve a production line for manufacturing CDs. You might make a checklist that included several concepts: ways the line could be used to produce items other than CDs, ways the line could be reconfigured, ways the line could be made smaller, and so on. Note that the checklist tabulates *ways* in which a device or system could be improved. Once the areas for improvement have been listed, then brainstorming can be used to generate ideas in each area. An example checklist for improving the fuel efficiency of military aircraft is shown in Figure 7.2.

Another technique that can be used by individuals or groups is to create a table of the attributes of the device you are improving and the possible values or solutions for each attribute. This process is called **attribute listing**. For example, in creating ideas for heating a "smart home," you might consider three attributes: energy source (gas, oil, wood, electric, or solar), method of heat transmission (radiant, convection, or forced), and heat transmission medium (air, water, or other liquid).

The attribute values can be combined in different ways to generate new ideas. For example, you could combine attribute values *randomly* to generate a new system (i.e., design a radiant wood-burning stove with no heat transmission medium). In addition, you could combine *every possible solution* for every attribute with every possible solution for every other attribute and create a theoretical listing of every possible solution. This is called **morphological analysis**. In the "smart home" example, the combination of alternatives would generate 5 energy sources × 3 heat transmission modes × 3 heat transmission media = 45 ideas. Each idea could undergo further evaluation.

The **random forced relationship** technique is especially useful if you are interested in generating totally new ideas. The idea here is to force a relationship between two normally unrelated objects or words. One of the objects may be your project (or an

---

Checklist: Improving the Fuel Efficiency of Military Aircraft

- Ways to reduce weight
- Ways to make engine more efficient
- Potential alternative fuel mixtures
- Ways to increase lift
- Ways to decrease

Figure 7.2. Example of a Checklist

attribute of your project) and the other object is a randomly selected word. This word is used to act as a trigger to change the patterns of thought when a mental roadblock occurs. The random word can be used to generate other words and stimulate the flow of new ideas.

Returning to the "smart home" heating system example, suppose you randomly select "automobile" as a random forced relationship word. This might lead you to consider using antifreeze for the heat transmission medium or putting the heater on wheels to be rolled from room to room or installing a sunroof for solar heat. Clearly, not all of these ideas are reasonable. However, the random forced relationship technique can be useful to generate new ideas when brainstorming is unproductive. An example of using idea-generating techniques in engineering design is given in Example 7.1.

EXAMPLE 7.1 GENERATING ALTERNATIVES

Your design team has been asked to develop a new generation of a personal communications device (the cell phone of tomorrow). Your brainstorming session has bogged down. Use checklists and attribute analysis to restart the creative process.

SOLUTION

There are many ways to use checklists and attribute analysis to generate new ideas. A checklist for how to improve the cell phone might include several questions:

Can a cell phone be used for functions other than audio communication?

How can people use a cell phone more effectively?

What needs to happen to wireless communication to eliminate landline telephones completely?

An example attribute listing is given below.

Information Streams	User Interface	Unit Configuration
Audio only	Keypad	Handheld
Still pictures	Detachable keyboard	Wearable
Full-motion video	Voice-activated	Hypodermic implant
Holographic images	Brain wave–activated	

A random sampling might include the following designs: (solid arrows in the table) an audio-only, voice-activated, wearable device or (dashed arrows in the table) a handheld unit with detachable keyboard capable of producing holographic images.

Information Streams	User Interface	Unit Configuration
Audio only	Keypad	Handheld
Still pictures	Detachable keyboard	Wearable
Full-motion video	Voice-activated	Hypodermic implant
Holographic images	Brain wave–activated	

A morphological analysis would include all $4 \times 4 \times 3 = 48$ combinations.

7.3 ANALYZING ALTERNATIVES AND SELECTING A SOLUTION

7.3.1 Analyzing Alternatives

Key idea: Evaluate alternatives by applying the engineering analysis method to each alternative.

At this point in the engineering design method, you have defined the problem, gathered the pertinent information, and identified a number of potential solutions. To select the best alternative, the potential solutions must be analyzed and their performance capability must be evaluated. In other words, *you must apply the engineering analysis method to each alternative*. Potential solutions that are not optimal must be discarded, or modified and reevaluated. In light of new information from the evaluation of potential solutions, it may be necessary to redefine the problem, change the constraints, or change the evaluation criteria.

To illustrate the process of analyzing alternatives, suppose you are working with an advocacy group to assist wheelchair-bound people. Your group is designing a portable wheelchair ramp. To meet the requirements of the Americans with Disabilities Act, the ramp must be adjustable to slopes between 1:20 and 1:12 and have a minimum width of 36 inches. In addition, the ramp must be as inexpensive as possible and wheelchair-transportable. Your brainstorming activities result in three candidate designs: an inflatable ramp, a foldable (hinged) stainless steel ramp, and a fiberglass ramp.

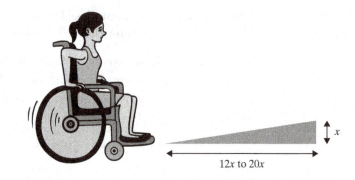

In this case, the analysis might consist of the following questions:

- Can the ramp be made to be adjustable to slopes between 1:20 and 1:12?
- Can the ramp be made with a width of at least 36 inches?
- Can the ramp with the above characteristics be transported on a wheelchair?
- What is the cost of a ramp with all of the above characteristics?

The analysis might result in modified alternatives. Suppose the fiberglass ramp is inexpensive, but too unwieldy to be moved by a wheelchair. You might steal the hinged idea from the second alternative and develop a foldable fiberglass ramp as another alternative to be analyzed.

The analysis of alternative solutions may be simple or very complex. In some cases, a preliminary sketch or a cursory analysis may show that an idea is not worthy of further consideration. In other cases, a component may need to be examined by laboratory tests. In still other cases, a comprehensive research program may be needed to determine the feasibility of a proposed solution. To facilitate the analysis, engineers often rely on models to evaluate the proposed solutions.

7.3.2 Selecting a Solution

Alternatives should be compared across a common set of criteria.

What criteria should you use to compare alternatives?

Key idea: Evaluate alternatives by and select assessing feasibility.

The answer lies in the types of feasibility, namely, technical (or engineering); economic; fiscal; and social, political, and environmental feasibility.

7.4   IMPLEMENTING THE SOLUTION

Key idea: Implement the solution through planning and action.

Implementation (literally, the act of filling up; from the Latin *in-* + *plere*, to fill) is the process of producing the product or system. Engineers participate in the implementation step by planning, supporting, and supervising the execution of the selected alternative. For many engineers, implementing the solution is the most satisfying step in the design phase. Finally seeing your ideas in concrete or as an accepted operating procedure on the shop floor is very gratifying and sets engineering apart from many other professions.

analysis paralysis: overanalysis of a problem to the extent that no action is taken

A common malady at this step in the engineering design process is **analysis paralysis**. People may become so involved in evaluating the alternatives that they never select or implement a solution. The analysis should end with a positive statement of the action to be taken to implement the selected alternative.

Implementing a solution to a design problem requires two steps: *planning* and *action*. The most important part of implementation is planning. In the planning stage, you must look at allocations of time and resources, anticipate bottlenecks, and identify a path to the finished product. Every step in the implementation process should be identified and documented. Using mathematical tools, implementation plans can be optimized and the effects of delays on the project schedule can be quantified. Two commonly used tools are the critical path method (CPM) and the program evaluation and review technique (PERT).

Finally, engineers and others involved in the problem-solving process must act on the plans they have formulated. They may execute a design alternative, fabricate a product, prepare a report, or conduct another planning activity. Great care should be taken with this phase, as good planning will not save a poor job of plan execution.

7.5 EVALUATING THE SOLUTION

continuous improvement: the process of continual reevaluation and redesign to continuously improve a product or process

Nearly all designs can be improved. Often, the deficiencies do not appear until months or years after the design is implemented. For many organizations, complex data systems are used to collect and analyze information from customers. The design team must be ready to repeat the entire design process to solve new problems as they appear. The repetitive design process, called ***continuous improvement***, is now recognized as essential for organizations of all types to achieve and maintain leadership positions in their fields.

7.6 DESIGN EXAMPLE

The world of the automobile has changed significantly since the first Model A rolled off Henry Ford's Detroit assembly line in 1903. However, the power plant (i.e., the internal combustion engine) and powertrain (i.e., transmission) used today would be understood by engineers of Ford's day.

The requirements for today's passenger automobiles are becoming increasingly stringent. Design goals are to meet near-zero emission standards and high fuel efficiency, while satisfying acceleration, safety, handling, comfort, carrying capacity, useful life, and maintenance cost constraints.

Internal combustion engines have three main problems. First, they are complex and therefore expensive to maintain. Second, they are inefficient and inflexible. For example, to satisfy acceleration requirements, their size must be 3 to 10 times greater than the size of engine that will provide the power required for cruising (i.e., power required for operation at constant velocity). Third, they produce air pollutants.

The design example concerns ways to improve vehicle performance.* The focus here is on the generation, evaluation, and selection of alternatives. Suppose your integrated project team conducted a brainstorming session. Three ideas emerged regarding how to optimize vehicle performance:

Alternative 1: Decrease the power requirements.

Alternative 2: Increase the efficiency of energy transfer from the fuel to the wheels.

Alternative 3: Employ complementary propulsion technologies.

In the usual design process, each alternative would be evaluated separately. This example will focus on the ways to decrease the power requirements. You know that automobiles must provide power to satisfy two demands. First, the power provided must overcome resisting forces. The power required to overcome the resisting forces is equal to the sum of the resisting forces multiplied by the velocity (power = force × velocity). Resisting forces include

- the drag force exerted by the car body on the air,
- the component of the gravitational force exerted in the direction of travel if the car is going up a hill,
- the friction forces from the tires meeting the road, and
- the force to accelerate the vehicle (from Newton's Second Law of Motion).

---

*For more information on this design example, see Hyman (2003) and Moore (1996).

Second, there should be sufficient power to satisfy the electrical systems (sound system, air conditioning, heating, and other accessories). Thus,

$$\text{power required} = (\text{sum of the resisting forces})(\text{velocity}) + \text{other power needs}$$

$$= (\text{drag force} + \text{gravitational force} + \text{friction force}$$

$$+ \text{ force to accelerate the vehicle})v + P_{\text{other}}$$

$$= \left[\tfrac{1}{2}\rho_a C_d A v^2 + mg(\sin \theta) + r_0 mg + ma \right]v + P_{\text{other}} \qquad (7.1)$$

where [standard values in brackets]

ρ_a = density of air [1.2 kg/m^3]

C_d = drag coefficient [0.3, dimensionless]

A = cross-sectional area of the vehicle [2.1 m^2]

v = vehicle velocity [compare alternatives at 90 km/h = 25 m/s]

m = vehicle mass [585 kg + two 68 kg passengers = 721 kg]

g = gravitational acceleration [9.8 m/s^2]

θ = road grade [compare alternatives at θ = 0 for level cruising]

r_0 = rolling friction coefficient [0.01 dimensionless]

a = vehicle acceleration [compare alternatives at constant velocity cruising, so a = 0]

P_{other} = other power requirements (heating, cooling, and accessories) [500 W]

This model includes five ways to reduce the power consumption: reduce the mass, reduce the cross-sectional area, reduce the drag coefficient, reduce the rolling friction coefficient, and/or reduce other power drains. Equation (7.1) is the analysis tool to assess the impact of changes in the mass, area, drag coefficient, rolling friction coefficient, and other power drains on the power requirements. Equation (7.1) must be coupled with reality: for example, the mass can be reduced only so much before safety constraints are violated.

The impact of potential changes on the power requirement are summarized in Table 7.1. With aluminum-based materials, mass reductions of 40% may be possible, with a resulting power savings of 7%. The greatest power savings can be obtained from a decrease in the drag coefficient. Why? At the test velocity (90 km/h), aerodynamic drag is the largest source of power demand. The relative contributions to power demand are shown in Figure 7.3.

TABLE 7.1 Impacts of Changes in Vehicle Design on Power Requirements

Parameter	Potential Change	New Value	Power Savings
Vehicle mass	10% reduction[a]	663 kg	1.7%
	40% reduction[a]	487 kg	7.0%
Cross-sectional area	minimum possible[b]	1.9 m^2	6.9%
Drag coefficient	minimum possible[b]	0.20	24%
Rolling friction coefficient	special tires	0.005	11%
Other power needs	50% reduction	250 W	3.1%

[a]Mass reduction in the vehicle without passengers.
[b]Minimum reasonable value consistent with safety, etc.

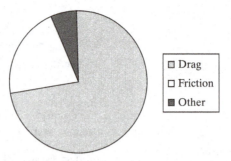

Figure 7.3. Contributions of Sources to the Power Requirements (calculated with the standard values following Eq. (7.1), so gravity and acceleration forces are zero)

In addition to reducing power requirements, complementary propulsion technologies (called auxiliary power units, APUs) have been proposed (alternative #3). Candidate APU technologies include fuel cells, gas turbines, and electric systems. To evaluate each APU, engineers have used the analysis process described in this chapter.

The hybrid–electric vehicle (HEV), a dual internal combustion engine/electric propulsion system, has emerged as a leading alternative, in part because it allows for energy recovery from braking. HEVs are favored over battery–electric vehicles (BEVs), because the usable power per unit mass of the batteries is hundreds of times less than the usable power per unit mass of gasoline; an additional pound of gas provides much more energy than an additional pound of battery. An increase in the mass will increase the power requirement (see the previous analysis).

Even if the choice is restricted to HEVs, the number of system configurations is large. Using morphological analysis (see Section 7.2.3), Steiber and Surampudi (2000) estimated that there are over 27,000 HEV combinations to be evaluated. Thus, the optimization of HEVs is likely to be a formidable design challenge.

Future generations of passenger vehicles must do more than transport people from point A to point B. The increased demands on automobiles require innovative uses of the analysis and design strategies. Relatively simple rules (such as providing sufficient power to meet demands) will be the basis of even the most complex analysis of the most innovative alternatives.

7.7   DESIGN PARAMETERS

7.7.1   Introduction

In this chapter, you have seen that analysis and design fit together. In a real sense, the way engineered systems are designed comes from repeated engineering analysis. One of the differences between engineers and engineering technicians is that engineers understand the analysis behind the design.

design parameters:
results of analysis that are used to determine the characteristics of a system

The results of analysis are sometimes summarized in easy-to-use collections of terms called ***design parameters***. Design parameters allow you to calculate key features of the system from known information. Some design parameters can become *code* (i.e., regulatory or legal requirements for design), design specification (i.e., design requirements in a project, also called *design specs*), or informal *rules-of-thumb* used by engineers in design. Using rules-of-thumb (or "seat-of-the-pants" engineering) is also called *heuristics* or *heuristic design* (from the Greek *heuriskein*: to discover).

7.7.2 Example

As an example, consider the design of a grit chamber. A grit chamber is a basin used to settle out small rocks so that downstream pumps can be protected. How big should such a chamber be? To answer this question, you must analyze the settling of particles. If the particle has a constant settling velocity, its trajectory in the chamber will be a straight line (see Figure 7.4). For simplicity, assume that if the particle impacts the right wall (path A in Figure 7.4), then it will be carried out of the chamber. Therefore, the critical trajectory is shown by path B. A particle settling as fast or faster than path B will be completely removed by the chamber. A particle settling slower than path B (e.g., path A) will not be removed by the chamber.

It is possible to calculate the settling velocity of a particle given its physical characteristics. The settling velocity is also related to the geometry of the chamber. Thus, it might be possible to devise a design parameter relating what you know (the settling velocity of the particle) to what you want to design (the dimensions of the chamber).

What is the settling velocity of a particle following path B? You already know that velocity = distance/time. So the settling velocity (v_s) of a particle following path B is

$$v_s = \text{distance/time} = D/(\text{time in the chamber})$$

where D is the chamber depth.

How much time does the particle spend in the chamber? It is pushed out of the chamber only by its horizontal velocity (v_h), so

$$v_h = L/(\text{time in the chamber})$$

where L is the length of the chamber in the direction of flow. Thus,

$$\text{time in the chamber} = L/v_h$$

Now, the horizontal velocity is given by the horizontal flow divided by the cross-sectional area of flow, or

$$v_h = \text{flow}/(WD)$$

where W is the width of the chamber. Thus,

$$\text{time in the chamber} = L/v_h = (LWD)/\text{flow}$$

and

$$\begin{aligned}
v_s &= D/(\text{time in the chamber}) \\
&= D/(LWD/\text{flow}) \\
&= \text{flow}/(LW) \\
&= \text{flow/plan area}
\end{aligned}$$

The plan area is the area of the floor of the chamber.

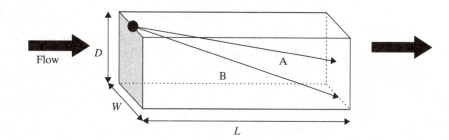

Figure 7.4. Possible Trajectories of a Settling Particle

Key idea: By specifying a value for the design parameter, you are specifying the performance of the system.

The term flow/plan area is a design parameter and is called the overflow rate.° *By specifying a value for the design parameter, you are specifying the performance of the system.* For example, suppose you want to settle out sand of 0.25 mm in diameter. You can calculate the settling velocity of the sand to be about 4.1 cm/s.[†] If you design a chamber with a flow/plan area = 4.1 cm/s, then all of the sand will be removed. The more common units of overflow rate are volume per time per area (gal/d-ft^2 or m^3/m^2/s). An overflow rate of 4.1 cm/s corresponds to about 87,000 gal/d-ft^2. (You might want to confirm the units conversion on your own.)

7.7.3 Uses of Design Parameters

Design parameters can be extremely useful for two reasons. First, they relate some measure of system performance with a system characteristic to be designed. In the settling chamber example, the plan area of the chamber was related to the removal of a certain sized sand particle. A design specification for a settling chamber might read: "Design the chamber with an overflow rate of 87,000 gal/d-ft^2." This is another way of saying, "Design the chamber to remove 0.25-mm-diameter sand completely."

Second, design parameters can be used easily to calculate system characteristics. In the settling chamber example, the plan area is calculated readily if the flow is known by plan area = flow/overflow rate. For example, removing sand from a flow of 10 million gallons per day would require $(1 \times 10^7 \text{ gal/day})/(87,000 \text{ gal/d-ft}^2) = 115$ square feet of area.

7.8   INNOVATIONS IN DESIGN

7.8.1   Introduction

In the engineering analysis and engineering design methods, you have been provided with lists of steps to be followed. Do not get the impression that analysis and design are cut-and-dried processes. Innovation is the key to success in both engineering analysis and design.

7.8.2   Need for Innovation

Although the general engineering analysis and engineering design approaches discussed in here have been successful, problems have arisen in using the standard approaches to bring products to the marketplace.

PONDER THIS

> **What problems do you see in using the standard design approach to design new products?**

Key idea: Design times can be very long for new products.

A major problem with the standard approach is that the *design time can be very long.* New products will take a long time to design if engineers and other professionals work independently. This might be called the "throw-it-over-the-wall" design approach:

- The engineering department designs a new product.
- They throw the design over the wall of the cubicle to the marketing department.
- Marketing changes the design to make it more customer friendly.

°It may seem strange that the settling performance appears to be independent of the chamber depth. In fact, if scouring of the particles off the chamber bottom can be ignored, then the depth is not important. Minimum depths usually are used to account for scouring.
[†]The settling velocity of a particle is given by $v_s = g(\rho_s - \rho)d^2/(18\mu)$, where g = gravitational acceleration = 980 cm/s^2, ρ_s = particle density = 2.65 g/cm^3 for sand, ρ = water density = 0.9997 g/cm^3 at 15°C, d = particle diameter = 0.025 cm, and μ = water viscosity = 1.37 $\times$ 10^{-2} g/cm-s at 15°C.

- Marketing throws design over the wall of the cubicle to the manufacturing unit
- Manufacturing changes the design to make the product easier and cheaper to make
- Manufacturing throws the revised design back to engineering
- Repeat *ad nauseum* until the design is complete

As you can see, this is a very inefficient way to bring a new product to market. It has been estimated that every blueprint page for the design of commercial aircraft in the past was revised 4.5 times, on average.

A lengthy design time is not the only problem if engineers and other professionals do not communicate.

PONDER THIS

What other problems arise if engineers develop new products without talking to others?

Key idea: Engineers must take into account the fact that products must be made, sold, used, and disposed of.

concurrent engineering: a design process where the manufacturing, quality control, end-user requirements, user support, and product disposal are considered in the original design

Engineers must take into account the fate of the design once it leaves their desks. First, the *product must be made.* In their design, engineers should take into account the difficulty in manufacturing or assembling products. In addition, the environmental impact of the manufacturing process must be considered. Second, *products must be purchased*, requiring the engineer to think about costs. Third, *products are used.* Thus, the service, maintenance, and support of the product must be considered in the initial design. Finally, the *product must be disposed of.* The environmental impact of products must be minimized. This includes post-consumer environmental impacts that occur when the consumer throws the product away.

7.8.3 Design Innovation by Concurrent Engineering

To address these problems, engineers have developed new design strategies, including concurrent engineering, reengineering (or redesign), and reverse engineering.

Concurrent engineering, CE, refers to a systematic approach to design, where all elements of product lifecycle are included. Elements include manufacturing, quality

green engineering: a design process where the impact on the environment of the raw materials, manufacturing process, product packing, product use, and product disposal are minimized

control, end-user requirements, user support, and product disposal. Design approaches to achieve CE include design for manufacturing (DFM) and design for the environment (DFE, also called **green engineering**).

A good example of DFM is a toy redesign executed by Mattel in the late 1980s.[*] In 1987, Mattel obtained the rights to a toy made by a Japanese toy company. The toy, called Color Spin, was designed to entertain and develop motor skills in infants. When the infant activated a roller (on the left in the accompanying picture), colored balls (on the right in the picture) would spin.

Unfortunately, there were several problems with the original Color Spin design. The main problem was that the toy cost too much to manufacture. Mattel's engineers and managers applied the principle of DFM to reduce the cost of the toy. An important aspect in DFM in this case was the idea of **design for assembly** (DFA): making the toy easier and less expensive to assemble. Assembly may not seem too important at first blush, but assembly costs can affect company profits significantly. For example, Mattel has estimated that over one-half million dollars per year could be saved by a *one-cent* reduction in the assembly of Barbie dolls.

There were several important changes in the assembly of the Color Spin toy. First, the total number of parts was reduced from 55 to 27. Second, parts requiring longer assembly times were reduced or eliminated. For example, fasteners (such as screws) require time to put together on an assembly line. The toy design engineers replaced two housing pieces and 10 screws with one piece of housing. (The housing pieces were attached by ultrasonic welding rather than fasteners, removing the construction task from the assembly line.) Third, parts were standardized to reduce the number of different pieces that the assemblers had to manipulate. Examples include the standardization of gears and the redesign of the roller so that the two roller halves were identical. The DFA approach resulted in a 38% reduction in assembly costs and a savings of $700,000 per year.

design for assembly: (DFA): a type of DFM where assembly requirements are considered in the original design

Two other notable changes improved the quality and marketability of the product. Throughput (the number of toys made per day) was limited because of back-ups on the assembly line. Back-ups occurred if an assembly line worker took too long to complete

---

[*]The background information for this example was taken from "The Mattel Color Spin: A Case Study in Design," created by the Engineering Systems Research Center at the University of California at Berkeley and part of the SYNTHESIS National Engineering Education Coalition (http://bits.me.berkeley.edu/develop/mattel2/welcome.html).

his or her job. Mattel installed storage bins between the workstations, allowing partially completed toys to accumulate. This, in turn, allowed workers to increase the quality of their work without the pressure of holding up the line. (This approach is called the "Pull System" of manufacturing.) Also, design engineers redesigned the packaging to save costs and allow parents to evaluate the toy in the store more effectively. It is believed that the packaging redesign by itself increased sales 5 to 10%.

7.8.4 Design Innovation by Reengineering

reengineering: the fundamental rethinking and radical redesign of a system

Reengineering refers to a fundamental rethinking and radical redesign of a system. The term "reengineering" is used to describe large changes in approaches to engineering, computer software, and business systems. Two examples in computer and electrical engineering can be found in recent "revolutions" proposed by Apple and Microsoft. In August 2000, Apple introduced a very different looking desktop computer, the Power-Mac G4 Cube. It was a very powerful computer. Through impressive engineering design, it fit into an $8'' \times 8'' \times 8''$ cube. Although elegant, the computer was criticized for a lack of expansion slots, limited audio inputs/outputs, and high cost. The design was not a success and was withdrawn in July 2001.

Also in 2000, Microsoft Corporation announced a new approach to business and personal computing: Microsoft .NET technology. The .NET initiative is an Internet-based computing platform promoting services distributed through the Web. Although Microsoft was not the first company to suggest that Web-based computing is the next wave, it has radically redesigned its software to take advantage of Web distribution of information and services. Will the .NET approach revolutionize computing? The answer, as with many initiatives in product engineering, is up to the consumers.

7.8.5 Design Innovation by Reverse Engineering

reverse engineering: the process of taking apart an object or system to see how it works

Reverse engineering refers to the process of taking apart an object or system to see how it works. You may have performed "reverse engineering" as a child on small household objects as your curiosity in technology grew. Reverse engineering is used in two ways. First, it can be used to obtain new ideas from competitors. Engineering ethics requires that applicable patent and copyright laws be strictly adhered to. Second, reverse engineering can be used to fabricate copies of parts for old equipment (sometimes called *legacy equipment*).

7.8.6 How to Innovate

How do engineers come up with completely new ideas? To reach a radically different design, you need the ability to think outside the normal constraints. This is frequently illustrated by asking the following:

PONDER THIS

Can you connect the nine dots in Figure 7.5 with four straight lines without lifting the pencil from the paper?

The solution, shown after the Problems section of this chapter, is to extend the lines *beyond* the constraining "box" formed by the dot array. Based on this puzzle, the ability to think beyond apparent constraints is known as thinking "outside the box."[*] Most people assume incorrectly that the lines may not be extended beyond the dots and are not able to solve the problem.

---

[*]The phrase "thinking outside the box" is overused and has been relegated to the dustbin of outdated business jargon. However, the phrase is a very clear illustration of the need to identify and question artificial constraints in engineering problem solving.

Figure 7.5. The Nine-Dot Problem

paradigm: a framework that defines boundaries around and successes in a discipline

paradigm paralysis: the often incorrect notion that past operating rules will always be successful in the future

Key idea: To solve a challenging engineering problem, you may have to go outside the existing paradigm and reengineer the system.

Thinking "outside the box" is described more formally as a *paradigm shift*. A *paradigm* is a model or pattern based on a set of rules that defines boundaries and specifies how to be successful within these boundaries (from the Greek *para-* + *deiknynai*, to show side by side). *Paradigm paralysis* occurs when a person or an organization is frozen with the idea that the rules successful in the past will *always* be successful in the future. Someday, you will hear a person in your organization say, "This is the way we have always done it. Everything seems to be going okay, so why should we change what we are doing?" This person is caught in paradigm paralysis. When a paradigm shifts, a new model based on a new set of rules replaces the old mode. The new rules establish new boundaries and allow solutions to problems that were previously unsolvable.

As an example of paradigm paralysis, American industry was stuck in the paradigm of "quality costs money" until Japanese industry demonstrated that design features that reduce complexity not only reduce cost but also improve quality. Major cost reductions and quality improvements occurred in American-built products in a few years after this new paradigm became accepted.

Similarly, industry has been trapped in the paradigm that "pollution reduction costs money." Only recently has industry recognized that money often can be *saved* by minimizing the amount of pollution discharged to the environment through the reuse and recycling of materials. A new paradigm, "pollution means inefficiency," has worked its way into the business culture.

Success in a new paradigm is only temporary. You must be open to the next new paradigm. For example, many companies that were once leaders in the rapidly changing technology of personal computers fell by the wayside when they failed to adapt to the next challenge.

7.8.7 Translating Failure into Success through Innovation

Key idea: Design is iterative; failure is an inherent part of engineering progress.

Truly innovative solutions that have significantly impacted our lives usually involved risk: risk that the solution will fail or will not be accepted. For example, it is said that Edison had hundreds of failures before finding a suitable filament material for his electric lamp. Even outright failure can result in opportunity if you are open to new ideas. A glue that did not stick well enough and was nearly abandoned became the solution for 3M's Post-it Notes®. As in these examples, engineers sometimes are called upon to translate failure into success.

Solving design problems is often an iterative process (see Figure 7.1). As the solution to a design problem evolves, you may find yourself continually refining the design. While implementing the solution to a design problem, you may discover that the solution you developed is unsafe, too expensive, or will not work. You then "go back to the drawing board" and change the solution until it works.

Despite the best efforts of engineering designers, designs occasionally fail. Bridges collapse, roofs fall in, and dams fail, potentially causing loss of life and property damage. Although the object of engineering design is to avoid failure, truly foolproof design is usually not economically feasible. One of the ironies of engineering is that failure is often one of the outcomes of success. In the interests of economy, there is a tendency to be more daring in design and to take greater risks. This eventually leads to failure. When a major engineering failure occurs, there is usually pressure to increase factors of safety and generally engage in more conservative engineering practices. Thus, engineering progresses slowly, with failure as an inherent (but sometimes tragic) component. For an example of the upward spiral of success and failure in engineering, see the *Focus on Design: What Comes Around, Goes Around.*

FOCUS ON DESIGN: WHAT COMES AROUND, GOES AROUND

BACKGROUND

Engineers seek to optimize the systems they design. This mentality is summarized in NASA's "faster, better, cheaper" management philosophy introduced in 1992. Sometimes, however, in seeking to make engineered systems less expensive or more attractive, reasonable limits of safety are violated. The result can be catastrophic failure. What happens next? The engineering community usually responds by *overdesigning* the next system (i.e., making the next system safer than necessary). The inevitable result is that the overdesigned systems are gradually reduced in safety to make them faster, better, cheaper, or more aesthetically pleasing—resulting eventually in an *underdesigned* system and failure once again. Like Sisyphus of Greek mythology (forced to eternally roll a stone block up a steep hill, only to have it roll back down), engineers can get locked into this cycle of failure, overdesign, gradual underdesign, and refailure. Engineering practice escapes this "infinite loop" only by a paradigm change to a new design approach.

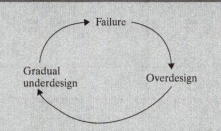

PARADIGM SHIFTS IN BRIDGE DESIGN

One example of how paradigm changes can break the success/failure cycle is in bridge design. It has been argued (Sibly and Walker, 1977) that bridge designs change radically about every 30 years, usually after a series of failures and conservative redesigns. By the 1930s, suspension bridges had become very popular in the United States. Two important features of suspension bridge design are (1) the ratio of the main span length to the bridge width and (2) the ratio of the main span length to the depth of the stiffener. The main span is the part of the bridge suspended between the main towers. The stiffener can be a truss (triangular arrangement of supports) or plate girder (solid structure resembling an I-beam) and is located under the roadway. The stiffener depth is its vertical dimension.

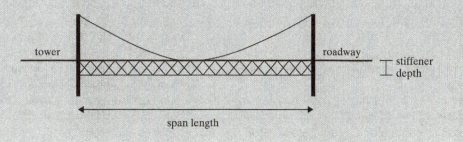

Tacoma Narrows Bridge after the 1940 collapse (note the narrow girder plates) (Photo courtesy of University of Washington Libraries.)

Rebuilt Tacoma Narrows Bridge with deep stiffening trusses (Photo courtesy of Corbis.)

A change in bridge design, the underdesign phase, was beginning in the 1930s. For example, the George Washington Bridge (opened in 1931) and Golden Gate Bridge (opened in 1937) both were wide with very deep (25 and 29 feet deep, respectively) stiffening trusses. However, some bridge designers, including Leon Moissieff, argued that stiff trusses did not dampen the deflections of a suspension bridge very much and that narrow long-span bridges could be successful. For example, the Bronx–Whitestone Bridge (opened in 1939) was built with relatively shallow (11-foot-deep) girder plates.

TACOMA NARROWS BRIDGE

Moissieff got the opportunity to put his ideas into concrete and wire with the Tacoma Narrows Bridge. The Tacoma Narrows is a beautiful, yet windy, passage between the Olympic Peninsula and the Washington mainland in western Washington state. The original design for the Tacoma Narrows span included 25-foot-deep stiffening trusses, but its cost ($11 million) far exceeded the cash available from the federal Public Works Administration and bonds. Moissieff proposed a bridge design that would come in under the available funds: a very narrow bridge with 8-foot-deep girder plates. The resulting bridge was extremely slender and graceful.

Most of the readers of this text already know the end of this sad story. On November 7, 1940, eight weeks after it opened to traffic, the Tacoma Narrows Bridge

collapsed catastrophically. Miraculously, the only fatality was a pet dog. The mechanisms behind the failure are complex, but fingers were pointed immediately at the narrow deck and shallow, solid girder plates.

What was the response of the engineering community to the failure? Overdesign. The Bronx–Whitestone Bridge was retrofitted with 14-foot-deep stiffening trusses, decreasing its length-to-stiffening-member depth ratio by a factor of two. The next suspension bridge built after the collapse was, ironically, the *second* Tacoma Narrows Bridge. In length-to-width ratio and length-to-stiffening-truss depth ratio (not plate girders!), it resembled the 1931 George Washington Bridge. The Mackinac Bridge (1957), although narrow relative to its length, also was built with a very conservative length-to-stiffening-truss depth ratio.

Was bridge design doomed to an endless cycle of underdesign, failure, and overdesign? No—a paradigm shift broke the cycle. The paradigm shift in bridge design was the introduction of cable stay bridges in the 1970s. If Sibly and Walker's idea that the bridge designs change radically about every 30 years is accurate, then we are due for a paradigm shift in bridge design soon. Will you be part of the next one?

Postscript: The Bronx–Whitestone Bridge is being retrofitted once again to recapture its original airy profile. The deep stiffening trusses will be removed and replaced with wedge-like structures to allow the wind to flow around the bridge and reduce twisting. And a twin span is planned across the Tacoma Narrows—the third Tacoma Narrows Bridge.

7.9   SUMMARY

Design results in new or improved devices or systems. In design problems, the system is poorly defined and more than one (or sometimes no) solution is possible. Design problems are solved by defining the problem, gathering information, generating multiple solutions, selecting a solution, implementing the solution, and evaluating the solution. Multiple solutions (i.e., alternative designs) are generated through brainstorming, making checklists, attribute listing (including morphological analysis), and using random forced relationships.

The alternatives are evaluated by assessing their feasibility using the engineering analysis method. Remember that a design is not completed until the solution is implemented (through planning and action) and monitored.

Design parameters are the result of analysis and are used to determine the characteristics of a system. They relate a measure of system performance with some system characteristic to be designed.

Standard analysis and design processes can create problems in product development if engineers work independently from other professionals. Design times can be very long, especially if manufacturing, assembly, marketing, and disposal of products are not considered early in the design process.

One technique for improving design is concurrent engineering, an approach to design where all elements of product lifecycle are included. Examples of concurrent engineering include design for manufacturing, design for assembly, and design for the environment (also called green engineering). A second technique, called reengineering, involves a radical redesign of a system or fundamental rethinking of an engineering problem. Reengineering requires a creative and fluid mind. Third, reverse engineering (the process of taking apart an object or system to see how it works) can be used in specialized situations to improve design.

Finally, watch for barriers to creativity in design, especially paradigm paralysis, where approaches used in the past are assumed to be valid in the present. Be aware, but not afraid, of risk. Remember that solving design problems is often an iterative process.

SUMMARY OF KEY IDEAS

- Solve engineering design problems by defining the problem, gathering information, generating multiple solutions, selecting a solution, implementing the solution, and evaluating the solution.
- Generate multiple solutions through brainstorming, making checklists, listing attributes, and employing forced random relationships.
- Attribute values can be combined in different ways to generate new ideas.
- Evaluate alternatives by applying the engineering analysis method to each alternative.
- Evaluate and select alternatives by assessing feasibility.
- Implement the solution through planning and action.
- By specifying a value for the design parameter, you are specifying the performance of the system.
- Design times can be very long for new products.
- Engineers must take into account the fact that products must be made, sold, used, and disposed of.
- To solve a challenging engineering problem, you may have to go outside the existing paradigm and reengineer the system.
- Design is iterative; failure is an inherent part of engineering progress.

Problems

7.1. How is the engineering design process different from the engineering analysis process?

7.2. Using each step in the engineering design process, generate alternatives and select a design for an elevated bed support for a dorm room.

7.3. Using each step in the engineering design process, generate alternatives and select a design for a fastener that can hold documents that are 2 to 150 pages long.

7.4. Using each step in the engineering design process, generate alternatives and select a design for a Web page for the online evaluation of engineering courses.

7.5. Another method for generating new ideas is bionics. The bionics technique uses analogies with the natural world to solve engineering problems. An example is the development of the hook and loop fastener. Swiss inventor George de Mestral observed that cockleburs attached to his wool pants and the fur of his dog; thus, he was inspired to develop Velcro® (named after the French *velour crochet*: velvet hooks). State a design problem and use bionics to generate solutions to the problem.

7.6. New ideas also can be generated by the method of *inversion*. In the inversion technique, you seek to achieve the opposite of the design goal and then invert the solution for the original design problem. For example, suppose your goal is to develop a very fast switch. Using inversion, you might consider how to make a slow switch (e.g., using high-resistance material) and then invert the solution (e.g., using low-resistance material) in your design. State a design problem and use inversion generate solutions to the problem.

7.7. Determine two design parameters for bridges from the *Focus on Design: What Comes Around, Goes Around*. Use the Internet to determine the values of the design parameters for two bridges.

7.8. Discuss an example of a local engineering project that had (or has) implementation challenges. How are engineers and other professions seeking to solve the implementation problems?

7.9. Write a short paper on how the development of Teflon is an example of a shortcoming turned into a success.

7.10. Practice thinking outside the proverbial box by solving the 16-dot problem. (Use a 4×4 matrix; Figure 7.5 shows a 3×3 matrix). Can you think of a strategy to solve the n^2-dot problem ($n \times n$ matrix)?

PART III
Engineering Problem-Solving Tools

I have no data yet. It is a capital mistake to theorise before one has data. Insensibly one begins to twist facts to suit theories, instead of theories to suit facts.
Arthur Conan Doyle

The purpose of models is not to fit the data but to sharpen the questions.

Samuel Karlin

8

Introduction to Engineering Problem-Solving Tools and Using Data

8.1 INTRODUCTION

8.1.1 Engineering Problem-Solving Tools

Engineers use four problem-solving tools. First, engineers collect *data* to test hypotheses, conduct analyses, and do design. Techniques to get the most out of your data are discussed in this chapter. Second, *models* are used. Models are conceptual, mathematical, or physical representations of the engineering system of interest. Engineers use models to verify system performance prior to design. Third, engineers use *computers* to perform calculations and visualize their results. Finally, engineers use feasibility concepts to evaluate alternative designs.

8.1.2 Using Data

All engineers generate data, use data, and frequently perform calculations involving data. In your engineering curriculum, you will take many classes in mathematics. It is likely that you will take a course or two in probability, statistics, and experimental design. In those courses, you will learn about the characteristics of data and how to manipulate data. In this chapter, a few basic concepts concerning experimental data will be introduced.

8.2 ACCURACY AND PRECISION

8.2.1 Introduction

Engineers measure characteristics of the real world. If life were perfect, you could collect data and determine *exactly* the parameter of interest to you. This is almost never the case. As an example, consider a human factors engineering study in which you must determine the distance from a computer user's eyes (modeled by a mannequin) to the computer monitor. The data collection requirements seem

OBJECTIVES

After reading this chapter, you will be able to:

- compare the concepts of accuracy and precision;
- round numbers appropriately;
- report numbers to the proper number of significant figures;
- list the important measures of central tendency and explain when to use them;
- list the important measures of variability and explain when to use them.

simple: you could easily measure the distance with a ruler or other measuring device. Common experience tells you that if you performed the measurement numerous times, you might get different results. If each one of your studymates made the measurement, you would likely get even more variety in the responses. In spite of the variability, there is one true distance (at least, one true distance when measuring at a fixed scale).

8.2.2 Accuracy

accuracy: a measure of closeness to the true value

How can you describe how closely your measurements are to the true answer? In this section, the relationship between measured and true values will be discussed in a general way. A more quantitative discussion may be found in Sections 8.4 and 8.5. The relationship between measurements and the true value is called **accuracy**. A measurement is said to be *accurate* if it is near the true value. For example, if the mannequin's eyes are set at 40.0 cm from the monitor, a measurement of 39.9 cm may be accurate and a measurement of 45.6 cm may be inaccurate (depending on the needs of the study).

8.2.3 Precision

precision: a measure of similarity in a set of values

The relationship between repeated measurements is called **precision**. A set of measurements is said to be *precise* if the measurements are similar in value. For example, suppose the measurements from the mannequin's eyes to the monitor are 31.6, 31.5, 31.6, and 31.4 cm. This set may be said to be precise (although inaccurate).* An example of accuracy and precision is shown in Example 8.1.

You may have seen the concepts of accuracy and precision illustrated with a dartboard or archery target. If the goal is to hit the bull's eye, then accurate shots are near the bull's eye and precise shots are clustered together (but not necessarily near the bull's eye). This is shown in Figure 8.1.

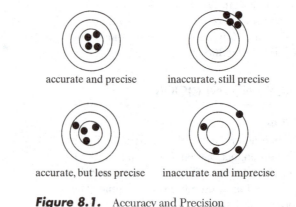

accurate and precise   inaccurate, still precise

accurate, but less precise   inaccurate and imprecise

Figure 8.1. Accuracy and Precision

---

*Be careful about the use of the word "precise." In common usage, "precise" is used to mean "exact" ("That is precisely my point."). In scientific work, use "precise" and "precision" only in reference to repeated measures.

EXAMPLE 8.1 ACCURACY AND PRECISION

For the human factors example, how would you label the following data sets with regard to accuracy and precision (if the "true" distance from the mannequin's eyes to the monitor is 40.0 cm)?

Set #1 = 40.1, 40.0, 39.8, and 40.0 cm

Set #2 = 39.8, 41.4, 39.4, and 40.9 cm

Set #3 = 35.2, 35.3, 35.3, and 35.1 cm

Set #4 = 36.7, 45.6, 46.2, and 34.9 cm

SOLUTION

The answer depends on the needs of the study. A reasonable answer is as follows:

Set #1 = accurate and precise

Set #2 = accurate, but less precise

Set #3 = less accurate, but precise

Set #4 = much less accurate and much less precise

The data are plotted in the following figure:

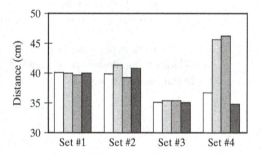

Note that the concepts of accuracy and precision are easy to see in this figure. More quantitative measures of accuracy and precision will be developed in Sections 8.4 and 8.5, respectively.

8.3   ROUNDING AND SIGNIFICANT DIGITS

8.3.1   Introduction

The concepts of precision and accuracy do not help you to *record* data and data calculations. A particularly troublesome area of data reporting and calculations concerns the number of decimal places to report. In the computer monitor problem, suppose you and a friend measure the distance from the mannequin's eye to the monitor. You use a meter stick and find the distance to be 40.6 cm. Your friend uses a yardstick and finds the distance to be 15 11/16 inches. Your friend realizes that the answer is supposed to be reported in centimeters and performs the following conversion:

$$\text{distance} = (15\ 11/16\ \text{in})(2.54\ \text{cm/in})$$
$$= (15.6875\ \text{in})(2.54\ \text{cm/in})$$
$$= 39.84625\ \text{cm}$$

Thus, your friend reports the distance as 39.84625 cm.

PONDER THIS

> **Is it really true that your measurement is only accurate to the nearest 0.1 cm, but your friend's measurement is accurate to 0.00001 cm?**

Key idea: Do not blindly report the numbers given to you by a calculator or spreadsheet—determine the proper number of digits to report.

The answer is, emphatically, *No! Just because your calculator reports five figures after the decimal point does not mean that the measurement is accurate to five figures after the decimal point.*°

8.3.2 Counting the Number of Significant Digits

significant digits: the number of digits justified by the precision of the data

If reporting all digits is incorrect, how *should* you report measurements and calculation results? To determine the number of decimal places, it is important to understand the idea of **significant digits** (or *significant figures*). Determine the number of significant digits of a number by the following procedure (for numbers containing a decimal place):

1. Starting on the *left* side of the number, move *right* until you encounter the first nonzero digit (ignoring the decimal place). Count this first nonzero digit as "one."

2. Continue moving to the right, counting each digit (still ignoring the decimal place). When you reach the last digit of the number on the right, you have counted the number of significant digits.

As an example, count the number of significant digits in the number 0.0504. It is a good idea to mentally separate each digit, ignoring the decimal place. One way to do this is to put each number in a box:

Counting the boxes from left to right, the first box containing a nonzero digit is the third box from the left. Number this box "1" and continue counting from left to right until you run out of digits:

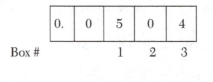

The last box is the third numbered box. Thus, 0.0504 has three significant digits. How many significant digits are in the number 120.0? Repeat the procedure by separating each digit (ignoring the decimal place):

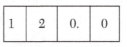

°By point of reference, 0.00001 cm = 100 nanometers, about the size of a virus. It is unlikely that a yardstick can measure a distance of about 40 cm to the accuracy of the size of a virus.

Counting the boxes from left to right, the first box containing a nonzero digit is the first box from the left. Number this box "1" and continue counting from left to right until you run out of digits:

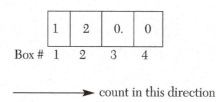

The last box is the fourth numbered box. Thus, 120.0 has four significant digits. See Example 8.2 for more examples of counting the number of significant digits.

8.3.3 Exceptions to the Rule: Numbers with No Decimal Point and Exact Numbers

The procedure in Section 8.3.2 for counting the number of significant digits works for most numbers. The rule implies that leading zeros to the left of the decimal point are ignored. What about numbers with *no* decimal point? Does the number "8" have the same number of significant digits as "8." or "8.0" or "8.00"? Numbers without decimal places are tricky. It is clear from Section 8.3.2 that "8.," "8.0," and "8.00" have one, two, and three significant digits, respectively. However, *you do not know how many significant digits there are in the number 8 when it is written without a decimal point.*

To avoid this uncertainty, refrain from writing numbers without decimal points. If you wish to indicate three significant figures with the number seven hundred, write it as "700." not "700" (i.e., write it *with* a decimal point). (In addition, *always* include a leading zero when reporting numbers between 1 and −1. It is much clearer if you write "−0.14" or "0.56" than if you write "−.14" or ".56".)

You also can use scientific notation to show the number of significant digits. Count the number of significant digits in a number in scientific notation by applying the procedure of Section 8.3.2 to the mantissa.[*] Thus, the numbers "$7. \times 10^2$", "7.0×10^2", and "7.00×10^2" have one, two, and three significant digits, respectively. (Again, mantissas without decimal places are tricky: avoid writing numbers such as "7×10^2".)

What about *exact* numbers? You may wish to do calculations that involve numbers that are exact. Exact numbers have no variability. For example, there are *exactly* 100 cm in a meter, *exactly* three feet in a yard, and *exactly* eight sides in an octagon. In engineering calculations, you may wish incorporate exact numbers into calculations, such as the number of roads in a city, the number of distillation columns in a factory, or the number of capacitors in a circuit.

PONDER THIS

How many significant digits are there in an exact number?

Exact numbers are treated as if they have an *infinite* number of significant digits. This may seem a little strange at first, but as you shall see in Section 8.3.5, the number of significant digits in exact numbers is ignored in calculations.

---

[*]Numbers in scientific notation have three parts: the base (usually 10), the mantissa (number before the "×" symbol), and the exponent (exponent on the base). Thus, in the number 1.71×10^5, the mantissa is 1.71, the base is 10, and the exponent is 5. The word *mantissa* means "an addition of little importance." In some software packages, "×10" is replaced with an uppercase letter "e." Thus, $1.71 \times 10^5 = 1.71E5$.

**EXAMPLE 8.2
SIGNIFICANT
DIGITS**

Count the number of significant digits in the following measurements: 43 cm, 4.3 kV, 0.43 Ω, and 0.043 microcurie (0.043 μCi). Count the number of significant digits in the numbers 691, 1.30, and 0.00000500.

SOLUTION

Start with the first nonzero digit from the left and count to the right until encountering the last digits. For example, the measurement "4.3 kV" can be counted as

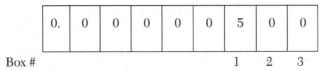

The number "0.00000500" can be counted as

0.	0	0	0	0	0	5	0	0

Box #   1   2   3

Thus, each number and unit in the set of measurements (43 cm, 4.3 kV, 0.43 Ω, and 0.043 μCi) has two significant digits. Each number in the set of numbers (691, 1.30, and 0.00000500) has three significant digits.

8.3.4   Reporting Measurements

Key idea: Report one more significant digit than the number of digits you are certain about. The last significant digit is understood to include some uncertainty.

You now know how to *determine* the number of significant digits in numbers. However, the question from Section 8.3.2 remains: how should you *report* measurements? The usual interpretation of significant digits is as follows: report *one more significant digit than the number of digits you are certain about*. In other words, the last significant digit is understood to be estimated and may include some uncertainty.

An example will help here. Suppose you are weighing concrete test specimens to test the strength of a new concrete formulation. The scale is marked in grams. You interpolate between gram markings to estimate tenths of grams. It would be proper to report a weight of 79.6 g. When another engineer reads this number, he or she will know that there is some uncertainty in the tenths of grams, because it is the last significant digit reported.

8.3.5   Rounding and Calculations

rounding: adjusting the value of certain digits to comply with the appropriate number of significant digits

Always report your calculations with the number of significant digits consistent with the data. Determining the number of significant digits involves two steps: deciding which digits to drop and deciding what to do with the digits you report. The latter process is called *rounding*.

There are two simple rules for rounding:

1. If the digit to be dropped is less than 5, then write the last digit retained as it is.
2. If the digit to be dropped is greater than or equal to 5, then increase the last digit retained by one.[*]

---

[*]Sometimes, a different convention is used if the digit to be dropped is equal to 5. Some people write the last digit retained as the nearest even digit if the digit to be dropped is equal to 5. For example, if you determine that three significant digits are appropriate, you would round 0.6225 to 0.622 and 1.235 to 1.24 in this system.

For example, if you determine that four significant digits are appropriate, you would round 95.673 to 95.67 (since the digit dropped, 3, is less than 5). Similarly, if you determine that five significant digits are appropriate, you would round 0.0124457 to 0.012446 (since the digit dropped, 7, is greater than 5).

How do you determine the appropriate number of significant digits in a calculation? Two rules will suffice here:

Key idea: The reported value is based on the smallest number of *significant digits* in the calculation for multiplication and division and on the smallest number of *decimal places* in the calculation for addition and subtraction. Exact numbers do not affect the number of digits reported.

1. When multiplying or dividing numbers, report the result to the number of *significant digits* of the value with the smallest number of significant digits.
2. When adding or subtracting numbers, report the result to the number of *decimal places* of the value with the smallest number of decimal places.

It is important to note the difference between how numbers are reported in different calculations. In multiplication and division, the reported value is based on the smallest number of *significant digits* in the calculation. This rule implies that the product or division of numbers cannot be more precise than the least precise number. For example, your calculator may report:

$$56.122/2.31 = 24.2952381 \ (\textit{Warning}: \text{Too many digits reported})$$

PONDER THIS

> **How would you round the results of the calculation of 56.122/2.31 = 24.2952381?**

You round the result to 24.3, because the smallest number of significant digits in the numbers on the left side of the equation is three ("2.31" has three significant digits).

In addition and subtraction, the second rule says that the reported value is based on the smallest number of *decimal places* (numbers to the right of the decimal place) in the calculation. As an example, your calculator may report

$$23.52 + 4.215 + 6.1 = 33.835 \qquad (\textit{Warning}: \text{Too many digits reported})$$

PONDER THIS

> **How would you round the results of the calculation of 23.52 + 4.215 + 6.1 = 33.835?**

You round the sum to 33.8, because the number "6.1" has only one digit to the right of the decimal point. Note that the sum is reported to three significant digits, even though one of the numbers on the left side (the number "6.1") has only two significant digits.

What about exact numbers in calculations? Exact numbers play no role in determining the number of digits reported.

PONDER THIS

> **Why should exact numbers not affect the number of reported digits in multiplication and division?**

Recall from Section 8.3.3 that *exact numbers are treated as having an infinite number of significant digits*. Thus, exact numbers do not affect the number of reported digits in multiplication and division. Why? Exact numbers can *never* have the smallest number of significant digits and can never control the number of digits reported.

In addition and subtraction, it also makes sense that exact numbers should not play a role in determining the number of digits reported. For example, suppose you are trying to convert a temperature reading from Kelvin to Celsius. A temperature of zero Kelvin (0 K) is defined to be exactly −273.16°C. Thus, 298.103 K is equal to −273.16 + 298.103 = 24.943 K. You report this temperature to three decimal places because the number "−273.16" is exact and does not affect the number of decimal places reported.

Finally, rounding is best performed on the final answer, not on intermediate calculations. If you round intermediate calculations, rounding errors may accumulate.

8.4   MEASURES OF CENTRAL TENDENCY

8.4.1   Introduction

Engineers usually take a more quantitative approach to the ideas of accuracy and precision than that presented in Section 8.2. To determine accuracy, you may wish to use one measure as representative of a set of data. This is called a measure of the *central tendency* of the data or the *average* (from the Arabic *'awariyah* meaning damaged merchandise, because the word "average" was originally applied to the process of proportionally distributing expenses for damaged goods during sea transport).

8.4.2   Arithmetic Mean

arithmetic mean: the sum of all values divided by the number of

$$\text{values} = \frac{1}{N}\sum_{i=1}^{N}x_i$$

There are many ways to take the average of a set of data. The most common measure is the **arithmetic mean** (often just called the *mean*). The arithmetic mean is calculated by summing all the values and dividing by the number of data points. For example, if the fuel efficiency of an innovative automotive engine was measured to be 56.2, 61.4, 55.2, and 60.9 miles per gallon (mpg), then the arithmetic mean would be

$$(56.2 + 61.4 + 55.2 + 60.9 \text{ mpg})/4 = (233.7 \text{ mpg})/4 = 58.4 \text{ mpg}$$

(Why was the answer reported to one decimal place? Each number to be summed was reported to one decimal place, so the sum should be reported to one decimal place. The number 4 is exact and does not affect how the result is reported.) If each data point is designated x_i and there are N data points $(x_1, x_2, x_3, \ldots, x_N)$, then the arithmetic mean is

$$\text{arithmetic mean} = \frac{x_1 + x_2 + x_3 + \cdots + x_N}{N} = \sum_{i=1}^{N}x_i$$

The uppercase sigma (see Appendix B) is read, "the sum from i equals 1 to i equals N of ...."

At first glance, it appears that the arithmetic mean is the *only* reasonable way to determine an average. However, the arithmetic mean can be misleading and is not always appropriate. Can you think of a situation where the arithmetic mean is **not** the best measure of central tendency?

THOUGHTFUL PAUSE

Take a guess at the mean wealth of eyeglass-wearing men in Washington state with the initials WHG who were born in 1951.

Now guess the mean wealth of this group if *one* member, Microsoft President Bill Gates, is excluded. The exclusion of Bill Gates would probably decrease the arithmetic mean of the wealth significantly.

Key idea: The arithmetic mean is sensitive to extreme values in the data set.

Similarly, note what happens to the arithmetic mean if you change *one* data point in the fuel efficiency data. If the fuel efficiencies were 56.2, 61.4, 55.2, and *20.9* mpg (rather than 56.2, 61.4, 55.2, and *60.9* mpg), then the arithmetic mean changes from 58.4 mpg to 48.4 mpg. These exercises demonstrate that the *arithmetic mean is sensitive to extreme values*. In other words, the largest and smallest values in the data set strongly affect the arithmetic mean.

8.4.3   Median

median: the middle data point when the values are listed in numerical order

To avoid the influence of extreme values, the median sometimes is used as a measure of central tendency. The **median** of a data set is the middle data point when the values are listed in numerical order. (For an odd number of data points, the median is the middle value when ordered. For an even number of data points, the median is the arithmetic mean of the two middle values when ordered.)

Say the engineering library has ten very old personal computers with hard-drive storage capacities of 1.2, 4.5, 6.4, 5.2, 6.4, 5.0, 2.3, 3.4, 6.3, and 8.2 gigabytes (1 gigabyte = 1 GB = 10^9 bytes). To determine the median storage capacity, order the values (8.2, 6.4, 6.4, 6.3, 5.2, 5.0, 4.5, 3.4, 2.3, and 1.2 GB). Since there is an even number of values, take the arithmetic mean of the middle two values (5.2 and 5.0 GB). The median amount of storage capacity is then 5.1 GB. (You may wish to confirm that the arithmetic mean amount of storage capacity is 4.9 GB. This is similar to the median, since there are no really extreme values here.) As another example, find the median value of 5.62, 4.1, and 6.2:

Values: 5.62, 4.1, 6.2

Ordered values:

4.1	5.62	6.2
smallest		largest

Median: 5.62

(use the middle value, since the number of values is odd)

8.4.4   Geometric Mean

geometric mean: the product of the values raised to the 1/N power = $(x_1 x_2 x_3 \ldots x_N)^{1/N}$ = $\left(\prod\limits_{i=1}^{N} x_i \right)^{1/N}$

There are several other types of less commonly used averages. The **geometric mean** is the product of the values raised to the 1/N power $(x_1 x_2 x_3 \ldots x_N)^{1/N}$, or, equivalently, the Nth root of the product of the values:

$$\text{geometric mean} = (x_1 x_2 x_3 \ldots x_N)^{1/N} = \sqrt[N]{x_1 x_2 x_3 \ldots x_N}$$

You can use uppercase pi (Π) to read "the product of ..." (just as Σ means "the sum of ...").

Thus,

$$\text{geometric mean} = \left(\prod_{i=1}^{N} x_i \right)^{1/N}$$

You can show that the logarithm of the geometric mean of a set of positive numbers is equal to the arithmetic mean of the logarithms of the numbers (see also Problem 8.10).

The geometric mean sometimes is used as a measure of central tendency with values that change over several orders of magnitude. For example, in environmental engineering, treated wastewater can contain no more than a specified number of a certain kind of microorganism. Since microorganism concentrations can vary greatly, the geometric mean is regulated. If the data for one week are 400, 100, 250, 100, 15, 20, and 15,000 organisms per 100 milliliters, then the seven-day geometric mean is as follows:

$$(400 \times 100 \times 250 \times 100 \times 15 \times 20 \times 15{,}000)^{1/7}$$

or 240 organisms per 100 milliliters.

8.4.5 Harmonic Mean

harmonic mean: the reciprocal of the arithmetic mean of the reciprocals of the values =

$$\dfrac{N}{\sum\limits_{i=1}^{N}\dfrac{1}{x_i}}$$

The **harmonic mean** is the reciprocal of the arithmetic mean of the reciprocals of the values:

$$\text{harmonic mean} = \frac{1}{\dfrac{1}{N}\sum\limits_{i=1}^{N}\dfrac{1}{x_i}} = \frac{N}{\sum\limits_{i=1}^{N}\dfrac{1}{x_i}}$$

The harmonic mean is used when the *reciprocals* of the data are important. For example, computer speeds often are assessed by benchmark tests, where the computation speeds (expressed in millions of instructions per second or MIPS) for several tasks are recorded. The computation *time* is more important than the computation *speed* for most computer users. The computation time is inversely proportional to the computation speed (speed = number of operations/time, so time = number of operations/speed). Thus, the appropriate measure of central tendency for computational speed is the harmonic mean.

As an example, suppose that five benchmark programs execute at 30, 700, 15, and 13,000 MIPS. The harmonic mean is

$$\cfrac{4}{\dfrac{1}{30\ \text{MIPS}} + \dfrac{1}{700\ \text{MIPS}} + \dfrac{1}{15\ \text{MIPS}} + \dfrac{1}{13{,}000\ \text{MIPS}}} = 39\ \text{MIPS}$$

The harmonic mean is influenced most strongly by the *smallest* values. For example, if the *largest* computation speed was doubled from 13,000 to 26,000 MIPS, the harmonic mean remains unchanged at 39 MIPS. However, if the *smallest* computational speed was doubled from 15 to 30 MIPS, the harmonic mean increases from 39 MIPS to 59 MIPS.

8.4.6 Quadratic Mean

quadratic mean (root mean square, RMS): the square root of the arithmetic mean of the squares of the values =

$$\sqrt{\frac{1}{N}\sum\limits_{i=1}^{N}x_i^2}$$

The **quadratic mean** (commonly called the **root mean square, RMS**) is the square root of the arithmetic mean of the squares of the values:

$$\text{quadratic mean} = \text{RMS} = \sqrt{\frac{1}{N}\sum\limits_{i=1}^{N}x_i^2}$$

The quadratic mean is used when an important property is proportional to the *square* of a measured value. For example, suppose a mechanical engineer bombards a surface with high-energy particles to learn more about the surface properties of the material. The information to be gathered depends on the energy of the particles. The engineer may be more interested in the quadratic mean velocity (RMS velocity) of the particles, rather than the arithmetic mean, because the energy of the particles is proportional to the velocity squared. (Recall that the kinetic energy = $\frac{1}{2}mv^2$, where m = mass and v = velocity.)

8.4.7 Mode

mode: the most frequently occurring value in a data set

Finally, the **mode** of a data set is the most frequently occurring value in a data set. In the hard-drive storage capacity example (Section 8.4.3), the mode is 6.4 GB, since that value is present at a higher frequency (2 out of 10 values) than any other (all others present at one out of ten values). An example of how to select the most appropriate measure of central tendency is given in Example 8.3.

**EXAMPLE 8.3
MEASURES OF
CENTRAL
TENDENCY**

Select and calculate the most appropriate measure of central tendency for the diameters of catalyst particles used in ammonia synthesis. A particle size analysis of 400 catalyst particles revealed 100 particles with diameter 5.4 μm, 100 particles with diameter 10.6 μm, 100 particles with diameter 7.5 μm, and 100 particles with diameter 8.4 μm. You are interested in the particle diameter, surface area, and surface-to-volume ratio (S/V).

SOLUTION

For the particle diameter, either the mean or median is appropriate, since the distribution is narrow. The mean of the particle diameters is $[(100)(5.4\,\mu\text{m}) + (100)(10.6\,\mu\text{m}) + (100)(7.5\,\mu\text{m}) + (100)(8.4\,\mu\text{m})]/400$ or **8.0 μm**. The median of the particle diameters is the arithmetic mean of 7.5 μm and 8.4 μm or **8.0 μm**. (Note the number of significant figures through this problem.)

Surface area is proportional to the diameter squared, so the quadratic mean of the particle diameters is the appropriate type of mean. The quadratic mean of the particle diameters is $\{[(100)(5.4\,\mu\text{m})^2 + (100)(10.6\,\mu\text{m})^2 + (100)(7.5\,\mu\text{m})^2 + (100)(8.4\,\mu\text{m})^2]/400\}^{1/2}$ or **8.2 μm**.

S/V is proportional to the reciprocal of the diameter, so the harmonic mean of the particle diameters is the appropriate type of mean. The harmonic mean of the particle diameters is $400/[(100)/(5.4\,\mu\text{m}) + (100)/(10.6\,\mu\text{m}) + (100)/(7.5\,\mu\text{m}) + (100)/(8.4\,\mu\text{m})]$ or **7.5 μm**.

(*Note:* The surface area and S/V also can be calculated directly and the arithmetic mean of their values calculated.)

8.5 MEASURES OF VARIABILITY

8.5.1 Introduction

Precision is a qualitative indicator of the variability of the data. There are three common *quantitative* measures of data variability: variance, standard deviation, and relative standard deviation.

Before developing the formulas for these measures, it is important to review two important types of data sets. If you examine all possible members of some group, then the measures of central tendency and variability are called **population** measures. For example, if you measured propulsion characteristics of all Space Shuttle engines, you could calculate the population mean of the propulsion characteristics.

population: all possible members of a group

In many cases in engineering, you can examine only a few members of the population. If so, label your measures **sample** measures. For example, you may determine the failure rate of a handful of circuit boards from a production line and calculate the sample mean of the failure rate. (Why test only a handful? If you tested *all* the boards to failure, there wouldn't be any left to sell.) If you have n members of the sample and N members of the population (where $n \leq N$), then you can define

sample: selected members of a group

$$\text{sample (arithmetic) mean} = \bar{x} = \frac{1}{n}\sum_{i=1}^{n} x_i, \text{ and}$$

$$\text{population (arithmetic) mean} = \mu = \frac{1}{N}\sum_{i=1}^{N}x_i$$

Note that different symbols are used for the sample and population means.

8.5.2 Variance

variance: a measure of data variability proportional to the sum of the squares of the differences between each data point and the mean

The formulas for the population mean and sample mean look similar. However, the differences between the sample and population measures of *variability* are more pronounced. The **variance** is one measure of data variability. The variance is proportional to the sum of the squares of the differences between each data point and the mean. The *sample variance* is given by

$$s^2 = \frac{1}{n-1}\sum_{i=1}^{n}(x_i - \bar{x})^2$$

The *population variance* is given by

$$\sigma^2 = \frac{1}{N}\sum_{i=1}^{N}(x_i - \mu)^2$$

(Note that the sample variance is calculated by dividing by $n-1$, while the population variance is calculated by dividing by N.) Why square the difference between the data and the mean? Squaring makes all the terms in the summation positive. Thus, contributions to the variance from data points *less* than the mean do not cancel out contributions from data points *greater* than the sample mean.

The variance has one big disadvantage as a measure of variability. To illustrate the problem, calculate the sample mean of the fuel efficiencies discussed in Section 8.4.2. The sample mean was 58.4 mpg. The sample variance is

$$s^2 = [(56.2 \text{ mpg} - 58.4 \text{ mpg})^2 + (61.4 \text{ mpg} - 58.4 \text{ mpg})^2 +$$
$$(55.2 \text{ mpg} - 58.4 \text{ mpg})^2 + (60.9 \text{ mpg} - 58.4 \text{ mpg})^2]/3$$
$$= 10.1 \text{ (mpg)}^2$$

Is 10.1 $(\text{mpg})^2$ big or small? It is hard to tell, because of the strange units of the variance. The units of the variance are the squares of the units of the observations.

8.5.3 Standard Deviation

standard deviation: a measure of data variability with the same units as each data point and equal to the square root of the variance

A more easily interpreted measure of data variability is the **standard deviation**. The standard deviation is the square root of the variance. Thus, the sample standard deviation is $s = (s^2)^{1/2}$ and the population standard deviation is $\sigma = (\sigma^2)^{1/2}$. For example, the sample standard deviation for the fuel efficiency example is

$$s = [10.1 \text{ miles}^2/\text{gallon}^2]^{1/2} = 3.2 \text{ mpg}$$

The standard deviation makes it clear that the variability is fairly low: the sample standard deviation is small compared with the sample mean.

8.5.4 Relative Standard Deviation

relative standard deviation (RSD or standard error): a dimensionless measure of data variability and equal to the standard deviation divided by the mean

The mean and standard deviation can be compared even more directly in the final common measure of variability: **relative standard deviation (RSD**, also called the **standard error**). The relative standard deviation is the standard deviation divided by the mean. It is usually expressed as a percentage. For the fuel efficiency data, the RSD is

$$(3.2 \text{ mpg})/(58.4 \text{ mpg}) = 0.055 \text{ or } 5.5\%$$

This reaffirms the observation that the variability in the data is fairly low. Another example of calculating the measures of variability is given in Example 8.4.

EXAMPLE 8.4
MEASURES OF
VARIABILITY

Calculate the relative standard deviations for the volumes of the Great Lakes and the lengths of all the pencils in the Western Hemisphere.

The volumes of the Great Lakes are as follows: Superior ($11,800$ km^3), Michigan ($4,800$ km^3), Huron ($3,500$ km^3), Erie (500 km^3), and Ontario ($1,600$ km^3). Using the pencils in my desk as examples, the lengths are 18.3 cm, 17.0 cm, 13.2 cm, 16.5 cm, and 18.5 cm.

SOLUTION

Since the data for all the Great Lakes are known, use population statistics. Thus, $\mu = 4,400$ km^3, $\sigma = 3,969$ km^3, and RSD $= \boldsymbol{\sigma/\mu} = \textbf{0.89 or 89\%}$. (Remember to divide by $N = 5$ for σ^2.)

Since the data for the pencils are samples, use sample statistics: $\bar{x} = 16.7$ cm and $s = 2.1$ cm, so RSD $= \boldsymbol{s/\bar{x}} = \textbf{0.13, or 13\%}$. (Remember to divide by $n - 1 = 4$ for s^2.)

8.5.5   Variability and Data Collection in Engineering

In engineering, variability leads to uncertainty. For example, increased variability in a parameter such as Young's modulus means more uncertainty about the stability of a structure. More uncertainty leads to more conservative designs. More conservative designs are more expensive. For example, if you are uncertain about the properties of a capacitor, then you may specify an overly large capacitor to account for the uncertainty. As a result, variability leads to higher costs.

Key idea: Variability increases uncertainty, leading to higher costs. Engineers collect data to reduce uncertainty.

How do engineers reduce uncertainty? Simply put, *engineers collect data to reduce uncertainty*. Two examples will demonstrate the relationship between data collection and uncertainty reduction. Suppose your firm is hired to design a waterproof cover for a baseball infield. You might design the cover with the assumption that the bases are spaced 90 feet apart. Are the bases exactly 90 feet apart? No, it is possible that they might be off by a tiny distance. However, the uncertainty regarding the base path distance is very small. In addition, the cost of overdesign (i.e., the cost of making the cover a little too big) is also very small. Therefore, the cost of actually measuring the base path distances probably is not outweighed by the reduction in uncertainty that the measurements would bring.

Now imagine you are designing the steel columns in a high-rise building. The columns transmit the force generated by the structure's mass to the foundation. Is it worthwhile to test the mechanical properties of the steel columns? The answer might very well be *yes*. While the cost of collecting the data is high, the potential payoff is very high. If you have to design for uncertain structural properties, then you might have to specify many additional columns. In addition, the cost of failure (in this case, the cost of collapse) is extremely high. Another example of how engineers pay to reduce uncertainty is shown in the *Focus on Variability: Paying to Reduce Uncertainty*.

safety factor: a multiplier of a design parameter used to account in part for uncertainty

Uncertainty in design sometimes is expressed by a *safety factor*. Design parameters sometimes are multiplied by safety factors to account for uncertainty. For example, steel buildings are designed with a safety factor of 2, while wooden buildings are designed with a safety factor of 6. Why the difference? The structural properties of wood are more variable than the structural properties of steel.[*]

---

[*]As the examples in the text show, safety factors are determined in part by the uncertainty in materials properties and the cost of overdesign. Safety factors also are influenced by the variability in loads and by the cost of failure. For example, the cables in high-speed elevators (which experience variable loads and for which the cost of failure is high) are designed with a safety factor of 11.9.

FOCUS ON VARIABILITY: PAYING TO REDUCE UNCERTAINTY

This chapter was devoted to how to *use* data. Perhaps a more fundamental question should be asked: why do engineers gather data in the first place? After all, data collection—whether a phone call or an hour on the Internet or two year's worth of laboratory research—costs money. The simple answer from the text is that engineers collect data to reduce uncertainty. When engineers pay for data collection, they are really paying to reduce uncertainty. So when should you collect data? *You collect data only when the benefits of reducing uncertainty outweigh the cost of data collection.*

You sometimes can quantify both the cost of collecting data and the degree of uncertainty reduction. Consider the case of simple random sampling. In many cases, it is common to collect a small number of samples relative to the total possible number of samples (using the symbols in Section 8.5.1: $n \ll N$). From the samples collected, it is possible to estimate the standard deviation of the sample mean. For example, if you collected five samples on ten occasions, you could calculate ten different sample means. Using those ten values, you could calculate the variance of the sample mean.

One result from statistics is that the variance of the sample mean $[\mathrm{var}(\bar{x})]$ is equal to the population variance divided by the number of samples (σ^2/n):

$$\mathrm{var}(\bar{x}) = \sigma^2/n$$

Rearranging terms yields

$$n = \sigma^2/\mathrm{var}(\bar{x})$$

This equation tells you that you need to collect twice as many samples in a given population to reduce the variance in the sample mean by half. In other words, decreasing $\mathrm{var}(\bar{x})$ by a factor of two means increasing n by a factor of two. In addition, the equation tells you that

an inherently more variable population (i.e., larger σ^2) will require more samples to maintain the same variance in the sample mean.

Suppose you are working on a new aircraft design and need to test the tensile strength of an aluminum alloy. How many alloy samples should you test? Say, from previous work, you know that $\sigma = 0.3$ ksi (tensile strength is measured in thousands of pounds per square inch, or ksi). Each test costs $25. By reducing the uncertainty in $\bar{x}$, you increase the benefits. Suppose that the relationship between benefits and variability is given by

$$\text{benefits (in \$)} = 2{,}000[0.3 - \mathrm{sd}(\bar{x})]$$

where $\mathrm{sd}(\bar{x})$ is the standard deviation of the sample mean in ksi. You know that

$$\mathrm{sd}(\bar{x}) = [\mathrm{var}(\bar{x})]^{1/2} = (\sigma^2/n)^{1/2} = \sigma/n^{1/2}$$

The cost of testing n samples (in $) is $25n$, since the tests cost $25 each. Thus,

$$\begin{aligned} \text{net benefits} &= \text{benefits} - \text{costs} \\ &= 2{,}000[0.3 - \mathrm{sd}(\bar{x})] - 25n \\ &= 2{,}000(0.3 - \sigma/n^{1/2}) - 25n. \end{aligned}$$

The net benefits are plotted as a function of n in the accompanying figure. Note that collecting a *small* number of samples is undesirable: the sample cost, although small, is not outweighed by the very small reduction in uncertainty. (A negative value for net benefits means that the costs are larger than the benefits.) Similarly, collecting a *large* number of samples is undesirable: the large sample cost is not offset by the corresponding reduction in uncertainty. Net benefits are maximized by collecting five samples in this case.

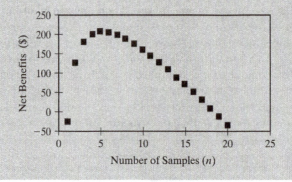

8.6 SUMMARY

Engineers generate and use data. Data may have errors, described qualitatively by the concepts of precision and accuracy. Calculated values should be presented with an appropriate number of significant digits (usually **not** the number given by your calculator or spreadsheet). Use measures of central tendency and variability to summarize your data and quantify data variability.

SUMMARY OF KEY IDEAS

- Do not blindly report the numbers given to you by a calculator or spreadsheet—determine the proper number of digits to report.
- Avoid writing numbers without decimal points, because such numbers have an indeterminate number of significant digits.
- Report one more significant digit than the number of digits you are certain about. The last significant digit is understood to include some uncertainty.
- The reported value is based on the smallest number of *significant digits* in the calculation for multiplication and division and on the smallest number of *decimal places* in the calculation for addition and subtraction. Exact numbers do not affect the number of digits reported.
- The arithmetic mean is sensitive to extreme values in the data set.
- Variability increases uncertainty, leading to higher costs. Engineers collect data to reduce uncertainty.

Problems

8.1.   Describe whether the following measures are more related to accuracy or precision:

   a. The range of scores on the midterm exam
   b. Free-throw shooting percentages
   c. Tolerance values on spark-plug gaps
   d. Length of a cold medication capsule
   e. Reproducibility of the lengths of cold medication capsules

8.2.   The subtraction of two values can result in a loss in the number of significant digits. Give an example of this phenomenon.

8.3.   Find three numbers in this text outside of this chapter. Determine the number of significant digits in each number and explain your reasoning.

8.4.   Consider two resistors having resistance R_1 and R_2. If the resistors are in series (i.e., connected end-to-end), the overall resistance R is given by $R = R_1 + R_2$. If the resistors are in parallel (i.e., the current is split between the resistors), then $1/R = 1/R_1 + 1/R_2$.

   a. What kind of mean of R_1 and R_2 would you use to calculate the overall resistance of resistors in series?

   b. What kind of mean of R_1 and R_2 would you use to calculate the overall resistance of resistors in parallel?

8.5.   Using the *Help* functions in your favorite spreadsheet software, find and report the spreadsheet functions used to calculate the arithmetic mean, geometric mean, harmonic mean, median, mode, sample standard deviation, and population standard deviation.

8.6. Using the data in Section 8.4.2, how does *median* change when one fuel efficiency value is changed? Does this make sense?

8.7. Measure the heights of ten students as samples of the larger student population. Calculate the variance in their heights.

8.8. For each of the following situations, state the most appropriate type of mean:

a. The mean of the interest rates for three years (i_1, i_2, and i_3), if you are interested in the interest rate i over the entire three-year period. [*Hint*: You want the most appropriate type of mean of i_1, i_2, and i_3 to describe i, where $(1 + i)^3 = (1 + i_1)(1 + i_2)(1 + i_3)$.]

b. The mean speed of four legs of an automobile trip, if you are interested in the mean speed of the entire trip.

c. The mean frequencies of the three A notes nearest to middle C on a piano. (*Hint*: The frequencies of the A notes nearest middle C are 220, 440, and 880 Hz, where 1 Hz = 1 hertz = 1 cycle/s.)

8.9. Can the geometric mean ever be greater than the arithmetic mean?

8.10. The geometric, harmonic, and quadratic means are related to the arithmetic mean in a way that helps in calculating their values. It is possible to find three functions so that

$$f(\text{geometric mean}) = \text{arithmetic mean of } f(x_i)$$
$$g(\text{harmonic mean}) = \text{arithmetic mean of } g(x_i)$$
$$h(\text{quadratic mean}) = \text{arithmetic mean of } h(x_i)$$

where f, g, and h are the functions and the x_i are the data. Another way to write this is

$$\text{geometric mean} = f^{-1}(\text{arithmetic mean of } f(x_i))$$
$$\text{harmonic mean} = g^{-1}(\text{arithmetic mean of } g(x_i))$$
$$\text{quadratic mean} = h^{-1}(\text{arithmetic mean of } h(x_i))$$

where f^{-1}, g^{-1}, and h^{-1} are the inverses of f, g, and h, respectively. The inverse of a function means that $f^{-1}(f(x)) = x$, $g^{-1}(g(x)) = x$, and $h^{-1}(h(x)) = x$. For example, if $f(x) = e^x$, then $f^{-1}(x) = \ln(x)$ because $\ln(e^x) = x$.

a. Find a function f so that geometric mean = $f^{-1}(\text{arithmetic mean of } f(x_i))$.

b. Find a function g so that harmonic mean = $g^{-1}(\text{arithmetic mean of } g(x_i))$.

c. Find a function h so that quadratic mean = $h^{-1}(\text{arithmetic mean of } h(x_i))$.

9

Engineering Models

9.1 INTRODUCTION

To facilitate analysis and design, engineers often rely on models. There are several different types of engineering **models**.

To some people, the phrase "engineering model" conjures up an image of a hastily drawn sketch of a Rube Goldberg machine on a napkin (see Figure 9.1). This is an example of a *conceptual model*. Conceptual models will be discussed in more detail in Section 9.3.2.

To others, the term "engineering model" evokes images of clay cars on pedestals or small airplanes in wind tunnels or miniature supertankers suspended in wave tanks. These are examples of *physical models* and will be discussed further in Section 9.3.3.

Finally, "engineering model" may bring to mind page after page of densely written mathematical formulas. This is an example of a *mathematical model*. More details on mathematical models may be found in Section 9.3.4. Each type of model will be illustrated with the following problem: predict the time required to reach the engineering building from your apartment.

9.2 WHY USE MODELS?

Why will you use models as an engineer? Models serve a number of roles in engineering. First, models aid in organizing ideas about engineered systems. In particular, conceptual models are a useful way to enumerate the important elements of a system.

Second, models can be used to simulate expensive or critical systems prior to construction. Physical models often are used prior to assembly. It is now possible to design even large engineered systems by computer.

OBJECTIVES

After reading this chapter, you will be able to:

- explain why engineers use models;
- list the types of models used by engineers;
- solve engineering problems using models and data;
- explain how models and data interact.

model: a conceptual, mathematical, or physical representation of an engineering system

Key idea: Types of engineering models include conceptual, physical, and mathematical models.

"what if" scenario: a question (usually probed by models) regarding how a system will respond to a set of conditions

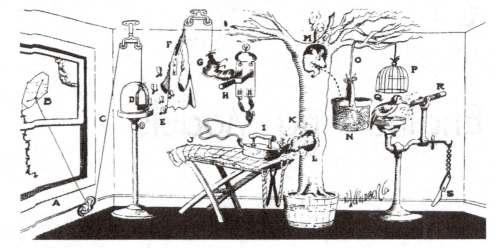

Figure 9.1. A Rube Goldberg Pencil Sharpener. Reuben (Rube) Lucius Goldberg (1883–1970) was a Pulitzer prize–winning cartoonist. He is known for his drawings of incredibly complicated machines designed to perform very simple jobs. The term "Rube Goldberg" refers to a complex solution to a simple problem. (Image courtesy of Rube Goldberg Inc.)

PONDER THIS

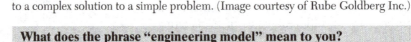

What does the phrase "engineering model" mean to you?

Key idea: Engineers use models to organize ideas, simulate expensive or critical systems prior to construction, and probe the response of a system to a large number of conditions.

Third, models aid in probing the response of a system to a large number of conditions. This use of models is sometimes referred to as a ***"what if" scenario***. Examples of "what if" scenarios include "What if the primary braking system fails on the Maglev (magnetically levitated) train?" and "What if a voltage spike occurs in a DVD player?" By way of another illustration, suppose you have a mathematical model for the steps involved in the construction of a high-rise apartment building. You could use the model to determine the effects of delays on the project completion time. (Delays may result from adverse weather, delivery delays, or labor strikes.) In this way, models can be used for prediction of future conditions.

9.3 TYPES OF MODELS

conceptual (descriptive) model: a model showing the main elements and how they interact, including *boundaries* (which define the system in space and time), *variables* (elements that may change), *parameters* (or constants: elements that do not change), and *forcing functions* (external processes that affect the system)

boundaries: the part of the model that defines the system in space and time

variables: elements of the system that may change

9.3.1 Introduction

As stated in Section 9.1, engineers use three types of models: conceptual, physical, and mathematical models. Each model type will be explored in more detail in this section.

9.3.2 Conceptual Models

A ***conceptual model*** (also called a ***descriptive model***) contains the main elements of the model and how they interact. Almost all modeling efforts begin with a conceptual model of the system. Conceptual models are often summarized in a sketch or diagram. A conceptual model should include the following elements of the system to be modeled: boundaries, variables, parameters, and forcing functions.

    Boundaries define the system. The system must be defined in both space and time. For example, the boundaries of a model for the movement of a pollutant through groundwater would include the area under investigation (spatial boundaries) and time period being modeled (temporal boundaries). The space and time domain described by the model boundaries is sometimes called the *control volume*.

    Variables are elements of the system that may change. For example, in modeling life support systems for the International Space Station (ISS), variables would include

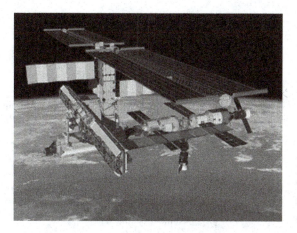

Artist's rendering of the International Space Station. The solar panel wings are 11.9 m × 34.2 m (39 ft × 112 ft) each. (Image courtesy of NASA/JPL.)

the size of the crew and the water usage rate. The crew size and water usage rates are expected to change over time.

Variables can be divided into two types. *Independent variables* serve as inputs to the model. In the ISS example, the crew size is an independent variable. *Dependent variables* are calculated by the model. For the ISS example, dependent variables include the size of the air cleaning system (called the atmosphere revitalization subsystem) and the capacity of the water treatment system (called the water recovery and management subsystem).

As the names imply, dependent variables *depend* on independent variables. For example, the sizes of the ISS subsystems depend on the crew size.

parameters: elements of the system that do not change

Parameters (also called *constants*) are system elements that do not change. If you were modeling the velocity of a water droplet in a decorative fountain, parameters would include gravitational acceleration, water density, and water viscosity (if density and viscosity are constant over the spatial and temporal domain of interest). In some types of mathematical models, the values of some parameters are changed to best fit the data (see Section 9.4). These parameters are called *adjustable parameters*: they are not functions of the variables, but are changed in the mathematical modeling process.

forcing functions: external processes that affect the system but are not modeled explicitly

Conceptual models also should include external processes that affect the system. These processes are called *forcing functions* (or *inputs*). Forcing functions are external to the model at hand and not modeled explicitly. If you were modeling the water level in a reservoir behind a hydroelectric dam, then the forcing functions might include rainfall and evaporation.

The boundaries, variables, parameters, and forcing functions combine to form the conceptual model. A conceptual model of the time-to-school problem is shown in Figure 9.2. In this case, a model is developed for commuting by bicycle.

The system boundaries are listed in the sketch title and include the spatial path (e.g., Main Street to University Avenue to Engineering Lane) and the time (year-round). The model is designed to calculate the speed at any time during the commute (also called the *instantaneous speed* and shown in the thickly outlined box in Figure 9.2). The total commuting time, the model output, will be calculated from the instantaneous speed. Thus, time is the independent variable and instantaneous speed and total commuting time are the dependent variables. Note that the dependent variable is expected to change over the course of the system boundaries and should be modeled.

Parameters include information about the route (i.e., the number of stop signs and the hill slope). Forcing functions include the bike condition, traffic, weather, and initial fatigue (i.e., fatigue at the beginning of your commute). The conceptual model shows

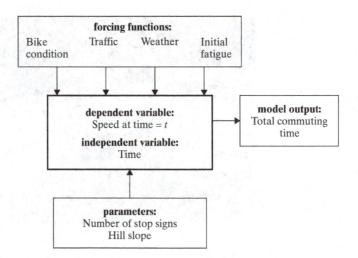

Figure 9.2. Conceptual Model for the Time-to-School Problem

that the forcing functions will influence the instantaneous speed. This, in turn, will affect the time required to complete the trip. Another example of a conceptual model is presented in Example 9.1.

9.3.3   Physical Models

physical model:
a (usually) smaller version of the full-scale system

mock-up: a full-size physical model

Physical models are often used for evaluating proposed solutions to engineering problems. A *physical model* usually is a smaller version of the full-scale system. (A full-size physical model is called a *mock-up*.) Physical models are typically used with large engineering projects, from the Great Pyramids to automobiles to the Space Shuttle.

**EXAMPLE 9.1
CONCEPTUAL
MODELS**

Develop a conceptual model for the design of a wooden pedestrian bridge over a river.

SOLUTION   A conceptual model for a wooden pedestrian bridge over a river would include the following items:

1. Boundaries:
   spatial boundaries (e.g., a river crossing a location), temporal boundaries (e.g., design life)
2. Independent variables:
   number of pedestrians crossing the bridge over time, properties of wood that change over time
3. Dependent variables:
   design elements (e.g., deck, trusses, railings, piers)
4. Parameters:
   wood properties that do not change over time, gravitational acceleration
5. Forcing functions:
   weather, cost constraints

A wind tunnel is an example of a physical engineering model. Conditions of an aircraft in flight can be simulated by placing a physical model of the aircraft in a wind tunnel and moving the air past it. Knowledge of fluid mechanics is used to correct for the size of the model and wind tunnel conditions to provide a good prediction of how the full-size aircraft will perform in flight. Other examples of physical models include rotating hydrodynamic laboratories (to study the effect of the Earth's rotation on the movement of water in large water bodies), laboratory robotic systems (to study interferences in automated materials-handling systems), and rapid mixing devices (to simulate chemical reactions and separations in the synthesis of plastics).

To develop a physical model for the commuting example, consider transportation to school by skateboard. You could build bench-scale physical models of the terrain and skateboard. This would allow you to perform experiments to estimate the travel time. The success of the model (i.e., its ability to predict the actual commuting time) would depend on how well the physical model mimicked friction, air resistance, and other elements of the system.

mathematical model:
a representation of the logical and quantitative relationships (usually mathematical expressions) between model components

deterministic model:
a mathematical model which provides a single answer for a given set of inputs

9.3.4   Mathematical Models

Mathematical models commonly are used in engineering evaluations. **Mathematical models** are built on the logical and quantitative relationships between model components. If the model is valid, then the real system can be probed by altering the independent variables and observing the model output.

Mathematical models may be further subdivided into deterministic and stochastic models. In a **deterministic model**, the input *determines* the output. Typically, a deterministic model provides a *single answer for a given set of inputs*. For example, the equation $t = d/v$ is a simple model describing the time t required for an object to travel a distance d at a constant velocity v. The model is deterministic: any combination of d and v determines a single value of t. The model $t = d/v$ is an accurate representation of the motion of a satellite in space.

PONDER THIS

> **Would a simple mathematical model (such as $t = d/v$) accurately predict the time required to commute to school by bus?**

This simple mathematical model would fail in many instances because it assumes a constant velocity. It does not take into account acceleration, deceleration, time spent at traffic lights or stop signs, or time spent at bus stops. A deterministic model sophisticated enough to include all factors may be so complex that it has little value.

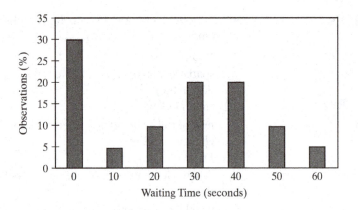

Figure 9.3. Hypothetical Distribution of Stoplight Waiting Times for a Time-to-School Stochastic Model

stochastic (or probabilistic) model: a mathematical model which provides a distribution of outputs for a given set of inputs

Stochastic models (from the Greek *stochos*, meaning target, aim, or guess) have different outputs, each with their own probability, for each set of inputs. Stochastic models have variables or parameters with probability distributions. For example, imagine that your commute to school passes through only one intersection and the intersection is controlled by a traffic light. You carry a stopwatch with you for a year and record your waiting time at the traffic light. Since you reach the traffic light at a random time in its cycle, your waiting time is expected to show a *distribution* of values, say between zero (if you hit a green light) and one minute (if you hit the light as it turns yellow). An example of a distribution of waiting times is shown in Figure 9.3. You could incorporate the distribution of waiting times into a stochastic model.

The stochastic model output would also be a distribution of commuting times, each with its own probability of occurrence. In other words, the output of the stochastic model could state, "There is a 50% probability that the commuting time will be greater than 20 minutes." Compare this statement with the output of a deterministic model, which might read, "The expected commuting time is 23 minutes." Other examples of mathematical models are given in Examples 9.2 and 9.3.

In engineering applications, mathematical models usually have a basis in theory. For example, suppose you are designing a new way to airlift food supplies to flood victims. You want to know how far the food crates will fall in a given time. Starting with basic definitions, you can derive the following common kinematic expression:

$$d = \tfrac{1}{2}gt^2$$

where d = distance fallen, g = gravitational acceleration, and t = time.

empirical model: a mathematical model based on observations, not theory

The equation $d = \tfrac{1}{2}gt^2$ is derived by theory. Suppose you had sought the relationship between the distance fallen and the time by a series of experiments. By analyzing the experimental data, you might come up with the following relationship: d is proportional to t^2, or $d = kt^2$, where k is a constant. The model $d = kt^2$ is an **empirical model**.

Trajectory of a falling object (vertical distance fallen is proportional to t^2)

Empirical models are based on observations, not theory. Most engineers are more comfortable using models with theoretical underpinnings. However, empirical models also are useful and may lead to research designed to provide the theory to support the observations.

**EXAMPLE 9.2
DETERMINISTIC
MATHEMATICAL
MODEL**

You are developing new polychromatic lenses for sunglasses. Polychromatic sunglasses darken when exposed to ultraviolet light. The glass in the lenses contains silver chloride (AgCl). You find from experiments that the degree of darkening of the lenses is proportional to both the thickness of the lens (d) and the concentration of AgCl in the layer (C). Further experiments show that when d and C are both doubled, the degree of darkening increases by fourfold. Develop a mathematical model for the degree of darkening of the lenses.

SOLUTION

The degree of darkening is proportional to the lens thickness d and AgCl concentration C. There are two possible models to describe this behavior:

$$\text{Model \#1: darkening} = ad + bC$$
$$\text{Model \#2: darkening} = edC$$

where a, b, and e are constants.

Both models describe the observation that the degree of darkening of the lenses is proportional to both d and C. However, recall that other experiments showed that the degree of darkening increased fourfold when d and C both were doubled. Model #1 predicts that the degree of darkening will only double when d and C both are doubled. Model #2 correctly predicts the behavior of the system when both d and C are doubled. Thus, **Model #2 correctly predicts the observations**.

Note: Model 2 is analogous to the Beer–Lambert Law of absorbance.

**EXAMPLE 9.3
STOCHASTIC
MATHEMATICAL
MODEL**

A nonprofit organization has asked you to help them evaluate the ticket price for a carnival ride at a charity event. For safety reasons, the riders must be at least 54 inches tall. You anticipate that 20% of the fair customers will be unable to ride the roller coaster because of the height constraint. The tickets for the ride cost $1 and the ride costs $7 per hour to operate. During a particular one-hour time period, 10 customers visited the charity event. What is the probability that the revenues collected during the hour will equal or exceed the operational costs?

SOLUTION

Systems of this sort follow a binomial distribution. If the proportion of eligible riders to the total population is p, then the probability (P) that n out of N people are eligible to ride is

$$P = \binom{N}{n} p^n (1 - p)^{N-n}$$

Here,

$$\binom{N}{n} = \frac{N!}{n!(N - n)!},$$

where $N! = N(N - 1)(N - 2)\cdots(2)(1)$. ($N!$ is read "N factorial," with $0! = 1$.)

In this example, $p = 1 - 0.20 = 0.80$ and $N = 10$. The number of eligible riders is n ($n = 0, 1, 2, \ldots, 10$). The revenue is

$$\text{revenue} = (\text{ticket price})(\text{number of eligible riders}) = (\$1)(n) = n \text{ (in dollars)}$$

The probability that there will be exactly n eligible riders (and therefore, exactly n dollars taken in during the hour) is

$$P = \binom{10}{n} 0.8^n \, 0.2^{10-n}$$

The revenue and probabilities of occurrence for each value of n are listed in the following table:

Number of Eligible Riders per Hour (n)	Revenue (dollars per hour)	Probability that Exactly n Riders Will Be Eligible
0	0	1.02×10^{-7}
1	1	4.10×10^{-6}
2	2	7.37×10^{-5}
3	3	0.000786
4	4	0.00551
5	5	0.0264
6	6	0.0881
7	7	0.201
8	8	0.302
9	9	0.268
10	10	0.107

Revenues will equal or exceed $7 per hour only when 7, 8, 9, or 10 riders are present each hour (i.e., only when $n = 7, 8, 9,$ or 10). The P values for $n = 7, 8, 9,$ and 10 are 0.201, 0.302, 0.268, and 0.107, respectively. The probability that $n = 7$ or 8 or 9 or 10 is the sum of the probabilities with $n = 7, 8, 9,$ and 10. Summing these values, **the probability that the revenue will equal or exceed the operational costs of $7 per hour is 0.879 or about 88%.**

Note: If there are only eight visitors per hour ($N = 8$), then the probability that the revenue will equal or exceed the operational costs decreases to about 50%. Of course, if only six or fewer people enter the charity event every hour, then the probability of generating $7 per hour drops to zero.

9.3.5   Other Kinds of Models

As computing power increases, the line between mathematical and physical models is being blurred. Computer-controlled milling machines now make it possible to quickly turn mathematical models into physical models. One name for this approach is "3-D printing." As the name implies, engineers may soon be able to "print" prototypes at their desktops as easily as they print reports.

In some cases, mathematical models coupled with computer-controlled milling machines allow engineers to bypass physical models completely. For example, the Boeing 777, first flown in June 1994, was the first aircraft to be designed without physical models. For more information, see *Focus on Models: Mathematical or Physical Model?*

FOCUS ON MODELS: MATHEMATICAL OR PHYSICAL MODEL?

BACKGROUND

Physical models often give engineers a sense of comfort. A model of a bridge or skyscraper sitting on a table can reduce the anxiety over whether the parts really can come together to make a whole. On the other hand, a large-scale physical model (especially a full-scale mock-up) can be very expensive. Engineers must decide whether a physical model is economically feasible; that is, whether the value of the information produced by the physical model exceeds its cost. For many years, engineers have sought other modeling tools that could provide the "comfort level" of a physical model at less expense.

Rapid changes in computing power have created opportunities for replacing some large-scale physical models with mathematical models. A good example is the design of the Boeing 777 series of jetliners. With a wingspan of 60 m and a length of about 64 m, the Boeing 777-200 is the world's largest twinjet airliner. Clearly, the design and assembly of such a large, sophisticated aircraft was a formidable challenge.

(Image courtesy of Boeing Commercial Airplane Group.)

COMPUTER-AIDED MANUFACTURING

Prior to the 777, aircraft parts were manufactured and assembled in large assembly buildings. If parts did not fit, change orders were issued to redesign and remanufacture the part. The Boeing 777 was preassembled digitally using solid, three-dimensional parts generated by computer. The computing power necessary for this effort was extremely high. Over 1,700 workstations were linked to a mainframe cluster of four IBM mainframe computers; this was the largest mainframe installation of its kind in the world at the time.

The innovative design approach had a number of advantages over traditional design/assembly methods.

The approach allowed engineers to identify parts-fitting problems without the extensive use of physical models. In addition, design and manufacturing operations could occur concurrently. In fact, a number of parts for the

A Boeing engineer designing a portion of the 777 (Image courtesy of Boeing Commercial Airplane Group.)

craft were made by metalworking machines controlled directly from the design software. This approach, called *computer-aided design/computer-aided manufacturing* (CAD/CAM), greatly sped up the time from design to construction of the first aircraft.

DIGITAL CLAY

The replacement of computer models for physical models is not limited to aircraft. Some automobile manufacturers have adopted "digital clay": computer representations of prototypes instead of clay models. The use of computer models allows designers all over the world to work on the same new car around the clock. The result is a cheaper, faster design. Volvo used digital clay to design a new station-wagon concept car. The company saved $100 million and cut the design time in half.

Returning to the CAD/CAM process, did Boeing use physical models or mathematical models for the 777? The answer is not readily apparent. Without a solid mock-up, you might say that the design used a very sophisticated mathematical model. However, when solid parts are generated digitally and allowed to interact, it is easy to argue that the design tool was as close to a physical model as you can get without the use of a milling machine. The design of the 777 represents the latest step in the blurring of the differences between mathematical and physical models. You can imagine a day where all physical models are holograms and the term "mathematical model" is no longer in use.

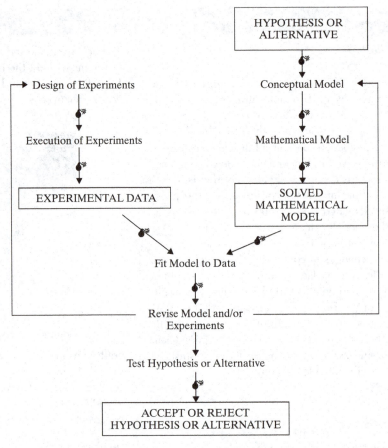

Figure 9.4. Interplay of Models and Data

9.4 USING MODELS AND DATA TO ANSWER ENGINEERING QUESTIONS

9.4.1 Interplay of Models and Data

Data collection is an important part of hypothesis testing and alternatives evaluation. The interplay of models and data is illustrated in Figure 9.4.

There are several steps to using models and data to answer questions. First, develop a conceptual model (see the upper right corner of Figure 9.4). An example conceptual model for the commuting problem was shown in Figure 9.2.

Second, translate the conceptual model into a mathematical model. A reasonable mathematical model might contain both deterministic elements (e.g., $t = d/v$) and stochastic elements (e.g., waiting time at traffic lights).

Third, solve the mathematical model. Note that experimental data may be needed to solve the mathematical equations. Why? Experiments may provide parameters needed for the model. As you may know from past experience, errors may occur during the solution of mathematical equations.

Key idea: Data are used to determine model parameters, and models influence the design of data collection activities.

On a parallel path with model development, you may gather data. Data may be used to determine model parameters (discussed in Section 9.4.3) or to compare model output with measured values. For example, you might measure the time required to reach school under a number of conditions and compare the measured values with the times predicted by the model. Note that *the model influences the design of experiments*.

For example, your model may include parameters such as hill slope and the number of stop signs. Thus, procedures must be developed to measure these parameters.

The interplay between modeling and data gathering is the engineer's friend. Data can point out errors in the model. A carefully constructed model may lead to the remeasurement of key parameters. The process shown in Figure 9.4 may have to be iterated several times until the model gives satisfactory results. The iteration process is the way that engineers refine their view of engineering problems and solutions.

If the model fit is not satisfactory (see Section 9.4.4), then revise the model. In some cases, the experiments may require revision as well. If the model is revised incorrectly, errors can occur on the next iteration. For example, imagine that the output of the commuting model does not fit the data. After some thought, you conclude that wind resistance must be included in the model to account for the discrepancies. If the real culprit is an incorrect formulation of the waiting time probabilities, then the model revisions may not help.

9.4.2   Potential Errors

Key idea: Models are limited by their scope (i.e., the elements included or excluded) and the input data.

The bombs in Figure 9.4 indicate where errors could occur. The first chance for error comes in the formulation of the conceptual model. If the conceptual model is incomplete (i.e., if important variables, parameters, or forcing functions are left out), then the model will be a poorer representation of reality. In general, *models are limited in their usefulness by their scope and input data.* If the commuting time is influenced significantly by road construction and road construction is *not* included in the model, then the model will likely not predict the commuting time accurately. Similarly, if you make an error in measuring the hill slope, then the model output may be useless. In other words, we apply the term **GIGO** (garbage in, garbage out).

GIGO (garbage in, garbage out): the idea that model output will be meaningless if the model input data are poor

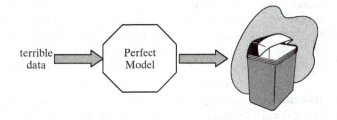

Errors also can creep in if the selected mathematical model is incompatible with the conceptual model. For example, if the conceptual model included the elements of acceleration and deceleration, then a constant-velocity model (such as $t = d/v$) would be inappropriate.

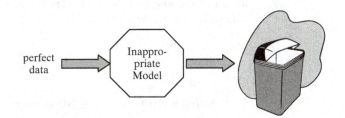

Errors can arise in experiments as well (see the bombs in Figure 9.4). Experiments can be improperly designed. For example, suppose you chose to determine the length of a road up a hill by measuring the hill height and hill length and by using basic trigonometry. This design is inappropriate if the hill is undulating.

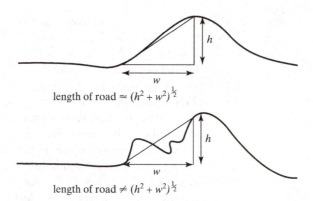

$$\text{length of road} \approx (h^2 + w^2)^{\frac{1}{2}}$$

$$\text{length of road} \neq (h^2 + w^2)^{\frac{1}{2}}$$

The next step is to execute the experiments to gather the data. Again, errors can occur through improper execution. In the bicycling-to-school example, the stopwatch may be slow or the surveying equipment may be incorrectly calibrated.

9.4.3 Model Fits

Key idea: Fit the model to the data; **never** fit the data to the model.

Assuming the model and experimental errors are small, you now can compare the model output with the measured values. It is critical at this stage to *fit the model to the data rather than the data to the model.* **Never** exclude data because they do not fit your preconceived notions of what the data should look like. In other words, do not reject data just because the data does not match the model.*

What is meant by "fitting the model to the data"? Fitting a model means finding the values of the adjustable parameters so that the model output matches the experimental data as closely as possible. This process is called *model calibration*.

model calibration: the process of finding the values of the adjustable parameters so that the model output matches the experimental data as closely as possible

There are a number of fitting tools used by engineers to calibrate models. In this text, a numerical approach will be introduced. Before discussing the fitting method, it is necessary to think about how you will know when the model output matches the data satisfactorily. A common approach is to formulate a function that describes the error in the model prediction and then pick adjustable parameter values to minimize the function. The objective then becomes to minimize the error. (The error function becomes the *objective function*.) Say, for example, that the deterministic model has one independent variable x, one dependent variable y, and one parameter m. From experiments, you have n pairs of x and y values. The x values are denoted $x_1, x_2, \ldots, x_n$ and the y values are denoted $y_1, y_2, \ldots, y_n$. Since the model is deterministic, each value of x (i.e., each x_i) will give one predicted value of y (usually denoted $\hat{y}_i$ and pronounced "why eye hat").

objective function: a mathematical statement of the success of the project

One possibility for an objective function is the sum of the differences between the model predictions and the data. This is called the *sum of the errors* (or SE). For the n data points (i.e., n pairs of x_i and y_i), SE is given by

$$\text{SE} = \sum_{i=1}^{n} (y_i - \hat{y}_i)$$

An example of computing SE is given in Figure 9.5. For the data in Figure 9.5:

$$\sum_{i=1}^{3} (y_i - \hat{y}_i) = -3 + 2 + 0 = -1.$$

---

*While you should not *reject* data that do not fit the preconceived model, sometimes it is appropriate to *question* data that do not fit the model. Models can be used to identify data that do not fit preconceived notions. If repeated measurements show that the original data are in error, then the original data are rejected. If the original data are *not* in error, then the model must be revised.

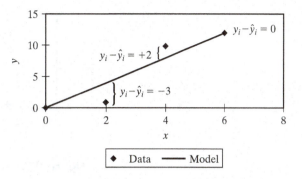

Figure 9.5. Example of Computing SE

PONDER THIS

Is SE a good measure of the differences between the model predictions and the experimental data?

SE is **not** a good measure of how well the model fits the data. Why? In SE, the "negative" errors (i.e., $y_i - \hat{y}_i$ less than zero) will cancel out "positive" errors (i.e., $y_i - \hat{y}_i$ greater than zero). For example, consider the data and model output in Figure 9.6. The model is $y_i = mx_i$. Model output is shown for $m = 2$. For this value of m, SE $= 0$ because the negative errors and positive errors cancel out. Even though SE $= 0$, it is clear from Figure 9.6 that the model $y_i = 2x_i$ is not a "perfect" model for the data.

There are many possible objective functions where the positive and negative errors do not cancel out (see Problem 9.2). A commonly used objective function is the *sum of the squares of the errors, **SSE***:

SSE: sum of the squares of the errors (square of the differences between the model output and data)

$$\text{SSE} = \sum_{i=1}^{n} (y_i - \hat{y}_i)^2 \tag{9.1}$$

To fit a model to data, the mathematical problem becomes "Find the set of adjustable parameter values that minimize the SSE."

To illustrate the use of SSE, consider an exciting new field of engineering: the use of *fractal geometry* to describe the dimensions of irregular objects. In common objects, the area (A) increases proportionally to the square of a characteristic length (l). Examples

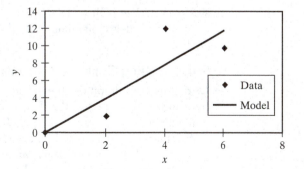

Figure 9.6. Example Model and Data to Illustrate Model Fit

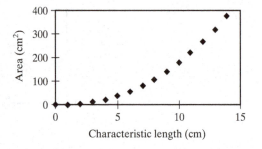

Figure 9.7. Data for Fractal Example

include circles (where $A = \pi r^2$; r = radius), spheres (where the surface area = $4\pi r^2$), squares (where $A = s^2$; s = side), cubes (where the surface area = $6s^2$), and equilateral triangles (where $A = \dfrac{\sqrt{3}}{4}s$). With fractal objects, A increases proportionally to l raised to the power n, where n is not necessarily equal to 2. Suppose you measure the area of a family of fractal objects and plot the area against the characteristic length (see Figure 9.7).

PONDER THIS

What is your model for the relationship between l and A?

The model is A = (proportionality constant)l^n. Suppose you know from other data that the proportionality constant is equal to 1. Thus, $A = l^n$. How would you find n? One approach would be to vary n and calculate the SSE. You can do this easily with a spreadsheet. Values of SSE are plotted against n in Figure 9.8. Note that the units of SSE are the units of the dependent variable squared.

PONDER THIS

What is your estimate of n from Figure 9.8?

Key idea: Select the values of the adjustable parameters to minimize the objective function (i.e., minimize SSE).

From Figure 9.8, SSE is minimized at $n = 2.2$–2.3. As shown in this example, the *values of the adjustable parameters should be selected so that SSE is minimized*.

The graphical approach of determining the values of the adjustable parameters that minimize SSE works well when your model has one adjustable parameter. The approach becomes more cumbersome with two adjustable parameters and virtually impossible to visualize with more than two adjustable parameters. A common approach to calibrating models containing more than one adjustable parameter is called *regression analysis*.

9.4.4 Using Calibrated Models

During model calibration, the values of the adjustable parameters are determined. In this process, the values of the dependent variables are calculated where data exist. For example, in the crate-drop example of Section 9.3.4, the values of the distance that the crate fell would be calculated for each time at which data were collected. These data are sometimes called the *calibration data set*. The model outputs for the calibration data set are called **model fits**. You *expect* the model fits to be close to the data, because you are using the data to fit the adjustable parameters.

model fits: comparison of model output to the calibration data set

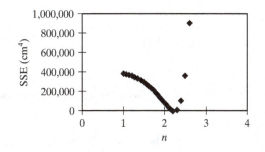

Figure 9.8. Variation of SSE with n in the Fractal Problem

model predictions:
comparison of model output to data outside the calibration data set

If you compare the calibrated model output to other data (i.e., data outside the calibration data set), the model outputs are called **model predictions**. Be sure to differentiate between model fits and model predictions in your engineering work. We want our models to be *predictive*; that is, they should predict data outside the calibration data set (but they have some limitations, as shown in Section 9.4.6).

9.4.5 Determining Model Fit

Another important issue in using models is determining how well the model fits the data. In fact, you already know one way to look at model fits: SSE. Unfortunately, SSE has a problem. It has units of $(\text{units of } y)^2$, so the magnitude of SSE depends on the units of y. Say you have a model relating the current in a photocircuit with light intensity. The photodiode produces a current in amps equal to a constant times the light intensity in watts. You calibrate the model twice with the same data set: once with current in amps and once with current in milliamps. The values of the SSE will be different even though the model is the same, as illustrated below:

Photodiode example: Data in watts and amps
model: current (A) = a (light intensity in watts)

Light Intensity (W)	Measured Current (A)	Predicted Current (in A, a = 0.3)	Error = measured − (A) predicted	Square of Error (A²)
0	0	0	0	0
1	0.2	0.3	−0.1	0.01
2	0.6	0.6	0	0
5	1.7	1.5	+0.2	0.04

SSE = 0.05 A²

Photodiode example: Data in mW and mA
model: current (mA) = a (light intensity in mW)

Light Intensity (mW)	Measured Current (mA)	Predicted Current (in mA; a = 0.3)	Error = measured − predicted (mA)	Square of Error (mA²)
0	0	0	0	0
1,000	200	300	−100	10,000
2,000	600	600	0	0
5,000	1,700	1,500	+200	40,000

SSE = 50,000 mA²

As this example shows, SSE would be more useful if it is dimensionless. One way to make SSE dimensionless is to compare your model with the simplest model for your dependent data.

What is the simplest possible model for *any* data?

The simplest model for a dependent variable y is y = constant. A reasonable value of the constant is the arithmetic mean of the y values. Thus, the simplest model is

$$y = \bar{y}$$

A more useful measure of model fit would be

(SSE for your model)/(SSE for the simplest model), or

(SSE for your model)/(SSE for the model y = mean y)

In mathematical terms, this new measure is

$$\frac{\sum_{i=1}^{n}(y_i - \hat{y}_i)^2}{\sum_{i=1}^{n}(y_i - \bar{y})^2}$$

***correlation coefficient* (r^2):** a dimensionless measure of the degree of fit of a model to data ($r^2 > 0.9$ is good)

This new measure is equal to zero when the model is perfect ($SSE = 0$) and is equal to one when the model is no better than the simplest model (y = mean y). This is okay, but it would be nice to have a measure that is equal to one when the model is perfect and is equal to zero when the model is no better than the simplest model. This is accomplished by defining the ***correlation coefficient*** r^2:

$$r^2 = 1 - \frac{\sum_{i=1}^{n}(y_i - \hat{y}_i)^2}{\sum_{i=1}^{n}(y_i - \bar{y})^2} \tag{9.2}$$

The correlation coefficient is a very valuable measure of the degree of fit of *any* model. It is dimensionless and near one if the model fits the data well. You can verify for the data in Figure 9.6 that r^2 is equal to 0.98, indicating a good fit ($r^2 > 0.9$ generally represents a good fit of the model to the data).

9.4.6 Are Engineering Models Real?

Key idea: When using models, remember that (1) the final model is only as good as the underlying conceptual model and resulting mathematical model, (2) models should not be used outside their calibration range, (3) engineers should not be misled by a good model fit, and (4) model output must be interpreted.

It is easy to become overly enamored with models and model output. Sometimes engineers come to believe that the calibrated model represents the truth and that data and model interpretation just get in the way.

You should remember four points about using engineering models. First, the final model is only as good as the underlying conceptual model and resulting mathematical model. As discussed in Section 9.4.2, models cannot predict behavior missing from the conceptual model.

Second, take great care not to use models outside the range of the independent variable for which the model has been calibrated. As an example, consider the prediction of the trajectory of a projectile (Figure 9.9). At short times, the height appears to be proportional with time. An extrapolation of the data at short time (≤ 6 seconds) is

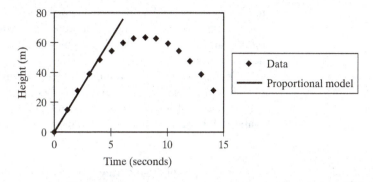

Figure 9.9. Example of an Extrapolated Model

shown by the line in Figure 9.9. The proportional model fits the data used to calibrate the model well. However, extrapolation of the model output *beyond* six seconds gives the wrong picture of the flight of the projectile. Predictions made by the proportional model outside the range of calibration could be disastrously wrong.

Third, do not be misled by good model fit. It is incorrect to assume that the model is "good" just because it fits the data. The classic example of an incorrect but good-fitting model is the Ptolemaic model of the galaxy. Claudius Ptolemy (ca. 100–ca. 170) proposed in the second century that the Earth was the center of the universe. The motion of the planets was described by circular orbits called *epicycles* as the planets orbited the Earth. More and more layers of epicycles were added to explain the observations as measurement techniques improved. In its final form, the Ptolemaic model was a mishmash of epicycles based on a flawed premise (i.e., an Earth-centered universe), but it fit the observed data remarkably well. In fact, the Copernican Sun-centered model (after Nicholas Copernicus, 1473–1543), although correct, did not fit the observations as well as the Ptolemaic model! The bottom line: do not assume that good-fitting models are "real."

Fourth, model output must be interpreted. Models are only part of the analysis or design process. It is important to realize that you must use the results of models *in conjunction with other information* to make conclusions about alternatives. The "other information" may be other types of feasibility.

9.5   SUMMARY

Engineers often rely on models to conduct analysis or to evaluate alternatives. Models are used to organize ideas, simulate expensive or critical systems prior to construction, and probe the response of a system to a large number of conditions ("what if" scenarios).

Three types of models are used in engineering. First, *conceptual models* contain the main elements of the model (boundaries, variables, parameters, and forcing functions) and how they interact. Second, a *physical model* is usually a smaller version of the full-scale system. Finally, *mathematical models* represent systems in terms of logical and quantitative relationships and are of two types. *Deterministic models* provide one output for each set of inputs. *Stochastic models* contain variables or parameters with probability distributions and therefore predict a probability of a certain outcome with a given set of inputs.

Models and data interact. As mentioned earlier, model parameters may be measured through experiments. In addition, models may identify key variables and parameters, thereby influencing the design of experiments. If the model predictions do not match the measured values satisfactorily, then both the model and experiments may need to be revised. Although numerous opportunities for errors exist, the interplay of models and data aids in developing models that adequately describe natural and engineered systems.

When using models, remember that the final model is only as good as the underlying conceptual model and resulting mathematical model. In addition, take great care in using models outside the range of the independent variable for which the model has been calibrated. Finally, do not be misled by good model fit in interpreting model results.

SUMMARY OF KEY IDEAS

- Types of engineering models include conceptual, physical, and mathematical models.
- Engineers use models to organize ideas, simulate expensive or critical systems prior to construction, and probe the response of a system to a large number of conditions.
- Data are used to determine model parameters, and models influence the design of data collection activities.
- Models are limited by their scope (i.e., the elements included or excluded) and the input data.
- Fit the model to the data; **never** fit the data to the model.
- Select the values of the adjustable parameters to minimize the objective function (i.e., minimize SSE).
- When using models, remember that (1) the final model is only as good as the underlying conceptual model and resulting mathematical model, (2) models should not be used outside their calibration range, (3) engineers should not be misled by a good model fit, and (4) model output must be interpreted.

Problems

9.1. Develop a conceptual model for a device to screen airline passengers for concealed weapons. Include the boundaries, variables, parameters, and forcing functions.

9.2. List two objective functions other than SSE where the positive and negative errors do not cancel out. Discuss the advantages and disadvantages of your objective functions compared with SE and SSE.

9.3. Use the Internet to find a picture and description of a bridge of interest to you. Build a physical model of the bridge using everyday materials (e.g., Popsicle sticks). What characteristics of the actual bridge does your bridge model well? What characteristics of the actual bridge does your bridge model poorly?

9.4. Develop a mathematical model to calculate how deeply a ship will sit in the water. (*Hint*: Perform a force balance, where the buoyancy force is proportional to the mass of water displaced by the ship. Develop your model for a specific geometry of the ship.)

9.5. Collect data of the height and weight of 10 friends. Develop an empirical model relating their height and weight.

9.6. For the data below, find the slope and intercept relating two hardness scales for steel, the Brinell number and the Vickers number. Calculate the correlation coefficient and comment on the applicability of the linear model for the data given.

Brinell Number	Vickers Number
780	1,150
712	960
653	820
601	717
555	633

9.7. Thermistors are semiconductors used for measuring temperature. The electrical resistance of a thermistor changes with temperature. One model for the effect of temperature on a specific thermistor is $R = 2,252e^{4000\left(\frac{1}{T} - \frac{1}{298.16}\right)}$, where R is the resistance in ohms and T is the temperature in K. You can verify that this thermistor has a resistance of $2,252\ \Omega$ at $25°C = 298.16$ K. For the data below, calculate the correlation coefficient for the model and decide whether the model fits the data well or not.

Temperature (°C)	Resistance (Ω)
0	7,850
10	4,400
20	2,900
30	1,500
40	1,000

9.8. The number of transistors on processor chips has doubled every 18 months or so for at least the last 30 years. This doubling sometimes is called *Moore's Law* (after Intel founder Gordon Moore). For the data on Intel processors below, determine the doubling time that minimizes the SSE. Decide whether the model (with your fitted doubling time) fits the data well or not.

Processor	Year of Introduction	Number of Transistors
4004	1971	2,250
8008	1972	2,500
8080	1974	5,000
8086	1978	29,000
286	1982	120,000
386	1985	275,000
486 DX	1989	1,180,000
Pentium	1993	3,100,000
Pentium II	1997	7,500,000
Pentium III	1999	24,000,000
Pentium 4	2000	42,000,000

9.9. One example of 3-D printing (Section 9.3.5) is printable prescription lenses. Printable lenses, developed by Saul Griffith (winner of the 2004 Lemelson–MIT Student Prize), may be used to provide low-cost eyeglasses for less-developed regions. Write a short report about printable prescription lenses.

9.10. What is the difference between stochastic and deterministic models? Give an example of each in the engineering field of your choice.

10

Computing Tools in Engineering

10.1 INTRODUCTION

Throughout history, engineers have used the most modern techniques of the day to perform engineering computations. For example, shortly after its development in the late 1600s, calculus was used to solve applied problems. (The words "calculus" and "calculation" come from the Latin *calx* meaning small stone, because pebbles were used as counters in older computation systems.) In modern times, engineers helped create and take advantage of the most up-to-date computing machines. Today, computers are part of the daily life of engineers.

In this chapter, you will be introduced to the myriad ways in which engineers use computers. In Section 10.2, the components that make up computers (i.e., hardware) are reviewed. Computing instructions (i.e., software) are presented in Sections 10.3 and 10.4. A specialized collection of hardware and software, the Internet, is discussed in Section 10.5. An important bit of advice is that you can *read* about engineering computing tools in this text, but to learn the tools, you must *put down the book* and fire up the computer.

10.2 COMPUTER HARDWARE

10.2.1 Computer Types

Computers are divided into types based on the number of users and the computing power of the machine (see Section 10.2.2 for a discussion of computing power). Machines used primarily by one user include ***personal computers***

OBJECTIVES

After reading this chapter, you will be able to:

- list the important types of computer hardware and how their performance is quantified;
- list general computer software types;
- explain what operating systems do;
- list the important types of software specific to engineering and science;
- discuss the structure of the Internet and how it can be used to solve engineering problems.

Key idea: Engineers helped create and take advantage of the most up-to-date computing machines.

A state-of-the-art computer from 1946:
ENIAC (Electronic Numerical Integrator and Computer)

personal computer: usually networked, the least-powerful computers employed by single users

workstation: always networked, the most-powerful computers employed by single users

and **workstations**. Personal computers[*] can be divided into desktop and laptop (notebook) types, with the former generally more powerful than the latter. In the good old days, personal computers and workstations were easy to identify because workstations were much more powerful. Another difference was that workstations were always networked, while personal computers often were stand-alone. In recent years, the gap between the computer types has narrowed. High-end personal computers may rival low-end workstations in computing power. In addition, virtually all personal computers are now networked.

server: a machine that provides computational services to personal computers and workstations in the network

supercomputer: the highest-performing computers, often a collection of other computers working together

Machines used primarily by more than one user include **servers** and **supercomputers**. Servers form the backbone of computer networks and provide computational services to networked personal computers and workstations. In the past, larger servers were called *mainframe computers*.

The name *supercomputer* is reserved for computers performing near the maximum computing speeds currently possible. Work done with supercomputers is often called *high-performance computing*. In many instances, supercomputers are a collection of less-powerful computers or processors that work together using a computing technique called *parallel processing*. In parallel processing, the computational work is shared among the processors or computers that make up the supercomputer.

Engineers also use low-power computing devices, such as calculators and personal data assistants (PDAs). It is worth your time to master the capabilities of your calculator or PDA so that problem solving in the future will be easier.

10.2.2 Microprocessors

microprocessor: a collection of transistors that control computer calculations

Computers are built up from component parts. Small pieces of semiconducting material are carved into transistors. Transistors are electrical components that amplify signals or open (or close) a circuit. Many millions of transistors are grouped together in personal computers and workstations on single silicon chips called **microprocessors**. The microprocessor (also called the *central processing unit* or *CPU*) controls the calculations performed by the computer.

---

[*]The phrase *personal computer* sometimes is abbreviated *PC*. The term *PC* also means *IBM-compatible personal computer* and is used to differentiate machines that use a Microsoft operating system and an Intel-like microprocessor from computers using the operating systems developed by Apple Corporation and microprocessors made by Motorola.

clock speed: the number of instructions that a microprocessor can execute per second (expressed in MHz or GHz)

Microprocessors process information in groups called *instructions*. Microprocessors are characterized by three aspects of instruction processing that determine the *computing power*. First, the microprocessors are rated by clock speed. The **clock speed** tells you how many instructions a microprocessor can execute per second. Thus, despite its name, a clock speed is a frequency. Microprocessor clock speeds usually are given in megahertz (MHz). (Recall that 1 hertz = 1 Hz = 1 per second. A clock speed of 1 MHz means 1×10^6 instructions per second, and a clock speed of 1 gigahertz, 1 GHz, means 1×10^9 instructions per second.) For example, a Pentium 4 microprocessor has about 55 million transistors and can operate at 1.4 to 2.2 GHz.

bandwidth: the number of bits of information that are processed in each instruction

The second element that affects computing power is the **bandwidth**. The bandwidth of a microprocessor is the number of bits of information that are processed in each instruction. (A **bit**, from *binary digit*, is the smallest unit of information on a machine. A bit can hold a value of 0 or 1. Eight bits form a **byte**, probably from *binary term*. A byte is large enough store a single character, e.g., the letter *A*.)

bit: the smallest unit of information, either 0 or 1

byte: eight bits

The third element that affects computing power is the *instruction set*. Some microprocessors (called *reduced instruction set computer* or *RISC* processors) use a small instruction set that they operate on very quickly. *Complex instruction set computer* (*CISC*) processors use a larger instruction set that they operate on more slowly.

Larger bandwidth (more bits per instruction) . . .

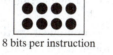

8 bits per instruction

16 bits per instruction

or larger clock speed (more instructions per second) . . .

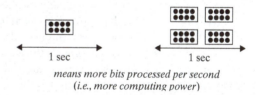

1 sec   1 sec

*means more bits processed per second
(i.e., more computing power)*

Key idea: Computing power is determined by the clock speed, bandwidth, and instruction set.

The computing power is determined by the clock speed, bandwidth, and instruction set. For example, a workstation with a 64-bit 3.2-GHz microprocessor is more powerful than an old personal computer with a 16-bit 50-MHz microprocessor.

10.2.3 Memory and Mass Storage

To perform calculations and archive information, computers store information in memory. Computer memory is divided into two types: main memory and mass storage devices. The units of memory are megabytes (MB) or gigabytes (GB).

Key idea: Computer memory (expressed in MB or GB) is made up of main memory (RAM) and mass storage.

Main memory is internal to the computer. It is commonly called *random-access memory* (or *RAM*). Computers often are rated by the amount of main memory. For example, a personal computer may have 128 MB of RAM. Computers with more main memory allow for more software to be run simultaneously and allow for more data to be easily available to software. Another characteristic of main memory is that it is *volatile*. This means that the information is erased when the power is turned off.

Mass storage (sometimes called *auxiliary storage*) is memory located on mass storage devices external to the RAM. Mass storage is *nonvolatile* (i.e., it remains when the power is turned off). Mass storage devices store information on magnetic or optical media. Magnetic media devices include hard drives, floppy disks, and other removable magnetic storage devices (flash memory cards, Zip disks, etc.). Optical storage devices include CDs and DVDs.

Computers also contain a small amount of nonvolatile, internal memory called *read-only memory* (ROM). ROM stores important programs such as the software responsible for system start-up (called boot programs; see Section 10.3.1). Integration of system components is explored in Example 10.1.

10.2.4   Input, Output, and Communication Devices

Computers are not very useful unless you can get data into the computer and acquire data from the computer. The most common input devices are the keyboard and the computer mouse. In addition, voice-activated computing is becoming more popular as the voice-processing algorithms improve.

digitizing tablet: an input device for graphical and spatial data, where each point on the tablet corresponds to a point on the monitor

Engineers also use digitizing tablets and pens to input graphical data. A *digitizing tablet* differs from a computer mouse. Each point on the tablet corresponds to a point on the monitor. By contrast, the relative motion of the mouse is translated to a movement of an object on the screen. With a digitizing tablet, data are inputted using a *cursor* (also called a puck—similar to a mouse) or a *pen* (also called a stylus).

The most common output device is the monitor. Most computer users also output data with speakers and printers. Engineering offices often house plotters capable of producing 2 ft × 3 ft plans* in color.

Most computers now communicate with other computers or other devices. The communication can be through direct wiring (also called *hard wiring*), through telephone lines (i.e., with conventional twisted-pair copper lines), through coaxial cable lines, by infrared light (as with wireless mice and keyboards), or by radio waves (called *wireless computing*).

Communication over telephone or coaxial cables or by radio waves is in the *analog* mode. In analog communication, a baseline sinusoidal signal is modified to transmit information. This process is called *modulation*. Computers use data in the *digital* mode, where information is communicated by bits that can have values of only zero or one. Therefore, it is necessary to process the analog signal to make it readable by digital computers. The signal processing devices are called *modems* (from *modulation–demodulation*). The digital signal from the computer is *modulated* to create an analog signal capable of transmission along analog communication lines. The analog signal is *demodulated* to be read by the computer (see Figure 10.1).

modem: a device that translates information between analog and digital modes

Communication speeds are determined by their bandwidth. Recall from Section 10.2.2 that bandwidth is the amount of information that can be transmitted per second. Thus, a modem capable of translating about 56,000 bits of information each second is called a 56-Kbps modem (bps = bits per second).

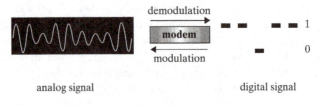

analog signal   digital signal

Figure 10.1.   Modem Operation

---

*Plans historically have been called *blueprints*, because the original printing process produced white lines on blue paper.

EXAMPLE 10.1 COMPUTER HARDWARE

To save money, you decide to build your own personal computer from a kit. The kit contains the following items: a case with spaces for disk drives (called drive bays); the power supply and cooling fan; and a motherboard. Using the information from Section 10.2, what other components should you buy to build your computer? (A motherboard is an integrated circuit board. It houses controllers for input/output devices and mass storage devices, as well as space for the CPU and memory.)

SOLUTION

The kit has the motherboard, but you still must purchase a **microprocessor** (CPU) and **RAM**. The motherboard has connections to mass storage devices, but you must buy the **hard drive** and whatever **removable media drive** you wish (e.g., a DVD drive). In addition, **input/output devices** must be obtained. At a minimum, you should buy a **monitor**, **keyboard**, and **mouse**.

10.3   GENERAL COMPUTER SOFTWARE

10.3.1   Introduction

Key idea: The three types of computer software are boot programs, operating systems, and application programs.

Computer software can be divided into three categories: boot programs, operating systems, and application software. *Boot* programs* run upon power up (or when the operating system is rebooted) and load the operating system into RAM. *Operating systems* are discussed in Section 10.3.2. All other programs are called *application software* (or *applications*). General application software is presented in Sections 10.3.3 and 10.3.4. The emphasis here is on how engineers use the generalized software. Application software specific to engineering is discussed in Section 10.4.

10.3.2   Operating Systems

operating system: software that provides services to the user and applications, including input/output functions, memory allocation, and processing time allocation

The *operating system* (OS) communicates with the user and with applications. (Computer engineers say that the OS *provides services* to the user and applications.) Communication with the user is accomplished through either a text-based interface (often called a *command line interface*) or, more commonly, through a *graphical user interface* (GUI). Communication with applications is provided through an application program interface (API).

What kinds of services does the operating system provide? The OS handles input and output from all the devices discussed in Section 10.2.4. For example, when you click the right mouse button, the operating system communicates to applications that the right mouse button has been pressed and released.

The operating system also allocates resources such as memory and processing time to applications that are running simultaneously (called *multitasking*). For example, the operating system allows you to continue to work on one document in a word processing program while another document is printing. In parallel processing systems (see Section 10.2.1), the OS will divide the program among processors.

Key idea: Common operating systems are Windows, Mac OS, and Linux for personal computers and Unix, Linux, and Windows-based operating systems for networks.

Many operating systems are available. The most common operating systems for personal computers are the Windows family, the Mac OS (for Macintosh computers), and Linux. Networks often use Unix, Linux, or Windows-based operating systems. The history of these operating systems is intertwined. Windows was derived from DOS (disk operating system), which was in turn derived from Unix. Linux, a form of Unix originally developed by Linus Torvalds, is *open-source software*; that is, it is software that can be modified by the user.

---

*The terms *boot* and *reboot* come from the word *bootstrap*. They refer to the idea that a small program (boot program) can launch a larger program (operating system), just as a small strap can be used to pull on a large boot.

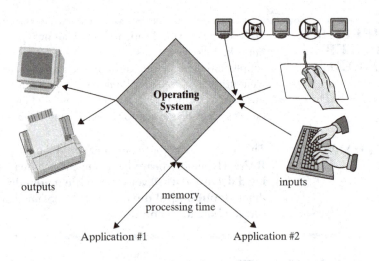

10.3.3 Communications Software

Engineering is not engineering until it has been *communicated*. Most of your engineering work will be communicated in written documents or oral presentations. In this section, the *software* needed for communicating engineering work will be discussed.

Key idea: Engineering is communicated through word processing software, email, and presentation software.

Engineering documents usually are created in *word processing software*. Documents are shared by printing or transmitted via email or posted on Web sites. Examples of word processing software include Microsoft Word and Corel's WordPerfect. It is highly recommended that you become very familiar with the use of at least one word processing program.

Engineering documents frequently include some of the more advanced features in word processing software. For example, engineers frequently use tables to summarize information. You should spend the time to learn how to create, format, and edit tables in the word processing software you use. In addition, engineers use equations. Word processing software usually comes with an equation editor. The editor allows you to write equations clearly, such as

$$y = \frac{1}{x}\sin\left(\frac{x}{2\pi}\right)$$

rather than in a more confusing format, such as

$$y = (1/x)[\sin(x/2\pi)]$$

Again, take the time to master the equation editor in the word processing software of your choice.

Key idea: Master a word processing program (including table and equation editors), email software, principles of Web design, and a presentation program.

It is also highly recommended that you master the art of creating, sending, and managing email (*electronic mail*). No engineering firm survives without the rapid communication provided by email. While it is not necessary to master the intricacies of Web page design, you may wish to become familiar with the general principles that make Web pages effective.

Oral presentations in engineering frequently make use of presentation software. A commonly used program for presentations is Microsoft's PowerPoint. An example of a PowerPoint slide is given in Figure 10.2.

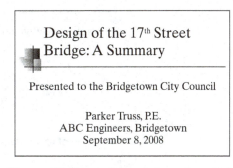

Figure 10.2. An Example PowerPoint Slide

10.3.4   Spreadsheet Software

Spreadsheet software allows you to manipulate a large amount of data at one time. Commonly used spreadsheet software includes Microsoft Excel, Corel's QuattroPro, and Lotus.

Most college freshmen know how to use a spreadsheet to perform simple calculations. For example, the sum of the values of cells A1 and B1 can be placed in cell C1 by typing

$$= A1 + B1$$

in cell C1. This is called *relative referencing*. If cell C1 is copied into cell D1, the contents of cell D1 become

$$= B1 + C1$$

The referencing cells have been shifted one column to the right as the cell contents were copied one cell to the right.

To refer to the *same* cell after copying, use *absolute referencing*. Absolute referencing is accomplished in Excel by placing a dollar sign ($) before the column indicator, row indicator, or both the column and row indicators. For example, to add the value of cell A1 to any cell, type

$$= \$A\$1 + B1$$

into cell C1. If cell C1 is copied into cell D1, the contents of cell D1 now will be

$$= \$A\$1 + C1$$

Note that the relative reference to cell B1 shifts to refer to cell C1, while the absolute reference to cell A1 remains. Another example of relative and absolute referencing is shown in Figure 10.3.

In addition to mastering the use of spreadsheets for simple calculations, engineers should learn how to use spreadsheets to perform complex calculations, plot data, and perform parameter estimation. Complex calculations are aided by built-in spreadsheet functions. Common built-in functions used in engineering are listed in Table 10.1.

The final spreadsheet skill that you should master is the use of the built-in nonlinear solver. Remember from Section 2.5 that engineers optimize. With a nonlinear solver, you can find the optimum value of an objective function by changing the values of adjustable parameters. In Microsoft Excel, the built-in optimization tool is called *Solver*. Instructions for using *Solver* are given in Appendix D.

Key idea: In a spreadsheet program, master the ability to perform simple calculations, use built-in functions, and use the nonlinear solver.

	A	B	C	D	E	F
1	**Example of relative referencing:**					
2					formula copied	
3						
4		What you see:	2	5	7	12
5						
6		Formulas:	2	5	=C3+D3	=D3+E3
7						
8						
9						
10	**Example of absolute referencing:**					
11					formula copied	
12						
13		What you see:	2	5	7	9
14						
15		Formulas:	2	5	=$C$13+D13	=$C$13+E13
16						

Figure 10.3. Relative and Absolute Referencing in Excel. The formulas in cells E4 and E13 were copied to cells F4 and F15, respectively

TABLE 10.1 Examples of Built-In Functions in Excel

Function Type	Examples	Engineering Example
Mathematical	trigonometric, exponential, logarithmic functions	Exponential decay $= \$A\$1*EXP(-1*B10)$
Financial	engineering economics calculations	—
Statistical	measures of central tendency and variability	Arithmetic mean $= AVERAGE(B10:B23)$
Logical	and, false, if, not, or, true	Conditional statement[1] $= IF(A1 = 1,2,0)$

[1] If this function is typed into cell B2, then cell B2 is equal to 2 if cell A1 is equal to 1. Cell B2 is equal to 0 otherwise.

10.4 ENGINEERING- AND SCIENCE-SPECIFIC SOFTWARE

10.4.1 Introduction

Key idea: Engineers also use programming software, symbolic math software, and engineering discipline-specific software.

Nearly every day, engineers use some of the general computer software discussed in Section 10.3. Of course, many people use word processing software and spreadsheets every day in their work. Engineers use additional software that is more specific to the engineering and scientific professions. In this section, three types of engineering- and science-specific software will be reviewed: programming software, symbolic math software, and engineering discipline-specific software.

10.4.2 Programming Software

programming language: software that can be used to write new software

Key idea: Programming software ranges from low-level (machine language) to intermediate-level (assembler language) to higher-level languages (e.g., FORTRAN).

Engineers often solve new problems or problems with new constraints. Unanticipated problems may be difficult to analyze with prewritten, commercially available software (sometimes called *canned software*). As a result, engineers sometimes must write their own software. New software is written with a ***programming language***.

In the days of yore, most engineers learned the programming language FORTRAN (named from *FORmula TRANslator*). While FORTRAN is still used in engineering programming, many other programming languages are now employed. Programming can be done on a number of levels, depending on the degree to which your program is interpreted by the processor before being executed. At the "lowest" level of programming, you must write programs in the native language of the microprocessor, also called *machine language* or *machine code*. Suppose you wish to add two forces (gravitational force and drag force) to obtain a net force. The machine code for adding two numbers together is*

```
000000 00001 00010 00110 00000 100000
```

As you can see, machine code is hard to read. The next level up in programming, called *assembler language* (or *assembly language*), is a little easier to read. The assembler language code for adding the two forces together is

```
lw $r0,gf  (copy gravitational force gf to register $r0)
lw $r1,df  (copy drag force df to register $r1)
add $r02,$r1,$r6  (add the forces; put result in register $r6)
```

Most engineers write programs in *higher-level languages*, where the commands resemble English but need to be interpreted before the processor can act on them. Examples of higher-level languages include FORTRAN, BASIC (from Beginner's All-Purpose Symbolic Instruction Code), and C. The BASIC command for adding the forces is (as implemented in Visual Basic)

```
NetForce = GravForce + DragForce
```

As you can see, higher-level languages are much easier to read.

The last category of programming languages is the software-generating capabilities inside of other applications software. For example, most word processors allow you to store and run small programs called *macros*. You can use a programming language called *Visual Basic for Applications* (VBA) and write programs to run inside most Microsoft products, such as Word and Excel. With macros or VBA programs, you can use the applications software itself as the user interface.

10.4.3   Trends in Programming Software

There have been three trends in programming in the last dozen years. First, programmers have sought graphical user interfaces (GUIs) that are at least as sophisticated as the GUIs of the operating systems (see Section 10.3.2). This has led to programming languages such as Visual Basic and Visual C++, which make the creation of a usable interface easier. Good interfaces for the user are critical if you are writing software to be used by a client.

One of the implications of a good user interface is that users will use it! As a result, the flow of most programs has changed. Old FORTRAN or BASIC code ran line by line: line 101 was executed immediately after line 100 (although some types of preprogrammed branching were allowed). Modern computer programs may run differently each time they are executed, depending on the actions of the user. This is called *event-driven programming*. The change in program execution requires the programmer to anticipate the needs and actions of the potential users.

Second, there has been a trend towards the development of software that can run over the Internet. Programming languages aimed at producing Net-based programs include Java and C# (read "C sharp").

---

*The command reads: add the contents of memory registers 1 and 2 and store the result in memory register 6. Interpretation: "000000" sets the type of instruction, "00001" refers to memory register 1, "00010" refers to memory register 2 ("00010" is "2" in binary), "00110" refers to register 6 ("00110" is "6" in binary), "00000" means "do not shift results to another register," and "100000" means "add."

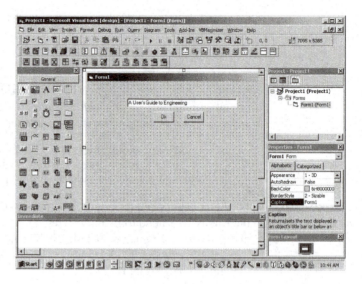

Design Environment for Microsoft's Visual Basic Version 6

A third trend is the emphasis on *object-oriented programming* (OOP). Traditional programming languages focus on *actions*. You can see this by noting that many programming commands are verbs (e.g., Do ... Loop or Goto). In OOP, the emphasis is on things (or nouns), mainly *objects* and *data*. The key to OOP is to identify *objects*, generalize them to *classes*, and define the *methods* by which the objects interact with each other and with the user.

As an example, suppose you wish to write a program to allow children to build devices on the computer out of simple machines. (The simple machines are the simplest devices with no moving parts: the screw, wheel and axle, wedge, pulley, inclined plane, and lever.) You might have a class called **SimpleMachine**. All members of the class share some properties (e.g., no moving parts, deliver mechanical advantage). Objects might be **Screw**, **WheelAxle**, etc., each with its own special characteristics. The methods are the ways that the simple machine objects interact (e.g., how the gears mesh).

Why is OOP popular? Among other advantages, OOP allows for the rapid generation of objects because the properties of the class are inherited by each new object in the class. Thus, if you have a class called **Beam**, then new beams (whether they are steel or concrete or wood) inherit properties of all beams (e.g., length, width, depth, Young's modulus, etc.).

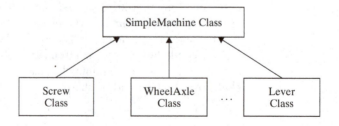

Do all engineers need to know how to program? You will find it useful through your engineering education and early career to be able to write short programs to solve engineering problems. As an example of the importance of programming in engineering, 6% of the Fundamentals of Engineering Examination (a step in professional licensing) is devoted to computers. (Topics include algorithm flowcharts, spreadsheets, pseudocode,° and data transmission and storage.)

10.4.4   Symbolic Math Software

symbolic math software: software to manipulate, simplify, and solve equations using symbols

Some programming languages are so sophisticated that they have become applications. An example in engineering is the **symbolic math software**. Symbolic math software allows you to manipulate, simplify, and solve equations using symbols. Examples of symbolic math software used in engineering include Maple, Mathcad, and Mathematica.

Symbolic math programs allow you to perform very sophisticated mathematical analyses. Many college courses in differential equations, engineering mechanics, and numerical methods are taught with symbolic math programs. Your path through your engineering education will be smoother if you master a symbolic math program.

10.4.5   Computer-Aided Design

Engineers must work visually. Thus, it is important that an engineer can represent objects in space with reasonable accuracy. As a result, most engineering curricula include a course on engineering drawing (or *drafting*).

computer-aided design (CAD): software used to create technical drawings

In today's world, drafting is done by computer. The computer software used to produce detailed drawings is called **computer-aided design**, or CAD, software. Common CAD software packages include AutoCAD and MicroStation.

Modern CAD programs are very sophisticated. They allow you to generate three-dimensional images and animations. CAD software can be integrated with computer-aided manufacturing (CAM) programs to allow for electronic design and manufacturing of objects.

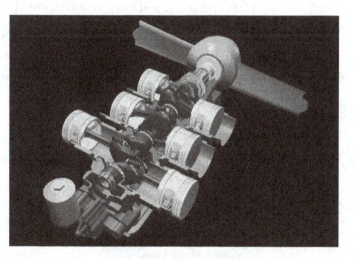

Example of an object drawn in AutoCAD

10.4.6   Discipline-Specific Software

In addition to the engineering software discussed Sections 10.4.2 through 10.4.5, each engineering discipline has its own specific software. Civil and mechanical engineers often use an analysis technique called *finite element analysis* (FEA). In FEA, a system is divided into parts (called *elements*) that are connected and interact. An example is a

---

°Pseudocode is a natural language description of a computer program. It is a useful tool for helping the programmer develop computer code. The pseudocode can be translated into the computer language of interest.

bridge. The elements of the bridge (trusses, decks, piers, etc.) are subject to forces, and the elements interact at the connecting points (called *nodes*). FEA problems can be very complicated and typically cannot be solved analytically (that is, you cannot find a solution where $y = $ some expression). The systems are solved numerically, using an approach called the *finite element method* (FEM). The solutions are approximations that can approach the true solution to your desired degree of accuracy. Commercially available FEM software includes ANSYS and ABAQUS. FEM can be used to solve problems in structural analysis, computational fluid dynamics (CFD), and process engineering.

Other engineering disciplines also have specific software. Electrical engineers use specialized software to design circuits. Industrial and chemical engineers employ software for the simulation of many systems, from production lines to chemical synthesis facilities. An example of how engineers use software is presented in Example 10.2.

EXAMPLE 10.2 COMPUTER SOFTWARE

You work at a company that makes video gaming hardware. Your division produces force-feedback joysticks. Your boss calls and asks you to put together a presentation for the board of directors' meeting that afternoon. She wants you to include detailed budget information from last month's progress report, structural analysis on the joystick that the Seattle office just finished, and a performance comparison with a competing company's product. What software would you use to put together the presentation?

SOLUTION

All the information likely would be imported into a **presentation program**. The budget information probably would be in tabular form and imported from **word processing** or **spreadsheet software**. You likely would call or email the Seattle office and have them email (or post to a secure Web site) the structural analysis results (probably generated with **finite element software**). The analysis results might be summarized visually in a drawing produced by a **CAD program**. The data concerning the competitor's product probably would be obtained from the **Internet**.

10.5 THE INTERNET

10.5.1 Introduction

Key idea: The Internet is composed of interconnected servers, a standardized transmission/routing language, and the information located on the servers.

The Internet is part of the daily lives of all engineers. The Internet grew out of a network of four government computers linked together by the Advanced Research Projects Agency (ARPA) in 1969. This first network was called ARPANET. The network expanded to include academic and commercial communication. In 1991, a milestone was reached when commercial Internet traffic first exceeded academic traffic. This led to the development of a portion of the network, called Internet2, to be devoted to government and academia.

10.5.2 Structure of the Internet

The Internet is a giant system for sharing information. It is made up of three elements. First, the physical portion of the Internet includes the servers that are networked together and the networking connections.

The second element of the Internet is the language established to allow communication between computers. This language is called TCP/IP and consists of a data transport protocol (Transmission Control Protocol, TCP) and a routing protocol (Internet Protocol, IP). Each computer (or central server) has an **IP address**. Through TCP/IP, all information is broken into pieces called *packets* and sent to the IP address of the computer requesting the information.

IP address: a standard format for locating a networked computer (or other device)

The third element of the Internet is the information on the servers. Through the physical network and transmission/routing protocols, information can be shared among users of the Internet.

A subset of the information on the Internet is the ***World Wide Web***. The World Wide Web is the portion of the Internet that uses a special protocol (called Hypertext Transfer Protocol or HTTP) to access text and graphics collections called W*eb pages*. A Web page is accessed by an address called a Uniform Resource Locator (URL). The URL consists of at least two parts that indicate

1. the *protocol service* and
2. the *domain name* where the information can be found.

World Wide Web: a portion of the Internet that uses a special protocol (HTTP) to access text and graphics collections

These two parts are separated by the characters ":/ /". For example, the URL http:// www.microsoft.com indicates that the Hypertext Transfer Protocol is to be used to access the information at the domain name www.microsoft.com.

Web pages are written in specialized computer languages. An example is Hypertext Markup Language (HTML). Programs called Web browsers interpret the HTML (or other Web page design languages) to re-create the Web page on your monitor. Examples of Web browsers include Internet Explorer, Netscape, and Opera.

10.5.3   Uses of the Internet

Key idea: Engineers use the Internet to communicate, to share files, and to gather information.

Engineers (along with most other people) use the Internet for three purposes. First, you can use the Internet for communication. A great deal of communication between practicing engineers now occurs by email or instant messaging.

Second, the Internet is used by engineers for sharing files. Files can be transferred rapidly between computers by a common protocol called File Transfer Protocol (FTP). This protocol can be accessed in Web browsers by using the command:

```
ftp://domain name
```

Note that the FTP protocol replaces the HTTP protocol in the URL.

Third, the Internet can be used to gather information from Web pages. Engineers frequently use the Web to gather information on technical specifications, regulations, products, and services. To obtain the information efficiently, you are advised to master the use of an Internet search engine. Examples include Google and Lycos. Much information on the Web is posted in a file format that preserves the formatting of the original documents. One such format is the *portable document format* or pdf. It is important as an engineer to use an Internet search engine that searches pdf files. The pdf files can be read with software such as Adobe's Acrobat Reader, which is available free of charge.

10.6   SUMMARY

Computers will become an integral part of your engineering education and professional engineering life. A basic understanding of computer hardware and mastery of several types of computer software are required of all engineers.

In terms of computer hardware, computers are divided into types based on the number of users and the computing power of the machine (from the least to the most powerful: personal computers, workstations, servers, and supercomputers). The basic hardware unit is the microprocessor. Computing power is determined by the microprocessor's clock speed (in MHz or GHz), bandwidth (in bits), and instruction set.

A second aspect of computer hardware is memory (expressed in megabytes or gigabytes). Main memory (commonly called RAM) exists inside the computer and is erased when the power is turned off. Mass storage devices (such as hard drives and floppy drives) are external to the main memory and are not erased when the power is turned off.

The third type of computer hardware includes input, output, and communications devices. Common input devices are the keyboard, computer mouse, and digitizing tablets (with pen or cursor). Common output devices are the monitor, printer, and plotter. Communication with other computers or devices is achieved via direct (hard) wiring, telephone lines, coaxial cables, infrared light, or by radio waves. (The latter is called wireless computing.)

The software used by engineers ranges from very general to discipline-specific. General software includes boot programs (used to load the operating system), operating systems (used to communicate with the computer user and application programs), and application programs. As an engineer, you should master the use of several types of communications software, including word processing (especially tables and equation editors), email, and presentation software.

Engineers use spreadsheet software on almost a daily basis. Learn to use simple formulas (by employing both relative and absolute addressing), more complex built-in functions, and nonlinear solving capabilities.

You will also become familiar with engineering- and science-specific software. Examples include programming languages, symbolic math software, and computer-aided design programs. A great deal of software has been written that is specific to certain engineering disciplines. You likely will learn to use software to perform finite element analysis, circuit design, or simulation of natural or engineered systems.

Finally, our world has changed because most computers are now networked. The Internet comprises the physical network, the communication protocols, and interesting content. Engineers use the Internet for communication, file sharing, and information gathering. Learn to use the Internet to your advantage by mastering the intricacies of a Web browser and search engines.

SUMMARY OF KEY IDEAS

- Engineers helped create and take advantage of the most up-to-date computing machines.
- Computing power is determined by the clock speed, bandwidth, and instruction set.
- Computer memory (expressed in MB or GB) is made up of main memory (RAM) and mass storage.
- The three types of computer software are boot programs, operating systems, and application programs.
- Common operating systems are Windows, Mac OS, and Linux for personal computers and Unix, Linux, and Windows-based operating systems for networks.
- Engineering is communicated through word processing software, email, and presentation software.
- Master a word processing program (including table and equation editors), email software, principles of Web design, and a presentation program.
- In a spreadsheet program, master the ability to perform simple calculations, use built-in functions, and use the nonlinear solver.
- Engineers also use programming software, symbolic math software, and engineering discipline-specific software.
- Programming software ranges from low-level (machine language) to intermediate-level (assembler language) to higher-level languages (e.g., FORTRAN).

- The Internet is composed of interconnected servers, a standardized transmission/routing language, and the information located on the servers.
- Engineers use the Internet to communicate, to share files, and to gather information.

Problems

10.1. Using spreadsheet software, prepare a table comparing the microprocessors in your calculator (or PDA), the computer that you use most frequently, and the most powerful computer at your institution. Include the clock speed, bandwidth, and number of processors.

10.2. Using word processing software, prepare a table comparing the RAM and mass storage devices (if any) in your calculator (or PDA), the computer that you use most frequently, and the most powerful computer at your institution.

10.3. List the input devices, output devices, and communications capabilities of your calculator or PDA.

10.4. Use the Internet to find out why the size of $3\frac{1}{2}$-inch floppy disks was selected.

10.5. Using spreadsheet software, prepare a table comparing the storage capacities of $5\frac{1}{4}$-inch floppy disks, $3\frac{1}{2}$-inch disks, CDs, and DVDs. In a column of your spreadsheet, normalize the values to the storage capacity of the $5\frac{1}{4}$-inch floppy disk.

10.6. List the software packages you would use to accomplish the following:

a. Convert one year's worth of daily flow data at a drinking-water treatment plant from units of millions of gallons per day to units of m^3/s.

b. Convince your boss that your company should create a research and development team.

c. Find a list of electrical engineering consulting firms in your state and contact them.

10.7. Find two protocol services other than FTP or HTTP.

10.8. Evaluate the cell phone as a computer. How does it compare with the computer you use most frequently? What technical obstacles must be overcome for the cell phone to replace the computer?

10.9. Write a short paragraph on how Internet search engines (such as Google) search and report hits. What implications do their search and reporting algorithms have for using their search engines in consulting engineering?

10.10. Discuss applications of blogging in managing a large engineering project.

11

Feasibility and Project Management

11.1 INTRODUCTION

An engineering solution that satisfies all appropriate criteria is called a *feasible* solution. There are four types of feasibility: technical (or engineering) feasibility; economic (or financial) feasibility; fiscal feasibility; and social, political, and environmental feasibility. Engineers use the four types of feasibility to evaluate and compare alternatives.

Technical feasibility addresses whether the proposed alternative satisfies its engineering criteria. Clearly, at a minimum, alternatives must *work*; that is, they must do the technical job required. Technical feasibility is discussed in Section 11.2.

Economic feasibility addresses whether benefits outweigh costs. If an alternative costs more over a given length of time than the benefits derived from it, then we say that the alternative is not economically feasible. Unfortunately, costs and benefits often accrue at different times. For example, an investment in a new facility this year may not provide benefits until several years in the future. Thus, to determine economic feasibility, it is necessary to understand how the value of money changes over time. Financial issues of interest to engineers are called *engineering economics*. Engineering economics is discussed in Section 11.3, and economic feasibility is explored in Section 11.4.

Many engineering projects are not built, even though they would generate more benefits than costs. One reason projects are not pursued is that the money needed to fund the project cannot be acquired. An alternative is said to be fiscally feasible if sufficient funds can be obtained to pay for it. *Fiscal feasibility* is discussed in Section 11.5.

Another reason why projects are not built even though they are technically, economically, and fiscally feasible is that they do not have support of the public and the politicians or

OBJECTIVES

After reading this chapter, you will be able to:

- list the types of feasibility used by engineers;
- calculate the future value of money;
- evaluate alternatives using technical; economic; fiscal; and social, political, and environmental feasibility;
- plan and schedule a simple engineering project.

Key idea: The four types of feasibility are technical (or engineering) feasibility; economic (or financial) feasibility; fiscal feasibility; and social, political, and environmental feasibility.

they place undue stress on the environment. These issues, called *social, political, and environmental feasibility*, are covered in Section 11.6.

After feasible projects are approved, the project benefits can be realized only if the project implementation is managed properly. An introduction to project management is provided in Section 11.7.

11.2 TECHNICAL FEASIBILITY

technical (engineering) feasibility: a test of whether the system satisfies its engineering criteria

Technical feasibility, also called engineering feasibility, addresses the question: *will it work?* An alternative is technically feasible if it satisfies the engineering criteria required of it. For example, suppose you required that an engine generate 300 horsepower to drive a certain system. Any engine designs that generate less than 300 horsepower are not technically feasible.

Technical feasibility also takes into account whether it is possible from an engineering standpoint to construct the project. Thus, the design must both satisfy the design requirements and be capable of being implemented. A space station design may work (i.e., achieve the technical objectives if built), but not be technically feasible because the required materials cannot be delivered into space with existing technology.

Key idea: An alternative is technically feasible if it satisfies the required engineering criteria and can be built.

Although technical feasibility is the easiest type of feasibility to understand, it takes a great deal of knowledge to be able to evaluate it. Most of the classes you take as an engineering student are devoted to (1) determining whether systems are technically feasible, or (2) designing technically feasible solutions. An example of assessing technical feasibility is given in Example 11.1.

EXAMPLE 11.1 TECHNICAL FEASIBILITY

You are designing an elevated bed for your dorm room. You need a platform to place under your mattress. The platform must be 7 feet long by 4 feet wide. You do not want the platform to sag more than $1/4$ inch. Its mass must also be less than 150 pounds to meet constraints on loading to the floor.

You come up with three alternatives: a $1/2$-inch thick steel platform, a 1-inch thick corrugated cardboard platform, and a $3/4$-inch thick plywood platform. Evaluate each choice for its technical feasibility.

SOLUTION

Each choice must sag less than $1/4$ inch and weigh less than 150 pounds. The platform mass can be calculated from its density multiplied by its volume (volume = length × width × thickness, where length L = 7 ft and width = 4 ft). Its maximum sag can be calculated from

$$\text{maximum sag} = \frac{5wL^4}{384EI}$$

where w = loading per unit length, E = Young's modulus, and I = moment of inertia = (width)(thickness)^3/3 in this case. If you, the mattress, and your bedding weigh 200 pounds, then w = (200 lb)/(7 ft) = 28.6 lb/ft.

The materials properties are listed in the table. The masses and maximum sag can be determined from these properties and mass and sag formulas.

Material	E (lb/in^2)	Density (lb/ft^3)	Thickness (in)	Maximum Deflection (in)	Weight (lb)
Steel	2.9×10^7	1,600	0.5	0.027	1,900
Cardboard	7.3×10^4	10	1	1.3	23
Plywood	2.0×10^6	38	0.75	0.11	67

The steel platform satisfies the maximum deflection requirement. However, it is not technically feasible because it weighs too much. The cardboard platform satisfies the weight requirement. However, it is not technically feasible because it exhibits too much deflection. **The plywood platform satisfies both the maximum deflection and weight requirements and is technically feasible.**

11.3 ENGINEERING ECONOMICS

11.3.1 Costs of Engineering Projects

Key idea: Engineering projects incur capital costs and operation and maintenance costs.

capital costs: one-time costs (e.g., for equipment and the physical plant)

operation and maintenance (O&M) costs: costs that accrue over time (e.g., for repair and personnel)

To appreciate the process of determining economic feasibility, a short introduction to engineering economics is in order. Three basic concepts are important. First, engineering projects incur two main kinds of costs. The initial costs to build a system are called *capital costs*. Capital costs are considered one-time costs typically accruing at the beginning of a project.

Annual upkeep costs are called operation, maintenance, and repair costs (often just called *operation and maintenance* or *O&M costs*). O&M costs accrue over time.

As an example, the initial price of a car is its capital cost. The annual costs of repairing, fueling, and insuring the car are O&M costs. O&M costs often increase over time (e.g., the repair costs on your car). In engineered systems, you often will trade off capital and O&M costs. For example, one type of pump may be less expensive initially (lower capital costs), but have higher upkeep expenses over time (higher O&M costs).

11.3.2 Time Value of Money

The second basic concept in engineering economics is that the value of money changes over time. To illustrate the time value of money, answer the question in *Ponder This*.

PONDER THIS

Would you rather have $100 today or $100 a year from now? Why?

Most people would rather have the money now. The value of money changes over time because of inflation. Due to inflation, $100 a year from now is worth less than $100 now. In other words, the $100 buys less one year in the future.

Inflation is reflected in part by the interest rate at which you borrow or loan money. If you had the money now, then you could deposit it in a savings account and have more than $100 one year from now.

$100 buys 20 widgets today…

but $100 buys only 19 widgets next year

Key idea: The value of money changes over time, so costs and benefits must be compared in constant dollars (i.e., dollars at the same year).

Why do you need to worry about the time value of money as an engineer? Capital costs and O&M costs accrue *at different times*. Thus, mathematical tools are needed to allow you to add capital and O&M costs so that costs and benefits can be compared and economic feasibility evaluated. Costs and benefits must be compared in *constant dollars* (i.e., dollars at the same year), because the value of money changes over time.

11.3.3 Calculating the Present and Future Value of Money

A few results from engineering economics will be presented here to give you some tools to determine economic feasibility. One common calculation is the interconversion of the *present value* of a system and the value of the system *in the future*. (In the language of engineering economics, we speak about the *equivalence* of the present and future values.) Suppose you put $100 in the bank at 3% annual interest. How much money would you have after one year? It is clear that after one year you would have:

$$
\begin{aligned}
\text{Money after one year} &= \text{initial investment} + \text{interest in one year} \\
&= \text{initial investment} + (\text{interest rate}) \times (\text{initial investment}) \\
&= \$100 + (0.03)(\$100) \\
&= \$100(1 + 0.03) \\
&= \$103
\end{aligned}
$$

After two years, you would have

$$
\begin{aligned}
\text{Money after two years} &= \text{money after one year} + \text{interest in second year} \\
&= (\text{money after one year}) + (\text{interest rate}) \\
&\quad \times (\text{money after one year}) \\
&= [\$100(1 + 0.03)] + (0.03)[(\$100) \times (1 + 0.03)] \\
&= \$100(1 + 0.03)^2 \\
&= \$106.09
\end{aligned}
$$

The value of money in the present is called the *present value* or *present worth* and is denoted by the symbol P. The value of money in the future is called the *future value* or *future worth* and is denoted by the symbol F. In general, after n years at an interest rate of i (expressed as a decimal) the present value will grow into a future value given by

$$F = P[1 + i]^n \tag{11.1}$$

The factor $[1 + i]^n$ is called the *single payment compound amount factor* and is denoted $(F/P, i\%, n)$. From this example, the future value after two years of $100 invested at an annual interest rate of 3% is $100(1 + 0.03)^2$ or $106.09. Working backwards, you can find the present value of some future cost. Solving Eq. (11.1) for P, we have

$$P = F[1 + i]^{-n} \tag{11.2}$$

The factor $[1 + i]^{-n}$ is called the *single payment present worth factor* and is denoted $(P/F, i\%, n)$. For example, suppose you want to have $100 four years in the future and the annual interest rate is 2.5%.

PONDER THIS

How much should you invest now to yield $100 four years in the future?

Then you now need to invest an amount equal to

$$P = (\$100)(1 + 0.025)^{-4} = \$90.60$$

11.3.4   Uniform Series

Another common calculation in engineering economics is converting a series of equal payments of A dollars over time into a present or future value. An example is to convert a series of constant O&M costs into an equivalent present value.

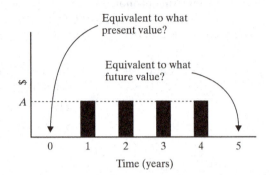

The appropriate equation is

$$P = A\left[\frac{(1 + i)^n - 1}{i(1 + i)^n}\right] \tag{11.3}$$

Key idea: To compare costs, the costs must be in the same units.

In Eq. (11.3), the term in square brackets is called the *uniform series present worth factor* and is sometimes written $(P/A, i\%, n)$.

Why convert a series of payments into a present value? To compare costs, *the costs must be in the same units*. A stream of costs over time has money expressed in different units: one year's costs may be year 2010 dollars and the next year's cost is in year 2011 dollars.

An example will make this point clearer. Suppose you just got your first engineering job and a financial planner calls you with two investment ideas. In the first plan, you pay him $1,000 today and receive $180 per year for six years. In the second plan, you pay him $1,000 today and receive $265 per year for four years. Which is the better deal? Without understanding the time value of money, you might make this simple calculation: you will receive ($180 per year)(6 years) or $1,080 with the first plan and ($265 per year)(4 years) or $1,040 with the second plan. Therefore, you might assume that the first plan is a better deal.

PONDER THIS

What is the fallacy in this approach to comparing costs?

The problem with this approach is that you are adding up costs with *different units* (i.e., different year's dollars). In fact, some of the funds with the six-year plan are being paid in dollars six years from now which are worth less than dollars four years from now. A better approach is to express your outlay in constant dollars. To calculate the present values of the plans, you need to know the interest rate. Assuming 2.5% annual interest, you obtain

present value of the first plan =

$$P = A\left[\frac{(1 + i)^n - 1}{i(1 + i)^n}\right] = (\$180)\left[\frac{(1 + 0.025)^6 - 1}{0.025(1 + 0.025)^6}\right] = \$991$$

present value of the second plan =

$$P = A\left[\frac{(1 + i)^n - 1}{i(1 + i)^n}\right] = (\$265)\left[\frac{(1 + 0.025)^4 - 1}{0.025(1 + 0.025)^4}\right] = \$997$$

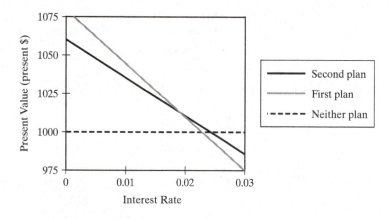

Figure 11.1. Present Value of Investment Options

What did you learn by this analysis? Due to the time value of money, the second plan had a higher present value, even though the total revenue (in mixed-year dollars) *appears* to be larger with the first plan.

Another important lesson from this analysis is that *both* plans have present values less than $1,000. This tells you that you will have a higher present value if you do not follow either plan, but just keep your $1,000! As with most engineering economics calculations, the answer depends strongly on the interest rate. As shown in Figure 11.1, you should choose the first plan if the interest rate is less than about 1.9%, choose the second plan if the interest rate is between about 1.9% and 2.4%, and choose neither plan (keep your $1,000) if the interest rate is greater than about 2.4%.

11.3.5 Engineering Economics Calculations

Key idea: Engineering economic factors can be determined by direct calculation, by tables, or by built-in spreadsheet functions.

Equations (11.1) through (11.3) are the basis for simple engineering economics calculations. The equations can be used in three ways. First, the terms in square brackets, often called *factors*, can be calculated directly using your calculator. With the factors calculated and recorded, you can easily interconvert present value, future value, and a uniform series of payments. Common engineering economics factors are listed in Table 11.1. You may wish to verify the formulas in Table 11.1 using Eqs. (11.1) through (11.3).

Second, the factors can be looked up in tables. A portion of a typical engineering economics table for $i = 4\%$ is given in Table 11.2. Again, you may wish to verify the values in Table 11.2 using the formulas in Table 11.1.

TABLE 11.1 Engineering Economics Factors

(Multiply the row value by the factor to get the column value. For example, $F = P[1 + i]^n$.)

	P =	F =	A =
P	—	$[1 + i]^n$ single payment compound amount factor (F/P)	$\left[\dfrac{i(1 + i)^n}{(1 + i)^n - 1}\right]$ capital recovery factor (A/P)
F	$[1 + i]^{-n}$ single payment present worth factor (P/F)	—	$\left[\dfrac{i}{(1 + i)^n - 1}\right]$ sinking fund factor (A/F)
A	$\left[\dfrac{(1 + i)^n - 1}{i(1 + i)^n}\right]$ uniform series present worth factor (P/A)	$\left[\dfrac{(1 + i)^n - 1}{i}\right]$ uniform series compound amount factor (F/A)	—

TABLE 11.2 Example of a Portion of a Compound Interest Table for Engineering Economics Calculations ($i = 4\%$)

	Single Payment		Uniform Series			
n	F/P	P/F	P/A	A/P	F/A	A/F
1	1.040000	0.961538	0.961538	1.040000	1.000000	1.000000
2	1.081600	0.924556	1.886095	0.530196	2.040000	0.490196
3	1.124864	0.888996	2.775091	0.360349	3.121600	0.320349
...						
28	2.998703	0.333477	16.663063	0.060013	49.967583	0.020013
29	3.118651	0.320651	16.983715	0.058880	52.966286	0.018880
30	3.243398	0.308319	17.292033	0.057830	56.084938	0.017830

TABLE 11.3 Useful Excel Functions for Performing Engineering Economics Calculations

Calculation	Excel Formula
Single Payment	
calculating F from P	$= \mathbf{FV}(i,n,,-1*P)$
calculating P from F	$= \mathbf{PV}(i,n,,-1*F)$
Uniform Series	
calculating P from A	$= \mathbf{PV}(i,n,-1*A)$
calculating A from P	$= \mathbf{PMT}(i,n,-1*P)$
calculating F from A	$= \mathbf{FV}(i,n,-1*A)$
calculating A from F	$= \mathbf{PMT}(i,n,,-1*F)$

Symbols: i = annual interest rate as a decimal (e.g., 0.02 for 2%), n = number of years, P = present value, F = future value, and A = annual payment.

Multiply the independent variable by -1 as indicated to obtain positive values for the dependent variables. Note the required double commas in three formulas.

Set the independent variables equal to 1 to obtain the factors in Table 11.2.

Third, you can use preexisting functions in most spreadsheet programs to perform engineering economics calculations. The built-in functions for Microsoft Excel are shown in Table 11.3.

11.4 ECONOMIC FEASIBILITY

11.4.1 Introduction

economic (financial) feasibility:
a test of whether the benefits of a system outweigh its costs

Economic feasibility, also called financial feasibility, addresses the question, *Are the benefits greater than the costs?* An alternative is economically feasible if its benefits are larger than its costs. The comparison of benefits and costs often is called a **benefit–cost analysis**. Benefit–cost analyses are required in almost every engineering project.

11.4.2 Comparing Alternatives

benefit–cost analysis:
quantification of the benefits and costs of a project

planning horizon:
the length of time over which alternatives are compared

As stated in Section 11.4.1, economic feasibility addresses whether benefits are greater than costs. The most common measures of economic feasibility are *net benefits* and *benefit–cost ratios*. With both approaches, it is necessary to compare costs and benefits fairly between alternatives. This means that the costs and benefits must be calculated using the *same dollars*. Either present value costs and benefits or future value costs and benefits can be used. In addition, each alternative must be compared over the same length of time. The time over which the alternatives are compared is called the **planning horizon**.

In the net benefit approach, the net benefits (total benefits – total costs) are calculated for each alternative. The alternative with the largest net benefits should be selected. In the benefit–cost ratio approach, the ratio of total benefits divided by total costs is calculated for each alternative. Alternatives with benefit–cost ratios less than 1.0 are rejected.

PONDER THIS

> **Why are alternatives with benefit–cost ratios less than 1 rejected?**

Alternatives with benefit–cost ratios less than 1.0 are rejected because they are not economically feasible (i.e., benefits do not exceed costs).

Key idea: Select the alternative that maximizes the net benefits, not the alternative that maximizes the benefit–cost ratio.

The net benefits and benefit–cost ratio approaches will give the same answer about economic feasibility. In some cases, more than one alternative may be economically feasible. In other words, the benefits may be greater than the costs for more than one alternative. To select from among economically feasible alternatives, choose the alternative with the largest net benefits (**not** the alternative with the largest benefit–cost ratio).

11.4.3   Example

As an example of using engineering economics, consider the following situation. A factory is deciding whether to upgrade its handheld inventory scanners. Two alternatives are being considered. In the first alternative, no new scanners will be bought now and the existing scanners will be sold (at $700 each) and replaced five years from now. The old scanners will require $80 each in annual maintenance costs over the next five years. In the second alternative, ten new scanners will be bought now for $1,000 each. The new scanners will allow each employee to bring in $200 more in revenues per year, with an annual maintenance cost of $50 each. After five years, the scanners will be sold at $800 each and replaced. Both alternatives replace scanners after five years, so a five-year planning horizon is appropriate.

Which alternative is better if the inflation rate is 3%? The best alternative will maximize net benefits and have a benefit–cost ratio greater than 1.0. For the first alternative, the costs and benefits are

> Capital costs: $0
> O&M costs: $10 \times \$80 = \$800$/year for five years
> Benefits: $10 \times \$700 = \$7,000$ five years from now from resale

For the second alternative, the costs and benefits are

> Capital costs: $10 \times \$1,000 = \$10,000$ now
> O&M costs: $10 \times \$50 = \$500$/year for five years
> Benefits: $10 \times \$200$ (= $2,000/year for five years from increased productivity) + $10 \times \$800$ from resale (= $8,000 five years from now)

To compare alternatives, convert each cost and benefit to present dollars (i.e., present values) and consider costs as negative and benefits as positive. For the first alternative, we have

> Present value capital costs: $0
> Present value O&M costs: $-\$800(P/A, 3\%, 5) = -\$3,664$
> Present value benefits: $\$7,000(P/F, 3\%, 5) = \$6,038$
> Net present value benefits: $\$0 - \$3,664 + \$6,038 = \$2,374$
> Present value benefit–cost ratio: $\$6,038/\$3,664 = 1.6$

For the second alternative, we have

Present value capital costs: $-\$10,000$
Present value O&M costs: $-\$500(P/A, 3\%, 5) = -\$2,290$
Present value benefits: $\$2,000(P/A, 3\%, 5) + \$8,000(P/F, 3\%, 5)$
$$= \$9,159 + \$6,901 = \$16,060$$
Net present value benefits: $-\$10,000 - \$2,290 + \$16,060 = \$3,770$
Present value benefit–cost ratio: $\$16,060/(\$10,000 + \$2,290) = 1.3$

Note that both alternatives are economically feasible. In other words, both alternatives have greater present value benefits than present value costs (and thus both have present value benefit–cost ratios greater than 1.0). However, the second alternative provides greater net present benefits ($3,770 versus $2,374). Thus, the second alternative is better (even though it has a smaller benefit–cost ratio). Remember, select the alternative that maximizes the net present benefits, **not** the alternative that maximizes the benefit–cost ratio. Another example of engineering economics is provided in Example 11.2.

EXAMPLE 11.2
ECONOMIC
FEASIBILITY

You are an industrial engineer in a large manufacturing company. The company has the opportunity to purchase a new production line for $1.5 million. The line will produce $600,000 in revenue in year 3 and $1 million in revenue in year 6. Would you recommend purchase of the new production line? Assume 2.5% interest.

SOLUTION

The costs and benefits are as follows:

Capital costs: $-\$1.5 \times 10^6$
Benefits: $\$600,000$ three years from now $+ \$1 \times 10^6$ six years from now

Converting to present values yields

Present value capital costs $= -\$1.5 \times 10^6$
Present value benefits $= \$600,000(P/F, 2.5\%, 3) + \$1 \times 10^6(P/F, 2.5\%, 6)$
$$= \$557,200 + \$862,300$$
$$= \$1.42 \times 10^6$$
Net present value benefits $= \$1.42 \times 10^6 - \$1.5 \times 10^6 = -\$80,500$
Present value benefit–cost ratio $= \$1.42 \times 10^6/\$1.5 \times 10^6 = 0.95$

Thus, present value benefits *do not* exceed present value costs and **the new production line should not be purchased**.

11.5   FISCAL FEASIBILITY

11.5.1   Introduction

fiscal feasibility: a test of whether sufficient funds can be obtained to pay for a project

Key idea: Fiscal feasibility addresses the question of whether you can obtain enough money to pay for the project.

Fiscal feasibility addresses the question, *Can you get enough money to pay for the project?* Many engineering projects and business deals are economically feasible. In other words, if you build the project (or seal the deal), then the expected benefits will exceed the expected costs. However, it may not be possible to raise enough money to start expensive projects. As an example, fast-food franchises typically are very profitable. Why doesn't everyone own such a franchise? Most people cannot obtain the start-up costs to purchase a fast-food franchise. Similarly, an economically attractive engineering project may remain unbuilt because start-up or operating funds cannot be secured. Such a project is fiscally unfeasible.

11.5.2 Bonds

How do companies or government agencies obtain millions of dollars to build projects? Large companies and government agencies often finance projects by borrowing money through the issuance of bonds. A **bond** is a loan, with the company or governmental authority as the borrower (called the *bond issuer*) and the public as a lender. The public buys certain denominations of bonds (called the *face value*). The bond issuer agrees to pay back the face value after a certain period of time (called the *maturity date*) and pay an annual amount at a specified interest rate (called the *coupon*).

bond: a loan from the public taken out by a governmental agency or company

11.5.3 Example

As an example, suppose that a municipality wishes to raise $10 million to upgrade a wastewater treatment plant. The city might issue 10,000 bonds each with a face value of $1,000, a 4% coupon, and a 10-year maturity. If you bought one bond, you would pay $1,000 and then receive $40 each year for ten years ($1,000 × 0.04 = $40) *and* get your $1,000 back at the end of 10 years.

Note that issuing the bond allows the city to raise cash quickly. Of course, the city must include the annual coupon payments and the final repayment funds in their benefit–cost analysis. Their costs are as follows:

$$\text{total costs} = \text{annual coupon payments} + \text{final repayment}$$

The annual coupon payments are i(funds received from bond sales) = (0.04) ($10,000,000) = $400,000 per year. The final repayment is $10,000,000.

PONDER THIS

> **How does the city make sure it has $10 million 10 years in the future?**

sinking fund: a special account to which deposits are made to pay for bonds or other debts

The city invests money every year in a special fund called a **sinking fund**. The payment to the sinking fund (called the *sinking fund payment*) is the uniform series of payments that will yield $10 million 10 years in the future. This is calculated using the sinking fund factor (A/F) in Table 11.1. The result is that a payment of $832,909 per year for 10 years at an interest rate of 4% will yield a future value of $10 million in 10 years. Thus, the equivalent annualized cost of the project is

$$\begin{aligned}
\text{annualized total costs} &= \text{annual coupon payments} + \text{annualized final repayment} \\
&= \$400{,}000 \text{ per year} + \$832{,}909 \text{ per year} \\
&= \$1{,}232{,}909 \text{ per year}
\end{aligned}$$

Thus, the project is fiscally feasible if the city can raise (say, by raising taxes or from cost savings generated by project benefits) $1,232,909 per year for 10 years to cover the cost of the bond. Another example of fiscal feasibility is shown in Example 11.3.

EXAMPLE 11.3 FISCAL FEASIBILITY

In the example of fiscal feasibility in the text, the interest rate on the bond (coupon) and interest rate on monies deposited in the sinking fund were the same. It is more common for bond interest rates and sinking fund rates to be different.

Suppose a city is considering the establishment of a sinking fund to repay a bond taken out to pay for bridge repair costs. The city will raise taxes to pay for interest payments on the bond and sinking fund payments. The $4 million bond is for 20 years at 6%. How much will taxes have to be increased if the sinking fund rate is 4%?

SOLUTION

The monies required to fund the sinking fund are

$$\text{tax increase} = \text{annual coupon payments} + \text{sinking fund payment}$$
$$= (i\%)(\text{bond amount}) + (\text{bond amount})(A/F, s\%, n)$$

where $i\%$ = bond coupon = 0.06, bond amount = \$4,000,000, $s\%$ = sinking fund rate = 0.04, and n = 20 years. Using the formulas from Table 11.1, the sinking fund factor $(A/F, 0.04, 20)$ is 0.033582. Thus, the tax increase is

$$\text{tax increase} = (i\%)(\text{bond amount}) + (\text{bond amount})(A/F, s\%, n)$$
$$= (0.06)(\$4,000,000) + (\$4,000,000)(0.033582)$$
$$= \$374,300 \text{ per year}$$

Taxes must be increased by **\$374,300 per year** to pay for the bond. The tax increase may be reduced by the benefits accrued from bridge maintenance.

11.6   SOCIAL, POLITICAL, AND ENVIRONMENTAL FEASIBILITY

social, political, and environmental feasibility:
a test of whether a project has sufficient backing of the public and political leaders and is compatible with the environment

Key idea: Social, political, and environmental feasibility addresses whether the project has sufficient backing of the public and political leaders and is compatible with the environment.

Social, political, and environmental feasibility addresses the question, *Does the project have sufficient backing of the public and political leaders and is the project compatible with the environment?* For a project to be implemented, support from stakeholders is essential. In the corporate world, this means that upper management must agree to the project. In large government projects (e.g., a new bridge or highway), support from the general public and politicians is required. Project teams that do not seek public input or ignore political realities risk not having their projects built.

In addition, there is an increasing awareness and concern about the impact of engineering decisions on the environment. Incidents such as the *Exxon Valdez* accident in Alaska's Prince William Sound have highlighted the potential cost of not adequately considering environmental impacts. Virtually all major government and private engineering projects now require a formal *environmental impact analysis* as part of the approval process. Engineering projects must be compatible with the environment to be considered environmentally feasible.

As an example of the increased attention to the environmental impacts of engineered projects, consider the New York Power Authority's Niagara Power Project. The Niagara Power Project generates 2.4 million kilowatts of electricity by diverting up to 375,000 gallons of water from the Niagara River *every second*. (This diversion rate would provide an 8-ounce glass of water to each person in the United States every 45 seconds.) When the Niagara Power Project was first licensed in 1957, **no** environmental regulations applied to its operation. In considering the renewal of its license 50 years later, approximately **40** international, national, and state environmental statutes apply.

New York Power Authority's Niagara Power Project

11.7 PROJECT MANAGEMENT

Key idea: Project
management activities
include planning,
scheduling, and supervising.

11.7.1 Introduction

Using the concept of feasibility discussed in this chapter, engineers generate, rank, and recommend alternatives. Many engineers also oversee the construction and implementation of projects. In general, project management involves three elements. First, the project is planned. In *planning* a project, an engineer breaks the project into tasks and determines the order of the tasks and how each task will be completed. Second, the project is scheduled. *Project scheduling* involves timing and coordinating the tasks to complete the project on time and on budget. Third, the project execution may be *supervised*. In this section, project planning and scheduling are emphasized.

Suppose a nursing home wishes to build a large storage shed on its property.[*] A reinforced concrete slab is required as a foundation. To construct the slab, the forms (i.e., the wooden framework) must be built and the reinforcement bars (rebar) installed. The concrete then will be poured into the form.

11.7.2 Project Planning

Few, if any, engineering projects are accomplished in one step. Engineers play a vital role in breaking projects into manageable tasks. Large projects often are broken down into logical areas, with each area containing hundreds or thousands of individual tasks. For example, the Department of Defense uses a format called a Work Breakdown Structure (WBS) to create the logical areas.

This project also must be broken down into tasks.

PONDER THIS

What tasks are necessary to construct the reinforced concrete slab on grade?

The project can be divided in several ways. One approach is as follows:

Excavate the site.
Purchase and deliver the wood for the form.
Purchase and deliver the rebar.
Build the form.
Place the rebar.
Pour the concrete.

11.7.3 Project Scheduling

In most projects, certain tasks must be completed before other tasks can commence. For example, in the development of the B-2 Bomber, over 900 new manufacturing processes had to be invented before the construction of the aircraft could begin.

One of the first project scheduling tasks is to develop an order of precedence of the tasks. In the example project, the tasks were listed in a logical order of precedence. However, it is necessary to determine *exactly* which tasks must be completed before a given task can be started.

PONDER THIS

What tasks must occur before the form is built?

---

[*]This example was modified from an example in Revelle et al. (2003).

TABLE 11.4 Precedence Table for the Example Problem

Task	Activity	Duration (days)	Predecessor Task(s)
A	Start	0	none
B	Excavate the site	2	A
C	Purchase and deliver the wood for the form	1	A
D	Purchase and deliver the rebar	3	A
E	Build the form	1	C
F	Place the rebar	1	B, D, E
G	Pour the concrete	2	F
H	Finish	0	G

Note: Task durations have been exaggerated in this example for illustrative purposes.

Clearly, the wood for the form must be purchased and delivered to the site before the form can be built. Ordering of the tasks sometimes is shown in a *precedence table*. A precedence table for the example project is shown in Table 11.4. In the precedence table, two "dummy" activities with zero duration have been added to the task list: a "Start" task and a "Finish" task. The predecessor tasks are the last tasks that must be completed before starting the task in question.

Make sure that the predecessor tasks in Table 11.4 make sense to you. Note that Table 11.4 tells you that Tasks B, C, and D can be started simultaneously. All three tasks can be started immediately after project initiation.

arrow diagram: a graphical way of showing task order in a project schedule

The order of the tasks can be shown graphically in an **arrow diagram**. An arrow diagram for the example project is shown in Figure 11.2. The form of the arrow diagram in Figure 11.2 sometimes is called an *activity-on-node network*, since the activity is indicated by the circles (called *nodes*).

11.7.4 Critical Path Method

In project management, you want to know how long the project will take and when work on each task can begin. One technique for determining the project duration and task start times is the **critical path method** or CPM. CPM was developed by the Sperry Rand Corporation and E.I. du Pont de Nemours & Co. in the mid-1950s to schedule chemical plant construction and process shutdown and maintenance activities.

critical path method (CPM): a scheduling technique used to determine the project duration and the timing of each task

At the heart of the method is the determination of the *critical path*. The critical path is the sequence of tasks from project start to project finish that requires the *longest* amount of time to complete. Thus, the time required to complete the critical path sets

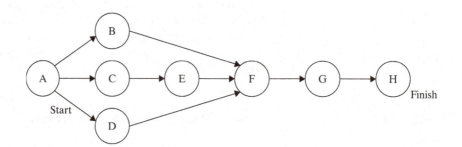

Figure 11.2. Arrow Diagram for the Example Project

the duration of the project. *Any delay on the critical path will increase the time required to complete the project.*

The mathematical approach for determining the critical path is beyond the scope of this text. However, in simple systems such as the example project, the critical path can be determined by calculating the times of all paths through the project.

PONDER THIS

How many paths are possible through the example project?

In the example project, three paths are possible:

$$\text{Path 1: A} \rightarrow \text{B} \rightarrow \text{F} \rightarrow \text{G} \rightarrow \text{H}$$
$$\text{Path 2: A} \rightarrow \text{C} \rightarrow \text{E} \rightarrow \text{F} \rightarrow \text{G} \rightarrow \text{H}$$
$$\text{Path 3: A} \rightarrow \text{D} \rightarrow \text{F} \rightarrow \text{G} \rightarrow \text{H}$$

By summing the task durations for each task on a path, the duration of each path can be computed. You can show that the path durations are five, five, and six days for Paths 1, 2, and 3, respectively. Thus, Path 3 is the critical path and the project will take six days to complete.

The start time of each task also can be determined. If each task is assumed to start as early as possible (i.e., immediately after all predecessor tasks are completed), then the task start time is called the *earliest start time* (or **EST**). The *earliest finish time* (**EFT**) is the EST plus the task duration. The EST and EFT values for each task are listed in Table 11.5.

Table 11.5 is an important guide for the project. It tells you, for example, that the concrete cannot be poured until Day 4 of the project. Thus, you know in advance when to arrange for the delivery of the concrete. For complex projects, CPM results in a large cost savings by allowing tasks to be scheduled in advance.

The EST and EFT information also can be presented graphically. A plot of the EST, EFT, and task durations for each project is called a ***Gantt chart*** (after Henry Laurence Gantt, 1861–1919). A Gantt chart for the example project (with tasks ordered by their EST values) is given in Figure 11.3. The critical path is highlighted by black bars.

The CPM approach is much more powerful than it may appear from this brief introduction. A modification of CPM called the *program evaluation and review technique* (**PERT**) incorporates uncertainty into the CPM framework. In this way, probabilities can be assigned to the project and task durations.

Gantt chart: a type of graph used to present the start and finish times of tasks in a project

TABLE 11.5 Earliest Start Time (EST) and Earliest Finish Time (EFT) for Each Task in the Example Project

Task	Activity	EST (days)	EFT (days)
A	Start	0	0
B	Excavate the site	0	2
C	Purchase and deliver the wood for the form	0	1
D	Purchase and deliver the rebar	0	3
E	Build the form	1	2
F	Place the rebar	3	4
G	Pour the concrete	4	6
H	Finish	6	6

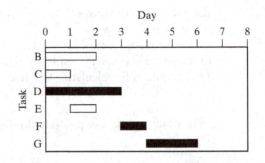

Figure 11.3. Gantt Chart for the Example Project (Tasks A and H omitted for clarity)

11.8   SUMMARY

Engineers often decide from among alternatives by using the concepts of technical (or engineering) feasibility; economic (or financial) feasibility; fiscal feasibility; and social, political, and environmental feasibility. *Technical feasibility* addresses whether the proposed alternative satisfies its engineering criteria and is technically implementable. Technical feasibility is the focus of most undergraduate engineering curricula.

Economic feasibility addresses whether benefits outweigh costs. The tools of *engineering economics* can be used to account for changes in the value of money over time. Engineering economics can be used to assess economic feasibility and also determine the alternative with maximum net benefits.

Fiscal feasibility addresses whether sufficient funds can be obtained to pay for an alternative. Government agencies and corporations typically raise money by floating bonds. Engineering economics tools can be used to determine the annual income stream needed to pay for a bond and thus determine fiscal feasibility.

Finally, engineering projects and design alternatives must have the support of the public and the politicians. In addition, they must exhibit an acceptable environmental impact. These aspects of design alternatives are assessed in *social, political, and environmental feasibility*.

Engineers also manage projects. They may plan, schedule, and supervise engineering projects. A common and powerful scheduling technique called the *critical path method* allows for calculation of the project duration and the start and finish times of each project task.

SUMMARY OF KEY IDEAS

- The four types of feasibility are technical (or engineering) feasibility; economic (or financial) feasibility; fiscal feasibility; and social, political, and environmental feasibility.
- An alternative is technically feasible if it satisfies the required engineering criteria and can be built.
- Engineering projects incur capital costs and operation and maintenance costs.
- The value of money changes over time, so costs and benefits must be compared in constant dollars (i.e., dollars at the same year).
- To compare costs, the costs must be in the same units.
- Engineering economic factors can be determined by direct calculation, by tables, or by built-in spreadsheet functions.

- Select the alternative that maximizes the net present benefits, not the alternative that maximizes the benefit–cost ratio.
- Fiscal feasibility addresses the question of whether you can obtain enough money to pay for the project.
- Social, political, and environmental feasibility addresses whether the project has sufficient backing of the public and political leaders and is compatible with the environment.
- Project management activities include planning, scheduling, and supervising.

Problems

11.1. List and define the types of feasibility. How do you decide if an alternative is feasible? How can you use measures of feasibility to select the best alternative from several feasible alternatives?

11.2. In Example 11.1, the conclusion about a technically feasible design for a mattress platform depended strongly on the thickness of the material. Repeat the calculations in Example 11.1 and find the following quantities:

a. The range of thicknesses in $\frac{1}{16}$-inch increments for which the steel platform is technically feasible.

b. The range of thicknesses in $\frac{1}{2}$-inch increments for which the cardboard platform is technically feasible.

c. The range of thicknesses in $\frac{1}{4}$-inch increments for which the plywood platform is technically feasible.

11.3. The Golden Gate Bridge cost $35 million to build in 1937. What is the value of $35 million in 2006 dollars, assuming the average interest rate between 1937 and 2006 is 5%?

11.4. A dam is to be constructed for flood control. Project costs and benefits vary with the reservoir volume as shown below. Assume a 50-year project life (planning horizon = 50 years) and 4% interest.

Reservoir Volume (acre-feet)	Initial Construction Costs ($)	Annual O&M Costs ($/year)	Annual Benefits ($/year)
50,000	4,500,000	32,000	317,000
100,000	5,000,000	79,000	761,000
150,000	8,000,000	127,000	1,078,000

a. Which dam size will maximize the net present value benefits?

b. What is the annualized net benefit of the 150,000 acre-feet project?

11.5. You just got your first engineering job. How much money would you have to save per month to buy a $25,000 car three years from now? Assume you can invest the money and get a monthly return rate of 0.25% per month.

11.6.   Wastewater treatment plants often are located in the lowest elevation of an area. The reason for this location is that wastewater usually flows by gravity to the treatment plant (rather than being pumped). In addition, wastewater treatment plants often discharge to rivers and rivers flow through the lowest elevations. However, low elevations typically are less desirable real estate and may be the neighborhoods of people with lower socioeconomic status.

Suppose your firm has recommended a site for a new wastewater treatment plant. You are asked to attend a public meeting, where many residents are upset about the proposed location. Prepare a set of notes for a public meeting to explain and defend the site recommendation. Role-play the public meeting with other students.

11.7.   In the text, the determination of economic feasibility was made by either the net benefits or benefit–cost ratio approach. Another approach is to calculate the *rate of return*. The rate of return is the hypothetical interest rate that makes the present value of an alternative equal to zero. An alternative is considered economically feasible if its rate of return is greater than a specified value (called the *minimum acceptable rate of return*, or MARR).

   a. You borrow $1,000 today and make ten annual payments of $110. Confirm that the rate of return is 1.77%.

   b. An engineering firm is debating whether to replace all their computers now (at a cost of $50,000) to receive an additional $12,500 in revenues each year for the next five years. Should they replace their computers if their MARR is 10%? (You can calculate the rate of return by iteration or using the **IRR** function in Excel.)

   c. Should the firm in Part B replace their computers if they will receive $1,400 more in revenues each year for the next five years than the value shown in part b.

11.8.   Write a short essay on an engineering project that faced social and/or political feasibility challenges. Examples may include NASA programs, urban renewal projects, and the use of the potential replacement structures for the World Trade Center in New York City.

11.9.   Engineers sometimes are criticized for asking *how* something can be built, but not asking *why* something should be built. Write a short paragraph on whether engineers should ask why projects should proceed. You may want to research the concepts of "Buddhist economics" and "intermediate technology" (Schumacher, 1973; www.smallisbeautiful.org; www.itdg.org).

11.10. Your student team has been assigned the task of building a catapult capable of launching a penny at least 50 feet.

   a. Divide the project into tasks, estimate the duration of each task, and construct a precedence table.

   b. Using the critical path method, determine the project duration.

   c. Create a Gantt chart for the project.

PART IV
Technical Communications

It usually takes more than three weeks to prepare a good impromptu speech.
Mark Twain (attributed)

If language is not correct, then what is said is not what is meant; if what is said is not what is meant, then what ought to be done remains undone.
Confucius

12

Introduction to Technical Communications

12.1 INTRODUCTION

Some people think the term *technical communications* is a contradiction in terms. Technical information, they say, is just numbers. They may snicker that engineers are not always natural after-dinner speakers. Why spend your time on the presentation side of things when some engineers are more comfortable grinding out the numbers?

People with these attitudes are sadly misinformed. Engineering often results in complex answers that need to be communicated simply and effectively. The truth is that *engineering work has no impact unless the message is delivered successfully*.

Technical presentations also must "tell a story." The conclusions of the story, of course, must be supported by data and solid reasoning. In evaluating your own technical writing or technical presentations, it is always important to ask yourself, Has the audience understood my story?

The purpose of this chapter is to introduce you to the importance of technical communications (Sections 12.2 and 12.3) and present ground rules common to all technical communication (Sections 12.4 through 12.9). In Section 12.4, the important questions you should answer before you start writing your report or technical talk will be discussed. Some techniques for organizing the presentation material will be presented in Section 12.5. In Sections 12.6, 12.7, and 12.8, you will learn in detail the ways that data are presented, including the design and construction of tables and figures. Section 12.9 discusses creativity in technical presentations.

You will notice some new terminology in this chapter. The recipients of the presentation will be referred to simply as the "audience," since the recipients could be either *readers* of your technical document or *listeners* of your technical talk. The word "presentation" will include both written documents and technical talks.

OBJECTIVES

After reading this chapter, you will be able to:

- explain why technical communication skills are important to engineers;
- list common misconceptions about technical communication;
- discuss how the presentation goals, the target audience, and the constraints shape technical communication;
- devise an outline for a technical presentation;
- use tables and figures to communicate technical information effectively.

12.2 ROLE OF TECHNICAL COMMUNICATION IN ENGINEERING

12.2.1 Technical Communication as a Professional Skill

Your interest in engineering may have been fueled by the important role of engineers in society and the challenges that engineers face every day. Take a moment to make a mental list of what engineers do.

PONDER THIS

What activities do engineers perform?

Key idea: Technical presentations must tell a story; always ask yourself whether the audience understood your story.

Your list may include activities such as designing, modeling, testing, building, and optimizing. While most engineers do at least one of these activities *some* of the time, all engineers communicate *all* the time. In a real sense, engineering is not engineering until you, the engineer, successfully communicate the results to someone else. Technical communication is not effective unless the audience understands the message you wish to deliver.

12.2.2 Technical Communication and Employment

Key idea: Strong technical presentation skills aid in obtaining a job and in advancing a career.

If you remain unconvinced of the importance of technical communication, consider a more practical reason to improve your communication skills. Engineering faculty frequently receive telephone calls requesting information about students (or former students) applying for jobs. Nearly every potential employer asks two questions: Can the person *write* effectively? Can he or she *speak* well? Potential employers ask these questions because they know that engineers spend a great deal of their time communicating. The result of a survey of graduates from the University at Buffalo's School of Engineering and Applied Sciences showed that respondents spent an average of *64% of their working hours* on written communication, oral presentations, and other oral discussions. To compete for employment opportunities, engineers must develop strong technical communication skills. Technical excellence is necessary (but not sufficient) to secure a good job in today's employment market.

Technical communication skills affect not only your ability to get a job, but also your ability to progress in your profession. In a survey cited by Paradis and Zimmerman (1997), over half of the research and development engineers and scientists polled (and 71% of the managers) knew of cases where technical communication skills had a serious impact on a person's career. Respondents to the University at Buffalo survey indicated that good technical communication skills can make the difference between receiving a raise and not receiving a raise. Good technical communication skills are prerequisites for success in your career.

12.3 MISCONCEPTIONS ABOUT TECHNICAL COMMUNICATIONS

Few areas of the engineering profession are more poorly understood or more underappreciated than technical communication. Common misconceptions are discussed in the next several sections.

12.3.1 Misconception #1: Technical Communication Is Inherently Boring

Key idea: Technical communication is a creative process.

Some people feel that engineers excel in dry facts and even drier numbers. How can an engineer possibly communicate creatively? The truth is that designing effective communication strategies is one of the most creative activities in engineering. Technical communication does not mean linking dull facts to form a sleep-inducing document or boring oral presentation. Today, engineers have many tools at their disposal for communicating ideas: everything from sketches on the back of a napkin to 3-D visualization techniques to Internet-based teleconferencing. Effectively communicating technical work is a challenging part of the optimization process that lies at the heart of engineering. Creativity in technical communication is discussed in more detail in Section 12.9.

Technical talks are *not* inherently boring.

12.3.2 Misconception #2: Engineering Communication Is Passive

Key idea: Technical communication is usually meant to be persuasive.

Many people think of technical communication as flat and one-sided. In this view, technical speakers and writers lay out a smorgasbord of facts that the audience records passively (as in a poorly designed lecture). In truth, much technical communication is both interactive and *persuasive*. Engineers often try to convince others of their point of view. Facts and figures rarely speak for themselves. They require thoughtful presentation to convince people of their worth.

12.3.3 Misconception #3: Technical Communication Is Best Left to Nonengineering Specialists

Key idea: Engineers can benefit from communication specialists, but the engineer must take responsibility for making sure the correct message is delivered.

In your career, you will benefit from working with many other professionals. Engineers often work collaboratively with communication specialists, such as technical writers and graphic designers. However, *you as the engineer are always responsible for making sure that the technical information is communicated clearly and concisely to the intended audience*. Remember, all your work (whether in a homework assignment or the design of a multimillion-dollar facility) is for naught if the intended audience does not understand your message. Taking control of the message is as important as taking control of the design calculations.

12.3.4 Misconception #4: Good Technical Communicators Are Born, Not Made

Key idea: All engineers can improve their technical communication skills.

It is true that not all of us will mesmerize° our audiences each time we stand before them or each time we put pen to paper. However, each of us can improve our speaking and writing skills *every time* we set out to communicate with our peers and others. Specific steps for honing your technical communication skills will be presented in Sections 12.4 through 12.9. Whatever level of comfort you have now with public speaking and technical writing, *know that you can improve your communication skills throughout this semester, throughout your university days, and throughout your career.*

°The word "mesmerize" comes from the Austrian-born physician Friedrich Anton Mesmer (1734–1815), who popularized the idea that doctors could induce a hypnotic state by manipulating a force he called "animal magnetism."

12.4 CRITICAL FIRST STEPS

Before you write a single word of a technical presentation, three elements must be identified clearly: the goals of the presentation, the target audience, and the constraints on the presentation. Each of these elements will be discussed in more detail in this section.

12.4.1 Presentation Goals

Key idea: Before preparing a technical presentation, write down the goals of the presentation.

One of the most important activities in the design of any technical talk or document is the identification of the *presentation goals*. It is absolutely critical to know what you are trying to accomplish in a presentation. The presentation will fail unless its goals are identified. Why? First, you cannot decide what information should be presented (or how to present it) unless you have described the objectives thoughtfully. Second, you need to know the goals to evaluate whether or not you have communicated the ideas successfully. In fact, *every* engineering project requires objectives so that the success of the project can be determined at its conclusion.

You should write out the presentation goals. For example, you might write, "The goal of the lab write-up is to tell the professor about the experimental methods employed, the results obtained, and the answers to the three discussion questions." This goal allows you to decide what should go in the lab write-up and how the material should be prioritized. Also, you now have a tool to judge whether your write-up was successful. You could compare the completed lab report with the goal to see if you met the goal. Remember, *a goal not written down is just a dream*.

12.4.2 Target Audience

target audience: the intended recipients of the information to be presented

The presentation goal should identify the **target audience**. The target audience consists of the intended recipients of the information you are presenting.

PONDER THIS

What is the target audience of this text?

Although professors order the text and professionals may read it, the target audience of this text is freshman engineering students.

As with presentation goals, identification of the target audience is critical to the success of your presentation. In your career, you will give oral and written presentations to many audiences, including colleagues (i.e., fellow engineers), managers, elected officials, students, and the general public. You must keep the background and technical sophistication of the target audience in mind when developing your presentation material. For example, you would not use the same approaches to communicate a bridge design to a city council as you would to communicate the same ideas to a professional engineering society.

Key idea: Identify the target audience (and their technical sophistication, interests, and backgrounds) before preparing a technical presentation.

The interests and backgrounds of the audience are as important as their technical sophistication. Each audience member will interpret the presentation through his or her own point of view. To engage the audience fully, you must know the backgrounds of its members. As an example, consider the choices available to an engineer presenting an idea for a new computer design. For an audience of managers and corporate executives, she may wish to emphasize the low cost and high profit margin of the new personal computer. For fellow engineers, she would likely focus on the technical specifications and performance data. Subtle changes often can make the presentation match the interests and background of the audience more closely.

12.4.3 Constraints

Identification of *constraints* on the presentation also is important. Engineering, like life, is a constrained optimization problem. Similarly, technical presentations almost always are

Know your audience

Key idea: Before preparing a technical presentation, quantify the constraints on the presentation (i.e., length limits, your time, and other resource limitations).

constrained. Common constraints are *presentation length* (page limits for written documents or time limits on oral presentations) and *resource limitations* (e.g., your time or money for photographs or specialized graphics). It is very important to heed the presentation length constraints. In oral presentations, going well over or under the allotted time limit is rude and unprofessional. With technical documents, many engineering proposals (and term papers) have gone unread because they exceeded the imposed page limit.

The resources required for technical presentations cannot be ignored. As a student, you know you must allocate time for *writing* a term paper as well as time for reading about the term paper topic. Similarly, practicing engineers learn to budget time for report preparation. Other resources required to produce a high-quality technical document (or oral presentation) include money for personnel, graphics creation, printing, reproduction, and distribution.

Common constraints on technical communication: time and money

12.5   ORGANIZATION

Once you have identified the goals, target audience, and constraints, you can begin to write the presentation. Technical documents and talks can be made or broken on their degree of organization. In a well-organized presentation, the audience always knows where in the presentation they are and where they are going. There are two keys to creating an effectively organized presentation: *structuring the material* and *showing your structure* to the audience.

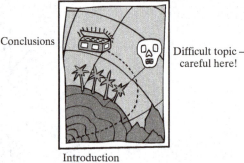

Organization is the map that guides your audience through the presentation.

12.5.1   Outlines

outline: a list of the major headings and subheadings in the presentation, showing the order of the main ideas and showing the secondary topics supporting the main ideas

The primary tool used to structure a presentation is the **outline**. An outline is a structured (or hierarchical) list showing the skeleton of the presentation. An example is shown in Example 12.1.

The purpose of the outline is to divide the presentation into manageable pieces. An outline shows three elements of the presentation:

- The main ideas (listed in the outline as major headings)
- The order of the main ideas
- The secondary topics (subheadings) that support and flesh out the main ideas

The main ideas, of course, depend on the goal of the presentation and the audience.

EXAMPLE 12.1 OUTLINE

Write an outline for a technical presentation on computer-aided manufacturing (CAM) in the production of aircraft.

SOLUTION

An example outline, with the parts of the outline labeled, is as follows:

Computer-Aided Manufacturing (CAM) in Aircraft Production

    I.   Introduction [major heading]
    II.   Background
         A.   History [subheading]
         B.   Contemporary examples
         C.   Current problems

III. Use of CAM in Aircraft Production

    A. CAM principles

    B. Applications

        1. potential barriers [subheading]

        2. examples

    C. Future trends

IV. Conclusions

Key idea: To organize a presentation, structure the material using an outline and show the structure to your audience.

The outline is a wonderful tool for organizing a presentation. It shows at a glance the relationships between parts of the document or talk. The outline helps you to see if the presentation is balanced; that is, whether the level of detail in a certain part of the presentation corresponds to the importance of that part in achieving your goals. The outline also helps determine the needs for more data or more presentation tools (i.e., more tables and figures). An outline can be changed easily as the presentation evolves. In fact, as the outline is annotated (that is, as more levels of subheadings are added), the document or oral presentation will nearly write itself.

signposting: indicators used to show the audience where they are in the presentation

12.5.2 Signposting

Organizing a presentation is only half the battle. You also must *let the audience know* that you are well organized. Showing the audience that you are organized is called **signposting**. An example of signposting is the headings used in this text. The consistency of the headings tells you where you are in the text:

Chapter title: 32 point Futura font, with the initial letters capitalized

Example: # Introduction to...

Section titles: 11-point Copperplate30ab font, all caps

Example: **12.5 ORGANIZATION**

Subsection titles: 11-point Futura Book font, initial letters capitalized

Example: 12.5.2 Signposting

12.6 USING TABLES AND FIGURES TO PRESENT DATA

Nearly every technical presentation you develop will contain data. The number of ways of presenting quantitative information is limited only by your imagination. However, some data presentation tools are more appropriate in a given situation than others.

12.6.1 Use of Tables and Figures

The two main ways to present numbers are *tables* and *figures*. Tables are used when the *actual values are important*. For example, a table would be an excellent way to show the estimated construction, operation, and maintenance costs for three polymer extruder designs. In this case, the exact costs are important and the audience wants to see the numbers.

Key idea: Use tables when actual values are important; use figures to show trends in the data.

On the other hand, figures are used to *show trends in the data*: that is, to show the relationships between variables. For example, suppose you collect data on the movement of an artificial limb in response to stimuli of varying voltage. A figure would be an appropriate way to show the trend in the dependent variable (here, the limb movement) as a function of the independent variable (here, the applied voltage).

12.6.2 Common Characteristics of Tables and Figures

Key idea: Tables and figures should have a number (by which they are referred to in the text) and a short, descriptive title.

While tables and figures are very different, they share several features. First, every table and figure in a technical document must have a number. Many numbering schemes are possible (e.g., "Table 1" or "Figure 4.2" or "Table II" or "Figure C"), but table and figure numbers are essential in technical writing. Why number your tables and figures? A number allows the figure or table to be *referred to* from the text. For example, in the text, you may write

> In Figure 2.3, the average wait time at the stoplight is plotted against the daily pedestrian traffic.

Remember, *do not include a table or figure in a technical document that is not referred to by number in the text*.

Second, every table and figure in a technical document must have a title. Titles are needed to give the audience a short description of the content of the table or figure. Titles should be concise and descriptive. They need not be complete sentences. Examples of table and figure titles are listed in Table 12.1. The numbers and titles appear together either at the top or bottom of the table or figure. Commonly (but not universally), table titles are placed at the *top* of tables and figure titles are placed at the *bottom* of figures. (Note that Table 12.1 has a number and title located together at the top of the table. Also, Table 12.1 was referred to in the text, so you knew when to look at it.)*

Key idea: Tables and figures must be interpreted in the text.

Third, tables and figures must be *interpreted*. This means that you should discuss the table or figure in the text. To continue the example at the beginning of Section 12.6.2, you may write

> In Figure 2.3, the average wait time at the stoplight is plotted against the daily pedestrian traffic. Note that the average wait time increases from baseline only when the pedestrian traffic exceeds 150 people per day.

Many inexperienced technical writers make the mistake of simply throwing the data at the audience rather than *presenting* the data. They write

> The data from the first study are shown in Figure 2.3. A second study was conducted in May 2005.

You included the table or figure for a reason. To satisfy that reason (and help you achieve your presentation goals), you need to guide the audience through the interpretation of the data in your tables and figures.

Key idea: Include units in the row or column headings of tables and the axes of figures.

Fourth, units must be listed for all data in tables and figures. In tables, units usually accompany the column or row headings. In figures, the axes must be labeled with units shown. You may want to take a moment and look through this text for examples of tables and figures with units in the headings or axis labels.

TABLE 12.1 Examples of Poor and Improved Table and Figure Titles

Poor Title	Problems with Poor Title	Improved Title
Table 2: Experimental Data	too vague: what data will the table contain?	Table 2: Ergonomic Data for Three Automobile Seat Designs
Figure 4.2: Problems with Acid Rain	insufficient detail: figure titles usually list the dependent and independent variables	Figure 4.2: Effects of pH on the Survivorship of Brown Trout in Lakes Receiving Acid Rain
Figure A.32: Current vs Voltage	insufficient detail: lists *only* the dependent and independent variables without putting the information in context	Figure A.32: Current–Voltage Curves for Four Electrode Configurations

---

*The astute reader will notice that some pictures in this text have no title and are not referred to in the text. An example is the cartoon labeled "Know your audience" in Section 12.4. The use of such pictures for illustrative purposes is common in textbooks and reflects the fact that the target audience of the text is students.

12.7 TABLES

Key idea: In tables, list the independent variables in the leftmost columns.

As stated previously, tables are used to present data when the actual values are important. Tables should be limited to the minimum number of columns needed to show the relevant data. In general, independent variables are listed in the first or leftmost columns, with dependent variables listed in the columns to the right.

With today's software, it is easy to create tables with myriad types of lines, shadings, colors, and font styles. However, these devices should be used sparingly and consistently. Each table has a goal; "bells and whistles" should be used only to make your point clearer.

An example table is given in Table 12.2.

PONDER THIS

Critique Table 12.2.

Table 12.2 is well constructed. Note that it is numbered and has a descriptive title. The independent variable (reinforcing bar type) is listed first. Units are given for all data (i.e., for every column). Lines are used minimally and mainly serve to separate the table from the surrounding text.

To demonstrate the importance of the order of the columns, examine Table 12.3. Table 12.3 contains the same data as Table 12.2, but the column order has been changed. Note how difficult it is to interpret Table 12.3. Even though the most important information probably is the weight, placing a dependent variable first does not communicate the information very effectively.

Table 12.4 demonstrates the potential for distractions in table design. The use of many fonts, lines, and types of shading adds little to the message and can be distracting.

TABLE 12.2 Characteristics of Standard Steel Reinforcing Bars

[*Caution:* Table may contain errors! See text for discussion.]

Type	Diameter (in)	Weight (lb/ft)
#2	0.250	0.167
#3	0.375	0.376
#4	0.500	0.668
#5	0.625	1.043
#6	0.750	1.502

TABLE 12.3 Characteristics of Standard Steel Reinforcing Bars

[*Caution:* Table may contain errors! See text for discussion.]

Weight (lb/ft)	Type	Diameter (in)
0.167	#2	0.250
0.376	#3	0.375
0.668	#4	0.500
1.043	#5	0.625
1.502	#6	0.750

TABLE 12.4 Characteristics of Standard Steel Reinforcing Bars

[*Caution*: Table may contain errors! See text for discussion.]

Type	Diameter (in.)	Weight (lb/ft)
#2	0.250	0.167
#3	0.375	0.376
#4	0.500	0.668
#5	0.625	1.043
#6	0.750	1.502

12.8   FIGURES

Key idea: Use scatter (x–y) plots when the independent variable is continuous.

scatter (x–y) plot: a type of plot using symbols or lines that is employed when the independent variable is continuous

Key idea: In general, use symbols for data and lines for calculated values (i.e., model output).

Recall that figures are used when the relationships between variables are important. There are three common types of figures used in technical presentations: scatter (or x–y) plots, bar charts, and pie charts.

12.8.1   Scatter Plots

The **scatter plot** (or **x–y plot**) is the most common type of graph in technical work. It is used when the *independent variable is continuous*; that is, when the independent variable could take any value. Examples of continuous variables are time, flow, and voltage. In the scatter plot, the independent variable is plotted on the x-axis (also called the *abscissa*) and the dependent variable is plotted on the y-axis (also called the *ordinate*). In general, symbols are used for data and lines are used for calculated values (i.e., for model fits or model predictions). An example of a scatter plot is shown in Figure 12.1.

PONDER THIS

Critique Figure 12.1.

In Figure 12.1, the independent variable (vapor pressure) is continuous. Thus, a scatter plot is appropriate. Note the important elements: figure title (here, at the bottom of the figure), axis titles with units, tick marks (small lines) near axis labels, and symbols that represent data. If more than one dependent variable were plotted, a legend would

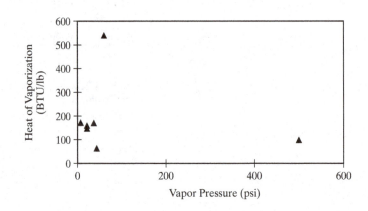

Figure 12.1.   Heat of Vaporization of Some Common Refrigerants [*Caution*: Figure may contain errors! See text for discussion.]

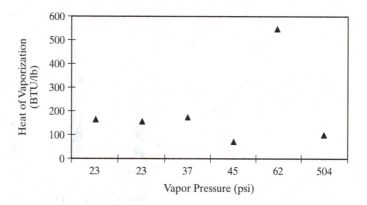

Figure 12.2.   Heat of Vaporization of Some Common Refrigerants [*Caution*: Figure may contain errors! See text for discussion.]

Key idea: Use the line chart type carefully in technical presentations (or, better yet, avoid it completely).

be necessary. Note that *a legend is not necessary if only one dependent variable is plotted*. (Legends are discussed with bar charts in Section 12.8.2.)

One final note on scatter plots. Most common graphing programs (including Microsoft Word, Microsoft Excel, Corel WordPerfect, and Corel QuattroPro) have a figure type (also called a *chart type*) called "line." With the line chart type, the *x* data points are spaced evenly, *regardless of their values*. The data in Figure 12.1 are replotted as a line chart in Figure 12.2. Notice that the relationship between heat of vaporization and vapor pressure appears to be distorted in the line chart. There are almost no cases where the line type is the *best* way to present technical data. It is recommended that *you avoid the line chart type completely*.

12.8.2   Bar Charts

bar chart: a type of plot using bars that is employed when the independent variable is not continuous

Bar charts are used when the independent variable is discrete (i.e., not continuous). Discontinuous independent variables are common in engineering. For example, you may wish to show how the properties of magnets vary with material type or how energy efficiency varies with industry category. The type of material or category of industry is a discrete variable and the use of a bar chart is appropriate.

Key idea: Use bar charts when the independent variable is not continuous.

An example of a bar chart* is given in Figure 12.3. Note the descriptive title, inclusion of units, and tick marks on the *y*-axis. Tick marks generally are not used on the *x*-axis in bar charts with vertical bars, since the tick marks would interfere with the bars. Note also in Figure 12.3 that two *y*-axes are used. Multiple *y*-axes are useful when the independent variables have different units or vastly different scales.

legend: a listing of the property represented by each symbol, bar, or line

In Figure 12.3, two variables are plotted; therefore, a legend is required. A **legend** tells the audience the meaning of each symbol, bar, or line. In this case, the legend tells you that the white bar represents the resistivity and the black bar represents tensile strength.

Key idea: In figures, select the axis ranges to encompass all the data without distorting the relative values.

In both scatter and bar charts, you must select the ranges of the axes carefully. Clearly, the ranges must be selected to encompass all data. In addition, it is generally a good idea to start the *y*-axis at zero.[†] Why? Starting at zero gives the audience a better view of the relative values of your data. In Figure 12.3, for example, it is obvious that the tensile strength of silver is about twice that of gold. If the data are replotted using smaller ranges for the *y*-axes, a skewed view of the relative resistivities and tensile strengths is created (see Figure 12.4). For example, the tensile strength of silver appears to be about five times that of gold in this figure.

---

*The common name for the plot in Figure 12.3 is a bar chart. Some software packages call it a *column chart* if the bars are vertical and a *bar chart* if the bars are horizontal.
[†]Do not, of course, start the *y*-axis at zero if you have negative *y* values. Also, avoid starting the *y*-axis at zero if the *y* values cluster around a large value.

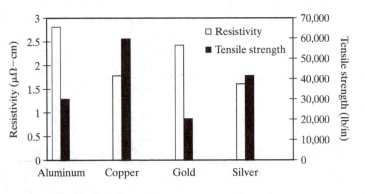

Figure 12.3. Physical Properties of Conductors

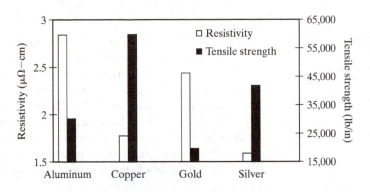

Figure 12.4. Physical Properties of Conductors [*Caution*: Figure may contain errors! See text for discussion.]

Key idea: Do not accept the default table or figure produced by the software without questioning whether it meets your objectives.

This lesson can be extrapolated. In general, *do not let the software pick the look of your tables and figures*. Always look critically at the default table or figure produced by the software package. Use your judgment: edit tables and figures to best meet your presentation goals.

12.8.3 Pie Charts

pie chart: a type of plot using pie slices that is employed to show the relative contributions of several factors to a whole

Pie charts are used to show the relative contributions of several factors to a whole. In most cases, pie charts are used to show percentages. Thus, pie charts have no independent variable. Although pie charts are not used very frequently in engineering, they can show the relative importance of discrete factors very effectively.

Key idea: Use pie charts to show relative contributions.

An example of a 3-D pie chart is shown in Figure 12.5. The slices of the pie may be defined with a legend or labels.

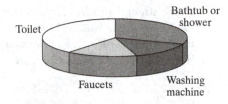

Figure 12.5. Water Use in the Home

An example of how the design of a figure may influence engineering decision making is shown in the *Focus on Figures: Of Plots and Space Shuttles*.

FOCUS ON FIGURES: OF PLOTS AND SPACE SHUTTLES

The explosion of the Space Shuttle *Challenger* on the cold morning of January 28, 1986, rocked the world. Subsequent investigation into the disaster pointed to the likely cause: hot gases from fuel combustion bypassed two seals, leading to the destruction of the booster segment and the loss of the lives of all seven astronauts. The booster segments were sealed with O-rings made out of a rubber-like material called Viton®. The two O-rings (primary and secondary) protected the segments from the combustion gases. The primary O-ring was closest to the fuel.

The *Challenger* disaster is often discussed as an example of engineering ethics. Although some facts are in dispute, it is clear that some of the engineers involved vigorously argued that the launch should be aborted. Why? The temperature at launch was forecasted to be much lower than previously experienced. Like typical rubber, the flexibility of Viton (and thus its ability to seal against the enormous pressures at launch) is dependent on temperature.

It has been argued (Tufte, 1993) that the available data, if plotted in the most meaningful way, would have provided overwhelming evidence for aborting the launch. According to this argument, the engineers were remiss in not presenting the data in the most powerful way. In other words, technical communication problems may have contributed to the launch and loss of *Challenger*. This point of view has been strongly challenged by the engineers involved (Robison et al., 2002). The arguments and counterarguments are complex and cannot be summarized in this short section. The interested reader is urged to read the cited papers. The purpose here is to show how data presentation can lead and mislead the engineer.

To appreciate the importance of how the data were plotted, it is necessary to understand what data were available at the time. There are several indicators of O-ring damage. One indicator is soot marks made by blackened grease as it blows through the primary O-ring. Soot is a very bad sign, indicating that the primary O-ring has been breached and the shuttle health depends only on the remaining secondary O-ring. The size of the soot marks (for shuttle launches with measurable

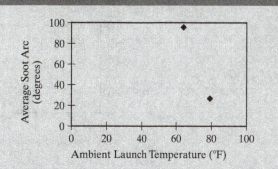

Figure 12.6. Influence of Temperature on Soot Including Only Data Where Soot Was Observed

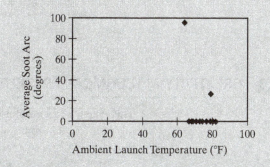

Figure 12.7. Influence of Temperature on Soot Including All Available Data

soot) are shown as a function of ambient temperature at launch in Figure 12.6. Based on these data, would you recommend launching at the launch temperature of 26°F on January 28, 1986?

Based on the data plotted in Figure 12.6, you *might* conclude that a launch at 26°F is inadvisable. Although there *appears* to be a trend that soot area increases with decreasing temperature, two data points are hardly enough to justify a quantitative relationship. The picture becomes even cloudier when all available data are included (Figure 12.7). Note that no soot was observed at many launch temperatures between the values shown in Figure 12.6. Does the trend appear weaker now?

Results of the testing of isolated rockets revealed no soot at O-ring temperatures between 47 and 50°F (see Figure 12.8). How would the rocket test data influence your decision to launch?

History proved that the advice not to launch was justified. It is impossible to know with certainty whether a plot such as Figure 12.7 or Figure 12.8 would have enhanced the argument of the engineers. Rather than a clear-cut lesson in technical communications, we are left with a tragedy.

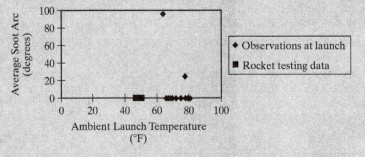

Figure 12.8.   Influence of Temperature on Soot Including All Available Data and Rocket Testing Results

12.9   CREATIVITY IN TECHNICAL PRESENTATIONS

This chapter has emphasized the need for structure in technical presentations and has introduced numerous rules. However, please do not forget that technical communication is a creative process. Much of the creativity in technical presentations is focused on two areas: conciseness and thinking visually.

12.9.1   Creative Conciseness

When in doubt, favor conciseness over verbosity in technical presentations. Simply filling the page or presentation time with words is always obvious and insulting to the audience. In addition, calculations, data, or analysis that are necessary, but secondary to the main points being made, can be very distracting. In a written document, they may be best placed in an appendix.

Finding the right degree of conciseness is not easy. Technical presentations, like homemade bread, are hard to digest if they are too dense. To use another food analogy: wine can be very pleasant. It can be distilled into a complex brandy. Overdistill and you end up with ethanol: harsh and undrinkable. Often a dense presentation can be made more palatable by building in repetition and explanatory text.

The idea of conciseness also applies to figures. Consider the three plots of the same data in Figure 12.9. The top panel is a typical figure produced by the built-in plotting software of a word processing program.

PONDER THIS   **What unnecessary elements do you see in the top panel of Figure 12.9?**

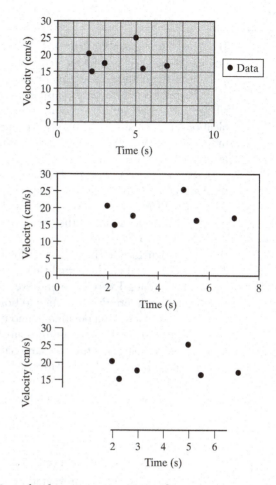

Figure 12.9. Example of Conciseness in Figures [*Caution*: Figure may contain errors! See text for discussion.]

The extraneous graphical elements in the top panel of Figure 12.9 include background color, grid lines, and legend. (No legend is needed, since there is only one set of symbols.) A clearer presentation (middle panel of Figure 12.9) is produced by eliminating the extraneous elements. In the bottom panel of Figure 12.9, nearly all extraneous lines have been removed. For most engineers, the bottom panel is on the verge of being too abstract: perhaps too much information has been removed. Figure 12.9 shows that some redundancy is needed to best communicate the information.

12.9.2   Thinking Visually

Another important creative element in technical presentations is the ability to *think visually*. The layout of the page or the slides can help make your points or distract the audience from your goal.

For more details (and fascinating examples), peruse the books by Edward Tufte listed in the references (Tufte, 1983, 1990). These are truly amazing and beautiful books that will greatly influence your thinking about the design of figures and tables.

12.10 SUMMARY

Technical communication is important in turning engineering ideas into reality. In addition, good technical communication skills are essential for obtaining an engineering job and advancing in the engineering profession.

Engineers must take responsibility for communicating their ideas and their work to a large and varied audience. As an engineer, you should think about technical communications as a creative and persuasive engineering tool. You must take control of the message, ask yourself if your message is understood, and seek to improve your communication skills at every opportunity.

Several general aspects of technical presentations (i.e., technical writing and technical speaking) were discussed in this chapter. Before putting pen to paper, you should take several steps. First, always identify the goals of the presentation, the target audience, and the constraints of the presentation. Second, organize the material to be presented. This can be done by using an outline to structure the information. Be sure to show your structure to the audience. Third, use the proper technique to present data. Tables are used when the actual values are important, while figures are used to show trends in the data. Every table and every figure in a technical document must have a number and a descriptive title. In addition, every table and figure must be referred to from the text of a written document, and its main points must be summarized.

Be sure to use the most appropriate type of figure: scatter $(x–y)$ plots when the independent variable is continuous, bar charts when the independent variable is not continuous, pie charts to show relative proportions, and line charts almost never. Look critically at the default table or figure produced by software and ask how it could be modified to best meet *your* presentation goals.

SUMMARY OF KEY IDEAS

- Technical presentations must tell a story; always ask yourself whether the audience understood your story.
- Strong technical presentation skills aid in obtaining a job and in advancing a career.
- Technical communication is a creative process.
- Technical communication is usually meant to be persuasive.
- Engineers can benefit from communication specialists, but the engineer must take responsibility for making sure the correct message is delivered.
- All engineers can improve their technical communication skills.
- Before preparing a technical presentation, write down the goals of the presentation.
- Identify the target audience (and their technical sophistication, interests, and backgrounds) before preparing a technical presentation.
- Before preparing a technical presentation, quantify the constraints on the presentation (i.e., length limits, your time, and other resource limitations).
- To organize a presentation, structure the material using an outline and show the structure to your audience.
- Use tables when actual values are important; use figures to show trends in the data.
- Tables and figures should have a number (by which they are referred to in the text) and a short, descriptive title.

- Tables and figures must be interpreted in the text.
- Include units in the row or column headings of tables and the axes of figures.
- In tables, list the independent variables in the leftmost columns.
- Use scatter (x–y) plots when the independent variable is continuous.
- In general, use symbols for data and lines for calculated values (i.e., model output).
- Use the line chart type carefully in technical presentations (or, better yet, avoid it completely).
- Use bar charts when the independent variable is not continuous.
- In figures, select the axis ranges to encompass all the data without distorting the relative values.
- Do not accept the default table or figure produced by the software without questioning whether it meets your objectives.
- Use pie charts to show relative contributions.

Problems

12.1. Identify the goals, target audience, and constraints for the following types of communication:

    a. Two roommates discussing how to divide the telephone bill

    b. A review article on the avian flu virus in a newsmagazine

    c. A NASA news briefing on the evidence of water on Mars

12.2. Write an outline for a research paper on career opportunities in the engineering field of your choice.

12.3. Discuss whether you would use a figure or a table to present the following data. If you choose a figure, state which type of figure you would use.

    a. The chemical composition (in percent by weight) of a concrete formulation

    b. Operation and maintenance costs of three pavement types

    c. Effect of fiber-optic cable length on the transmission of photons

    d. Percentage of zebra mussels killed under a specified treatment regime

12.4. Figure titles often are missing or incomplete in the popular press. Find two data figures in a newspaper or newsmagazine. Critique the figure titles and then write your own.

12.5. Find two data tables in a newspaper or newsmagazine. Critique and write your own table titles. Edit the table, if necessary, following the principles discussed in this chapter.

12.6. Some people refer to line charts as "bar charts with symbols." Explain this definition of line charts.

12.7. Write Newton's Second Law of Motion in a concise form for a technical audience and in a more expansive form for a general audience.

12.8.   Pick a figure in this text, critique it, and improve upon its design. State why your design is an improvement.

12.9.   Interview a practicing engineer and write a paragraph about the importance of technical communication in his or her professional life.

12.10. Explain the differences and similarities between technical communication and written or oral presentations you did in high school in nontechnical courses.

13

Written Technical Communications

13.1 INTRODUCTION

In this chapter, written technical communications will be discussed in much detail. Organization is the key to good technical communication. Thus, most of the chapter (Sections 13.2 and 13.3) is devoted to the organization of written documents. Grammar and spelling issues are reviewed in Section 13.4. Section 13.5 provides details on the types of engineering documents you will write, from formal reports to casual email.

13.2 OVERALL ORGANIZATION OF TECHNICAL DOCUMENTS

13.2.1 Introduction

The key to good written and oral presentations is organization. Technical documents must be organized on several levels. In this section, the general organization of technical documents will be discussed. Organization at the paragraph, sentence, and word levels is the subject of Section 13.3.

13.2.2 General Organization Schemes

Outlines should be used to develop organized presentations. What headings and subheadings should be employed? Clearly, the details of the outline will depend on the goal of the presentation and nature of the technical work. Although every technical report is different, several elements are common to many technical presentations. Important elements found in many technical presentations are given in Table 13.1. The common elements are as follows:

OBJECTIVES

After reading this chapter, you will be able to:

- list the elements of technical documents;
- organize a technical document;
- identify common grammatical and spelling errors in technical documents;
- proofread technical documents;
- write an effective technical document.

Key idea: Organize technical documents from the largest to smallest scale: outline level, paragraph level, sentence level, and word level.

Key idea: Common elements of technical documents include the abstract (or executive summary), introduction/background/literature review, methods, results, discussion, conclusions/recommendations, and references.

Key idea: The abstract should contain a summary of each element of the report.

- Abstract
- Introduction/Background/Literature Review
- Methods/Modeling
- Results
- Discussion
- Conclusions/Recommendations
- References

Each of the common elements will be illustrated with a report on a laboratory exercise conducted to test the conservation of momentum.

13.2.3 Abstract

Technical documents typically begin with an *abstract*. The purpose of the abstract is to provide a brief summary of the remainder of the document. The abstract should include the important points from each element in the document. An extended abstract (often written for nontechnical audiences) is sometimes called an *executive summary*.

A properly written abstract should be a miniature version of the entire technical document. The word *abstract* comes from the Latin *abstractus*, meaning drawn off. In a true sense, think of the abstract as being *drawn off of the whole document*. Thus, an abstract should include the following sections:

- An introduction (with enough background material to show the importance of the work),
- A statement on the methods or models employed,
- A short summary of the results and their meaning, and
- Conclusions and recommendations.

For the lab report on the conservation of momentum, the abstract might read as follows:

Abstract

The purpose of this lab was to test the law of conservation of momentum. Experiments were conducted with disks designed to remain together after collision. The masses and velocities of the disks were measured before and after collision. On average, the total momentum of the system after the collision was 101% of the total momentum before the collision. The calculated momentums were interpreted to be consistent with the conservation of momentum law.

TABLE 13.1 Elements in a General Technical Document

Section Title	Purpose
Abstract or Executive Summary	Summarizes the entire report, including all other elements
Introduction or Background or Literature Review	Brings the reader to the topic of the report; may give project history and/or a review of the appropriate technical literature
Methods or Modeling	Describes study approach, methods used, and model development (if any)
Results	Presents the results, including "raw" data with trends indicated but little interpretation of the data
Discussion	Interprets of the results
Conclusions and Recommendations	Summarizes main points and gives suggestions for further work, often in a list format
References	Lists references cited (may be in an appendix)

Note that the abstract contains all the elements of the full report: introduction (first sentence), methods (second and third sentences), results (fourth sentence), and conclusion (last sentence).

13.2.4 Introduction

The next element is the *introduction*. In writing the introduction section, assume that the reader knows only the information in the title of the report. After reading the introduction, the reader should have a good idea of the *motivation* for the report (i.e., why the report was written).

Key idea: The introduction should take the reader from the report title to an understanding of why the report was written.

In some cases, the introduction section may be fairly long. It may include a discussion of the project history, a review of pertinent technical literature, and a presentation of the goals and objectives of the work. On other occasions, the introduction is short and the other material is placed in separate sections (i.e., a background or a literature review section or a goals/objectives section).

The introduction section takes the reader from the title to an appreciation of why the document was written.

For the lab report on the conservation of momentum, the introduction might read as follows:

Introduction

Science and engineering are founded on a number of conservation laws. One example is the conservation of momentum. Momentum is the product of the mass of an object and its velocity. The law of conservation of momentum states that the momentum of a closed system remains unchanged.

The conservation laws are impossible to prove experimentally because of error. However, the data collected in a well-planned experiment should be consistent with the conservation laws. In this lab, a comparison was made between the momentum calculations from laboratory data and the law of conservation of momentum.

Key idea: In the methods section, justify the study approach, present data collection techniques, and discuss data analysis methods.

13.2.5 Methods

The introduction is usually followed by a section on the *methods* employed in the study. The methods section should describe three elements of the work. First, the methods section should justify the *study approach*. In most engineering studies, there are many ways to achieve the study goals.

PONDER THIS

> **How many ways can you think of to "test" the law of conservation of momentum?**

For example, you could explore the conservation of momentum law under controlled conditions with billiard balls or model cars or hockey pucks. You also could collect data in the real world. For example, a visit to a county fair would allow you to make measurements using bumper cars or the demolition derby. Even in this simple example, there are many ways to test the hypothesis of interest. As an engineer, you *choose* to follow a certain approach. It is important to justify your choice.

Second, the methods section should discuss the techniques involved in data collection. For experimental work, this means describing the measurement methods. For modeling studies, this means presenting the models developed specifically for your study.

Third, the methods section should discuss the approaches used to analyze the data. For example, suppose you measured temperature using a thermistor. A thermistor is a resistor that has a resistance related to temperature in a known fashion. In a study using a thermistor, it may be necessary in the methods section to describe how the temperature was calculated from electrical measurements.

The three parts of the methods section can be summarized as follows:

- *Why* did you do the work? (study approach)
- *How* did you do the work? (experimental procedure)
- *What* did you do with the work? (data analysis)

Often, the information about experimental set-up can be communicated most effectively by drawings or photographs. It should be noted that in some technical fields, information on methods is placed in an appendix rather than in the body of a report.

Elements of a methods section

For the lab report on the conservation of momentum, the methods section might read as follows:

Methods

Data collection was performed in a laboratory setting to enhance reproducibility. Tests were conducted on an air table to minimize friction.

Six experiments were conducted. For each experiment, the masses of two plastic disks were recorded. The disks were 5 cm in diameter and 0.5 cm thick. The rims of the disks were covered with a strip of Velcro tape to allow the disks to stick together upon impact. The disks were positioned about 2 m apart. One disk was propelled by hand towards the other disk. Disk velocities were measured immediately before and after collision.

Masses were determined with a Model 501 balance. To measure disk velocities, a digital video camera (VideoCon Model 75) capable of recording images at 30 frames per second was positioned above the initially stationary disk. The sides of the air table were marked in 0.1 cm increments. Images were examined frame by frame, with the instantaneous velocity calculated as (distance traveled between frames) divided by (time between frames). The velocities of the disks were averaged over one second prior to and after collision.

The average momentum was calculated as $p = mv$, where m represents mass and v denotes velocity.

In this example, the study approach is presented and justified in the first paragraph. The second paragraph gives the overall experimental procedure, with the measurement details in the third paragraph. The fourth paragraph outlines the data analysis approach.

13.2.6   Results and Discussion

Key idea: In the results section, present the results and note the general trends.

The results section comes next. In this section of the typical engineering report, the results are *presented* but not *interpreted*. The general trends shown by the data in tables and figures should be highlighted as the data are presented.

Key idea: Interpret data in the discussion section.

In the results section, there is generally little interpretation of the data. Data interpretation comes in the *discussion* section. Here, elements of the results section are combined and interpreted to reach the main conclusions of the engineering study. Often, data are compared with predictions from models in the discussion section. In a design report, alternative designs may be compared and a final alternative selected in this section.

In a results section:

The measured force values are shown against mass in Figure 1. The measured force (in newtons) increased nearly linearly with the increase in mass (in kg).

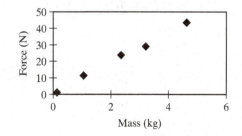

Figure 1: Dependency of Measured Force Values on Mass

In a discussion section:

Measured forces are compared with model predictions in Figure 2. The experimental results were consistent with the model, $F = ma$.

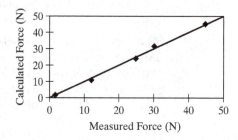

Figure 2: Comparison of Measured Forces and Model Results
(Line is calculated force = measured force)

The division between the results and discussion sections is not always clear cut. In fact, the results and discussion sections usually are combined in short reports. To illustrate the difference between the results and discussion sections, consider a large report on the effects of fatigue on the performance of assembly line workers. In the results section, the data on fatigue measures and performance measures might be reported. General trends (for example, that time to completion of critical tasks decreased as the level of fatigue increased) may be noted. More detailed interpretation of the data would be placed in the discussion section, where, for example, the predictions of a performance model might be compared with the data collected in the study.

For the conservation of momentum lab report, the results and discussion sections probably would be combined because the scope of the report is small. Example results and discussion sections are separated here for illustrative purposes.

Results

Measured masses and mean velocities from the six experiments are shown in Table 1. Note that the measured masses of the single disk (before collision) are similar, as expected. In addition, the measured masses of the coupled disks (after collision) are nearly double the masses of the single disk. By inspection of the data in Table 1, it appears that the velocity decreased by nearly a factor of two as the mass increased by about a factor of two.

TABLE 1 Measured Mass and Mean Velocity Data

	Before Collision		**After Collision**	
Experiment	**Mass (g)**	**Mean Velocity (cm/s)**	**Mass (g)**	**Mean Velocity (cm/s)**
1	2.5	99	5.0	51
2	2.5	102	5.1	48
3	2.4	96	4.9	48
4	2.5	93	5.0	45
5	2.6	102	5.1	51
6	2.5	105	5.1	54

Discussion

The calculated momentum values before and after collision are listed in Table 2. Note that the calculated momentum values before and after collision are nearly equal. As shown in the fourth column of Table 2, the momentum after collision averaged 101% of the momentum before the collision.

TABLE 2   Calculated Momentum Values (p) Before and After Collision

Expt.	p **Before Collision** (g-cm/s)	p **After Collision** (g-cm/s)	(p **After**)/(p **Before**) (%)
1	250	260	104
2	260	240	92
3	230	240	104
4	230	230	100
5	270	260	96
6	260	280	108
mean			101

The approach used in this lab was to compare two momentum values. Therefore, it is important to estimate the uncertainty in the mass and velocity measurements. The precision of the mass measurements can be estimated by the precision of the balance (given by the manufacturer as ±0.01 g). The instantaneous velocity was calculated as (distance traveled between frames) divided by (1/30 second per frame). The distance traveled was rounded to the nearest 0.1 cm, since the scale of the side of the table was marked in 0.1-cm increments. A difference of 0.1 cm over 1/30 s represents $(0.1 \text{ cm})/(1/30 \text{ s}) = 3$ cm/s. The uncertainty in velocity (3 cm/s) represents about 1.2% of the average velocity of 250 cm/s. Thus, the difference of 1% between momentum values before and after collision is not unreasonable. Given the uncertainty, the data collected are consistent with conservation of momentum during the collision of two disks.

13.2.7   Conclusions and Recommendations

Key idea: Conclusions and recommendations are often in list form and should be written very carefully.

The last main section of a typical engineering report is the *conclusions and recommendations* section. The conclusions and recommendations must be among the most carefully worded sections of an engineering report, since many readers may turn here first. Conclusions and recommendations often appear in a list format. The conclusions should stem directly from the discussion. In other words, no *new* information should be presented in the conclusions.

The recommendations section is a critical part of an engineering report. Why? Recall that engineers often select intelligently from among alternatives. The preferred alternative often is highlighted in the recommendations section.

An example of a conclusions section is given next.

Conclusions

An experimental study was conducted to explore the conservation of momentum law as applied to a collision of two discs on an air table. The momentum after the collision averaged 101% of the momentum before the collision. The experimental results were consistent with the conservation of momentum.

13.2.8   References

Key idea: Although many reference formats are acceptable, the references must be complete and consistent.

The last section of a technical report (often found in an appendix) is a list of references. There are many acceptable formats for listing references in technical material. The guiding rules are that the references should be *complete* (so that the reader can find the referenced material easily) and *consistent* (i.e., use the same format for all books or journals cited). Following are some examples of reference formats:

For books:

Author's last name, author's initials (for second authors, initials followed by name), book title in bold, publisher's name, publisher's location, publication date.

Example: Keller, H. **The Story of My Life**. Doubleday, Page & Co., New York, NY, 1903.

For journal articles:

Author's last name, author's initials (for second authors, initials followed by name), article title, journal name in bold italic, volume number, issue number in parentheses, page range, date.

Example: Dallard, P., A. J. Fitzpatrick, and A. Flint. The London Millennium Footbridge. *Structural Engineer*, **79**(22), 17–35, 2001.

For Web pages:

Author's last name, author's initials (for second authors, initials followed by name), article title, URL, date visited in parentheses.

Example: Anon., Standard Contract Documents, http://www.nspe.org/ejcdc/home.asp (visited March 21, 2005).

bibliography: a list of useful sources of information, including sources not cited in the text (as contrasted with the references, in which only cited material is listed)

Be careful about the differences between a list of references and a *bibliography*. A reference list consists only of the material cited in the text. A bibliography lists all useful sources of information, even if they are not specifically cited in the text. For examples, please see the References appendix of this text.

13.2.9 Signposting in Technical Writing

As discussed in Section 13.2.2, a good technical presentation is well organized and the organization is clear to the audience. The idea of showing the audience where you are in a technical presentation is called *signposting*. A common mistake in technical writing is to give the reader page after page of text with no guide to the content of the text.

Key idea: Use section headings or numbering schemes as signposts in technical documents.

In technical documents, signposting is usually accomplished in one of two ways. First, you may use *section headings* to show the readers where they are in the document. The divisions described in Section 13.2.1 (e.g., Introduction, Methods, Results, and so on) may be good section headings. Be sure to use a consistent theme to show the hierarchy of the headings. For example, major headings might be left aligned, while subheadings are indented. Or major headings might be all in capital letters, while subheadings are in initial caps.

Second, you can signpost with a *numbering scheme*. Numbers are an excellent way to show the hierarchy of headings. For example, a major section may be given a number (e.g., "3. Assessment of Alternatives"), with subheadings listed as sections under the number (e.g., "3.1 Soldered Joints Alternative"). Hierarchy can be shown by the numbering scheme (4, 4.1, 4.1.1, or I, I.A, I.A.1, or others), indentation, or use of boldface fonts.

Regardless of the system used, signposting *must be applied consistently*. If you use a bold font with initial caps with second-level subheadings, then use a bold font with initial caps with *all* second-level subheadings. Your readers will rely on your signals. Do not confuse the reader with inconsistent signposting.

13.3 ORGANIZING PARTS OF TECHNICAL DOCUMENTS

13.3.1 Paragraph Organization

Beyond organizing the overall presentation, each paragraph also should be structured. Each paragraph should tell a complete story and be structured by sentence. The paragraph should begin with a **topic sentence**. The topic sentence states the purpose of the paragraph. Each following sentence should *support the topic sentence*. Paragraphs should end with a *concluding sentence*, which summarizes the main points of the paragraph. Thus, each sentence in the paragraph has a specific purpose.

topic sentence: the first sentence in a paragraph in which the purpose of the paragraph is stated

PONDER THIS

Reread the previous paragraph and evaluate whether it is structured correctly.

13.3.2 Sentence Organization

Key idea: Each sentence should express a single idea.

A sentence is a grammatical structure containing a subject and a verb. Sentences should express a *single idea*. There are two common problems with sentences in technical documents: overly long sentences (with more than one idea) and too short sentences (lacking a subject or verb). Avoid using conjunctions (e.g., *and*, *but*, *or*, *nor*, *for*, *so*, or *yet*) to combine disparate ideas into one sentence. Consider the following sentence:

> Design parameters were calculated by standard procedures and all results were rounded to three significant figures.

This sentence contains two ideas. It should be split into two sentences at the word *and*:

> Design parameters were calculated by standard procedures. All results were rounded to three significant figures.

sentence fragment: an incomplete sentence (usually lacking a subject or verb)

Sentences can be *too short* if they do not include both a subject and a verb. Incomplete sentences are called **sentence fragments**. A common sentence fragment in technical writing creeps in when stating trends. For example,

> The higher the temperature, the shorter the annealing time.

PONDER THIS

Why is the sentence fragment "The higher the temperature, the shorter the annealing time" *not* a sentence?

This fragment has no verb and thus is not a sentence. Try to avoid such constructions in your technical writing. Say instead: "Annealing time decreased as the temperature was increased."

13.3.3 Word Choice

Key idea: Choose words to make your writing concise, simple, and specific.

The lowest level of organization is the choice of words. In choosing words to form sentences, try to be as *concise*, *simple*, and *specific* as possible.

Concise writing means that you should use the minimum number of words to express the thought clearly. To write concisely, avoid long prepositional phrases. Examples of common wordy phrases and suggested substitutions are listed in Table 13.2.

TABLE 13.2 Examples of Long Prepositional Phrases to Be Avoided
(adapted from Smith and Vesiland, 1996)

Wordy Prepositional Phrase	Possible Substitute
due to the fact that …	because …
in order to …	to …
in terms of …	reword sentence and delete phrase[a]
in the event that …	if …
in the process of …	delete, or use "while" or "during"
it just so happens that …	because …
on the order of [b] …	about …

[a] *Example*: The sentence "In terms of energy use, Alternative 3 was lowest" could be rewritten as "Alternative 3 had the lowest energy use."

[b] This phrase sometimes is used to indicate an order of magnitude (i.e., a power of 10), as in the following sentence: "On the order of 10,000 bolts were employed in the construction project."

For example, instead of writing

In order to find the optimum temperature, we conducted experiments.

it is preferable to write

To find the optimum temperature, we conducted experiments.

Although the general public sometimes feels that technical writing is impenetrable, written technical communication should be *simple*. In other words, use simple words to express your ideas as clearly as possible. Avoid sentences such as

System failure mode was encountered on three sundry occasions.

Instead, write more clearly:

The system failed three times.

Key idea: Avoid making up new words or new uses of words in conventional technical writing.

A common and annoying device used to make writing sound more technical is the use of nouns as verbs. One way this is accomplished is by adding the suffix *-ize* to almost any noun (e.g., initialize, prioritize, customize, and the like). Writers are converting nouns into verbs with increasing frequency. For example, a nationwide company offering photocopying services used to advertise itself as "The new way to office." (What does "to office" mean?) In your writing, avoid making up new verbs from nouns.

The heart of technical writing is its *specificity*. Make your writing specific by avoiding general adjectives such as *many*, *several*, *much*, and *a few*. Quantify your statements when you can:

Engine temperatures were 5°C above normal.

not

Engine temperatures were several degrees above normal.

13.4 GRAMMAR AND SPELLING

Important ideas about spelling and grammar will be reviewed in this section. The purpose of this discussion is not to provide you with a comprehensive list of the rules of grammar, but rather to identify common trouble spots in technical writing.

There are no excuses for errors in grammar or spelling in technical writing. The most important rules are reviewed here. Problem words will be discussed at the end of this section. For more details, please examine any of the excellent books listed in the bibliography at the end of the text.

You should be aware that there is some disagreement on several grammatical rules. It is important to differentiate firm rules from one writer's opinion. It is frustrating to learn and use one approach from a mentor, only to have it totally dismantled by another mentor. When in doubt about the feedback you have received, always ask questions.

13.4.1 Subject–verb match

Key idea: Make sure that the subject and verb agree in number (i.e., they must be both singular or both plural).

The subject and verb must match in number. In other words, use plural forms of verbs with plural nouns and singular forms of verbs with singular nouns. For example, you should write

The contacts of the integrated circuit were corroded.

not

The contacts of the integrated circuit was corroded.

The subject of the sentence is plural ("contacts") and thus a plural verb ("were") is required.

In most cases, the "subject–verb match" rule is simple. However, some sticky situations arise. For example, is the noun "data" plural or singular? In most technical literature, the word *data* is considered to be a plural noun. (Formally, it is the plural of the noun *datum*.) A growing number of technical writers consider *data* to be singular when referring to a specified set of data. The safe bet is to treat *data* as a plural noun:

The data fall within two standard deviations of the mean.

not

The data falls within two standard deviations of the mean.

If you wish to use a singular noun, use *data set*:

The data set was larger last year.

13.4.2   Voice

In grammar, **voice** refers to the person (people) or things doing the action. There are two general voices: active and passive. In the active voice, the subject is identified. In the passive voice, the person performing the action is not identified (either directly or by category).

There is some difference of opinion about which voice is best for technical writing. In general, the active voice is preferred. Why? In engineering, you usually want to know who did the action. You should write

Field technicians backfilled the soil.

not

The soil was backfilled.

The passive voice is appropriate when the identity of the person doing the action is obvious or unimportant. Thus, you will find the passive voice used in many engineering reports where the subject already has been identified. For example, the passive voice was used in the conservation of momentum example in Section 13.3. Regardless of the voice used, be consistent and use the same voice throughout.

Although the active voice is preferred, you should always avoid the use of the first person in technical writing. For example, write

XYZ Engineering personnel developed an ergonomic design.

or

We developed an ergonomic design.

not

I developed an ergonomic design.

13.4.3   Tense

Tense refers to when the action occurred. In technical writing, use the present tense unless describing work done in the past. Thus, write

Values were calculated by a nonlinear optimization algorithm.

This is in the past tense, since the calculation took place in the past. On the other hand, you might write

The results indicate the importance of the new quality assurance procedures.

Use the present tense here, since the results and their interpretation exist now.

13.4.4   Pronouns

Pronouns are substitutes for nouns. Examples of pronouns include *he*, *she*, *it*, *they*, and *them*. An all-too-common problem with pronouns in technical (and nontechnical) writing is *gender bias*. In the older technical literature, scientists and engineers were identified as males. It used to be common to write

An engineer must trust his abilities. [***incorrect**]

This construction is **not** proper, as it implies that all engineers are men.

One approach to remedying this situation is the use of the pronouns *they* or *their* in place of *he* and *his*. This leads to the statement

An engineer must trust their abilities. [***incorrect**]

Unfortunately, the solution is grammatically incorrect.

What is wrong with the statement, "An engineer must trust their abilities"?

In this case, the subject ("an engineer") is singular and the pronoun ("their") is plural. A much better solution to gender-specific pronouns is to rework the sentence completely so that the subject and pronouns match in number:

Engineers must trust their abilities.

Key idea: Use *who* as a pronoun for human subjects, *that* for specific nonhuman subjects, and *which* for nonhuman subjects in clauses set off by commas.

Another common problem is the proper use of the pronouns *who*, *that*, and *which*. When in doubt, use *who* for human subjects and *that* or *which* for nonhuman subjects. The pronoun *that* is used in reference to a specific noun, while *which* adds information about a noun and usually is used in clauses set off by commas. Thus,

The engineer *who* was on site had the contract documents.

The human subject takes the pronoun *who*. On the other hand,

The bolt *that* ruptured was installed improperly.

Here, use *that* in referring to a specific bolt. (We are discussing a particular bolt: the bolt that ruptured.) Finally,

The submitted proposal, *which* was missing page three, was thrown in the garbage can.

In this case, use *which* to add information in a separate clause. Often, you can avoid *who*, *that*, and *which* problems by incorporating the information as an adjective. For the examples just presented, you could write

The on-site engineer had the contract documents.

The ruptured bolt was installed improperly.

The submitted proposal was missing page three. It was thrown in the garbage can.

13.4.5   Adjectives and Adverbs

Adjectives modify nouns and adverbs modify verbs. Avoid using long lists of adjectives, sometimes called **adjective chains**. In adjective chains, it is often difficult to identify the noun. Consider the sentence:

adjective chain: a long list of modifiers to a noun (to be avoided)

High-grade precut stainless steel beams were specified.

The beam characteristics are clearer if the sentence is rewritten:

Precut beams made of high-grade stainless steel were specified.

split infinitive: insertion of a word between *to* and the verb (to be avoided)

Adjective and adverb *placement* also can be problematic. You should avoid placing an adverb between the word *to* and the verb. This construction is called a **split infinitive**. For example, you should write

The gear ratio was designed to drive the system efficiently.

not

The gear ratio was designed to efficiently drive the system.

Key idea: With adjectives and adverbs, avoid adjective chains and make sure the adverb or adjective modifies only the verb or noun you intend to modify.

Having railed against them, it should be noted that split infinitives are a tricky construction. The rule against split infinitives appears to stem from Latin, where splitting an infinitive is impossible. Place an adverb between *to* and the verb only to emphasize the adverb or to produce a sentence that sounds better. For example, there is a split infinitive in the phrase

To boldly go where no one has gone before …

However, it sounds better (at least to many people) than

To go boldly where no one has gone before …

Adjectives should be located next to the nouns they modify. Consider the following two sentences:

They only constructed three prototypes.

They constructed only three prototypes.

In the first sentence, *only* modifies *constructed*: they only *constructed* the prototypes, they did not construct and test the prototypes. In the second sentence, *only* modifies *three* (which, in turn, modifies *prototypes*): they constructed only *three* prototypes, rather than four prototypes. Make sure that the adverb (or adjective) modifies only the verb (or noun) you intend to modify.

13.4.6 Capitalization and Punctuation

Many neophyte technical writers find it necessary to use nonstandard capitalization and abbreviations. Please do not give in to this temptation. Few words in technical writing are capitalized. As suggested by Smith and Vesiland (1996), you usually capitalize the names of organizations, firms, cities, counties, districts, agencies, and states. In general, do not capitalize general references to these entities. Thus, you can write

The City of Rochester contracted for engineering services.

Here, capitalize "city" because it is specific to Rochester, New York, but

The city council met for three hours.

or

The federal government will meet the deadline.

Do not capitalize "city" and "federal" because they are general adjectives here. In addition, the titles of engineering reports are capitalized, and official titles of people are capitalized when they precede the names of the people (but not when they follow the name).

Key idea: Avoid nonstandard capitalization and abbreviations.

Standard abbreviations for scientific and engineering units and parameters should be used. If in doubt, define an abbreviation *the first time it is used*. There is no need to capitalize the words as an abbreviation is defined. Thus, write

The standard operating procedure (SOP) was followed.

not

The Standard Operating Procedure (SOP) was followed.

and not

The SOP was followed.

The last construction is improper if using the abbreviation for the first time, but desirable if the abbreviation *SOP* already has been defined in the document.

Common nontechnical abbreviations include the following:

e.g. (*exempli gratia* = for example)

i.e. (*id est* = that is)

etc. (*et cetera* = and so forth), and

et al. (*et alia* = and others)

(*Note*: The last term is **not** abbreviated et. al or et. al.) These common abbreviations sometimes are italicized (e.g., *e.g.*) to indicate their non-English origins.

Commas should be used to define clauses and separate items in a list. Usually, a comma is used even before the last item in a list. For example,

Materials included wood, steel, and concrete.

If the items in a list are long (or the items include commas or conjunctions), use semi-colons to separate them:

Materials included wood, natural materials, and fiber; steel and concrete; and thermoplastic resins.

13.4.7 Spelling

Key idea: Never assume a document is free of errors because it passes the spell checker.

There is no room for spelling errors in technical documents. One misspelled word could destroy an otherwise strong document. The fundamental rule of spelling is *Never, never, never trust your spell checker*. Spell-checking software is a good first start, but you must learn to proofread your writing very carefully. Spell checkers miss misspellings that result in another word (e.g., house/horse, dear/deer). A proofreading example is given in Section 13.4.10.

13.4.8 Citation

plagiarism: using someone else's words or ideas without proper credit

You are professionally and morally obligated to give credit when you use ideas from other people. Taking someone else's words or ideas without credit is called **plagiarism**. Plagiarism is defined in the University at Buffalo's University Standards and Administrative Regulations as

copying or receiving material from a source or sources and submitting this material as one's own without acknowledging the particular debts to the source (quotations, paraphrases, basic ideas), or otherwise representing the work of another as one's own.

Key idea: Make sure you give credit (by use of a citation) when presenting someone else's words or ideas.

Students who plagiarize are subject to disciplinary action. Engineers who plagiarize can lose their professional licenses.

Plagiarism is not just copying *words* from someone else. Plagiarism also means taking another person's *ideas* without giving them due credit. Always read your work carefully to make sure that you have not inadvertently included someone else's ideas "without acknowledging the particular debts to the source."

Credit is shown by *citing the work from which the material was taken*. There are many citation styles. One style (employed in this text) is to list the author's name and publication date in parentheses following the material cited, as in "Smith (2002)." Numbers, usually written as superscripts (e.g., Smith[2]), may be used with a numbered reference list.

paraphrase: to rewrite an idea in your own words

In nearly all cases, you **paraphrase** the material (i.e., rewrite it in your own words). On rare occasions (when the original words are required), it may be necessary to quote the words exactly. Quotation should be done sparingly and the citation always must be given. Indicate a direct quotation by the use of quotation marks or by doubly indenting the material. Examples of paraphrasing and quotation are shown in Example 13.1.

EXAMPLE 13.1 PARAPHRASING AND QUOTATION

Use the following material, from Paradis and Zimmerman (1997), in a paragraph, and cite it properly:

"Long sentences, often amounting to more than 30 words, are usually too complicated. Determine the main actions of the sentence. Then sort these into two or more shorter sentences."

SOLUTION   Here are several citation options:

1. Paraphrase with citation (preferred approach):

    Long sentences should be broken up into smaller sections according to their main actions (Paradis and Zimmerman, 1997).

2. Quotation using quotation marks with citation:

    Overly long sentences can be problematic. According to Paradis and Zimmerman (1997): "Long sentences, often amounting to more than 30 words, are usually too complicated. Determine the main actions of the sentence. Then sort these into two or more shorter sentences."

3. Quotation using indentation with citation:

    Overly long sentences are confusing to the reader. Several approaches have been developed to identify and eliminate run-on sentences. For example,

    > Long sentences, often amounting to more than 30 words, are usually too complicated. Determine the main actions of the sentence. Then sort these into two or more shorter sentences. (Paradis and Zimmerman, 1997).

The following approach is plagiarism because the work is paraphrased but no citation is given [*Warning*: **This material is not cited properly!**]

Long sentences—some can be up to 30 words long—should be subdivided. To do this, find its main actions and create a shorter sentence for each main action.

13.4.9   Other Problem Areas

In addition to the rules discussed previously, several other words and phrases cause problems in technical writing. Most of the words and phrases listed below were found in Strunk and White (1979) or Smith and Vesiland (1996):

affect/effect: These two words cause many difficulties in technical writing, but the rule regarding their use is simple. The word *affect* is almost always a *verb*. The word *effect* is almost always a *noun*. Thus, write "The effects of temperature were noted" (*effects* is a noun) and "Temperature affected the results" ("affected" is the verb).*

among/between: Use *between* when two people or things are involved and *among* when more than two or more people or things are involved. For example, write "The voltage was split between two capacitors," but "The work was divided among four engineers."

comprise: *Comprise* means *to consist of*: "The frame comprises four steel rods" (i.e., the frame *consists of* four steel rods) and "Four steel rods make up the frame" (*not* "Four steel rods comprise the frame").

Key idea: Avoid double negatives in formal writing.

double negatives: Avoid the **use** of two or more negatives (*not* or words starting with *un*) in the same sentence. Rewrite by canceling out pairs of negatives: "The project was like our previous work" (*not* "The project was not unlike our previous work.")

farther/further: *Farther* refers to distance, while *further* refers to time or quantity. Thus, "The ultrahigh-mileage vehicle went farther on a tank of gas," while "Further negotiations are necessary to seal the contract."

---

*While *affect* is usually a verb, it is used in psychology as a noun (for example, the Jones affect). The word *effect* almost always is a noun, but it is used *very rarely* as a verb, as in "Temperature effected a change in elasticity." (This means temperature *brought about* a change in elasticity.)

fewer/less: *Fewer* is used in reference to the *number* of things, while *less* refers to the *quantity* (or amount) of an object. For example, "Our model has fewer adjustable parameters" (i.e., fewer number of parameters), and "The high-efficiency engine used less gasoline" (i.e., a lesser amount of gasoline).

irregardless: *Irregardless* is an example of a double negative. The prefix *ir-* and the suffix *-less* both negate *regard*. Please write *regardless* anytime you are tempted to write *irregardless*.

its/it's: Here is a nagging exception to the rule that you add an apostrophe to indicate the possessive form. The word "its" is the possessive form: "Its color was red." The word *it's* is a contraction of *it is*: "It's hot today." In general, *avoid contractions in formal writing*.

Key idea: Avoid contractions in formal writing.

personification: Personification (also called anthropomorphism) is the assignment of human characteristics to nonhuman objects, as in "The day smiled on me." Personification should be avoided in technical writing. Some people dislike the assignment of any active verb to any inanimate objects. In this view, some say you should avoid statements such as "The data show ..." or "The experiments demonstrate ...." Although there is a difference of opinion on this matter, it is best to avoid egregious examples of personification in your technical writing (such as "The data really grabbed me by the throat," which is too informal as well).

precede/proceed: *Precede* means *to come before*, while *proceed* means *to continue or move forward*. Thus, "The air-conditioning study preceded the heating study" (meaning that the air-conditioning study was conducted first) and "The work proceeded without interruption" (meaning that the work continued without interruption).

presently: *Presently* means both *soon* and *currently*. Strunk and White (1979) suggest that *presently* be used only in the sense of *soon*.

13.4.10 Proofreading

Key idea: Always proofread your work.

The secret to good proofreading is practice. You can check your proofreading skills by asking others to read your work and give you feedback. An example of proofreading is given in Example 13.3.

EXAMPLE 13.3: PROOF-READING

Read the following paragraph and list the errors you encounter. Allow 60 seconds for this exercise. Rewrite the paragraph to eliminate the errors. [*Warning:* **The following text may contain errors!**]

Abstract

Project personnel conducted a laboratory study to definitively determine the engineeering feasability of polychlorinated biphenyl (PCB) removal by granular activated carbon. The study used an expanded bed granular activated carbon reactor in the upflow mode. PCB concentrations in the column effluent was measured by standard techniques. Study data is consistent with surface diffusion as the rate-limiting step, although much scatter in the data is observed. Columns were sacrificed at the conclusion of the study and carbon analysis revealed PCB saturation is the first 50% of the bed. Future studies will be conducted on the affect of the recycle rate on column performance.

SOLUTION

A list of errors (with the corresponding section numbers in parentheses) is given in Table 13.3.

TABLE 13.3 Errors in Proofreading Example

Sentence	Error(s)
First sentence	"to definitively determine" is a split infinitive (13.4.5), "engineeering" (engineering) and "feasability" (feasibility) are misspelled (13.4.7)
Second sentence	"...expanded bed granular activated carbon reactor..." contains an adjective chain (13.4.5)
Third sentence	"...concentrations... was..." is a subject/verb mismatch (13.4.1). The use of the passive voice (13.4.2) is discouraged, unless it is clear who analyzed the samples from other parts of the report.
Fourth sentence	"...data is..." is a subject/verb mismatch (13.4.1). *Note*: The sentence "...much scatter in the data is..." is fine, since the subject, "scatter," is singular.
Fifth sentence	This sentence is a long sentence (13.3.2). Also, the sentence should read, "...saturation *in* the first..." (rather than "...saturation *is* the first...").
Sixth sentence	Use of passive voice is inconsistent with the active voice used elsewhere in the paragraph (13.4.2). Also, "affect" should be "effect" (13.4.9).

Here is an improved version of the abstract:

Abstract

Project personnel conducted a laboratory study to determine definitively the engineering feasibility of polychlorinated biphenyl (PCB) removal by granular activated carbon (GAC). The study used an expanded bed GAC reactor in the upflow mode. A contract laboratory measured PCB concentrations in the column effluent by standard techniques. Study data are consistent with surface diffusion as the rate-limiting step, although much scatter in the data is observed. Columns were sacrificed at the conclusion of the study. Carbon analysis revealed PCB saturation in the first 50% of the bed. We plan to conduct future studies on the effect of the recycle rate on column performance.

13.5 TYPES OF ENGINEERING DOCUMENTS

13.5.1 Introduction

Thus far in this chapter, you have been exposed to the organization of engineering reports. Reports are used to present the results of a study. A report may transmit the results of the entire project (called a *final* or *full report*), transmit the results of a portion of the project (called a *progress report*), or transmit a small piece of a report in a short form (often called a *letter report*).

In addition to reports, engineers write several other kinds of documents. Common document types include letters, memorandums, and email.

13.5.2 Reports

Key idea: Reports should include a cover page and a transmittal letter.

The general outline of an engineering report was discussed in Section 13.2. Two other elements of a report deserve mention. First, every report should have a cover page. A cover page includes the names of the authors (and their professional titles), the names of the recipients (and their professional titles), the report or project title, a project identifier, and the date. Many formats are possible, as long as this information is included. Locate the required information for a cover page in the example cover page in Figure 13.1.

Second, most reports have a *transmittal letter* (also called a *cover letter*). The transmittal letter is a short letter that accompanies the report. The format of letters is presented in Section 13.5.3.

**Pumping Options for Stormwater Management
in Rivertown, Ohio**

Draft Final Report for
Rivertown DPW Project #2005-5-1214

Submitted to:

Mary J. Bremer, PE
Director of Public Works
Rivertown Public Works Department
1120 Bank Road
Rivertown, Ohio

Submitted by:

John H. Seal, PE
Senior Associate Engineer
AZA Engineering
12 Cunningham Parkway, Suite 114
Warsaw, Ohio

February 7, 2007

Figure 13.1.   An Example of a Cover Page

Key idea: Letters should have a heading (including the date, recipient's name, and title), closing (including your name, title, and signature), and structured paragraphs (the first paragraph should summarize previous correspondence and state the purpose of the letter, the next paragraphs should present supporting information, and the last paragraph should summarize the main point and state the required actions or follow-up communication).

memorandum: a short note used to document engineering work

Key idea: Memos should have a heading (including to whom the memo is written, who wrote the memo, the memo topic, the date, and the word *Memorandum*), and the same structured paragraphs as a letter.

13.5.3   Letters

Engineers use letters to document the transmission of ideas to the client or other agency. Letters must have structure. The heading of a letter includes the date and recipient's name and title. The first paragraph of a letter should summarize previous correspondence and state the purpose of the letter. In the next paragraph or paragraphs, supporting information should be presented. The last paragraph of the letter should summarize the main points and state the required actions or follow-up communication. In the closing information of a letter, include your name, title, and signature. An example letter is shown in Figure 13.2. Note the heading information; introductory, supporting, and concluding paragraphs; and closing information.

13.5.4   Memorandums

A **memorandum** (plural: memorandums or memoranda) is a short note. Similar to letters, memorandums are used for short documentation of engineering work. In fact, the word *memorandum* is a shortened form of the phrase *memorandum est*—Latin for "it is to be remembered." Memorandums are frequently used for messages inside an organization (called *internal memorandums*).

Memorandums (or memos) are structured similarly to letters (see Section 13.5.3), but without the heading and closing information of a letter. Heading information in a memo tells you to whom the memo is written, who wrote the memo, the memo topic, the date, and the word *Memorandum*.

AZA *Engineers*
Warsaw • Milton • Cleveland

March 10, 2006

Mary J. Bremer, PE
Director of Public Works
Rivertown Public Works Department
1120 Bank Road
Rivertown, Ohio

Dear Ms. Bremer,

As per our telephone conservation of March 9th, I am writing to summarize your comments on the draft stormwater report. Our responses to your comments also are included in this letter.

My notes indicate that your staff had three main comments on the draft report. First, the name of the Bilmore Pump Station was misspelled on page 6-2. Second, the flow calculations for the West Branch were based on 1980-2000 rainfall data, while all other system design calculations were based on 1970-2000 rainfall data. Third, your staff requested that the cradle design for Option 4 use a smaller factor of safety than the 2.5 safety factor in the report (p. 7-7).

We will correct the spelling error on page 6-2 and update the design calculations for the West Branch with rainfall data from 1970-2004. However, we feel best engineering practice requires the safety factor of 2.5 in the pump cradle design. Based on conversations with the pump manufacturer, lower safety factors will increase the chance of catastrophic failure. Therefore, we wish to retain the 2.5 safety factor in the design of Option 4.

To summarize, we plan to resubmit the report before March 31, 2006 with the spelling error corrected and with the design calculations for the West Branch updated to use rainfall data from 1970-2000. We will retain the safety factor of 2.5 in the pump cradle in Option 4.

Thank you for your thoughtful comments. I will call you next week to confirm the changes. We look forward to delivering you the final report on this project.

Sincerely,

J H Seal

John H. Seal, PE
Senior Associate Engineer

Figure 13.2.   Example of a Technical Letter

MEMORANDUM

To: Yvonne Ringland
From: J.H. Seal, PE
Re: Comments on Rivertown stormwater report
Date: March 9, 2006

I spoke with Mary Bremer at the Rivertown DPW today about the draft stormwater report. She requested that we use the same rainfall data for the West Branch design calculations as we did for the rest of the report. We used 1970-2000 rainfall data for the majority of the report.

Please redo the West Branch design with 1970-2000 rainfall data. The final report is due by March 31st. Please have the revisions to me by March 25th so we can get the changes to the word processing staff.

If you have questions about the requested changes, please call me at extension 36.

Figure 13.3. Example of a Memorandum

Memo paragraphs are similar to those of letters: previous correspondence and memo purpose should be summarized in the first paragraph, supporting information in the following paragraphs, and main points summarized in the last paragraph. An example of a memo is given in Figure 13.3. Note the heading information and purpose of each of the three paragraphs. A copy of this memo likely would be placed in the project file to document the internal communication of the consulting firm.

13.5.5 Email

Key idea: When writing business email, avoid contractions and emoticons, proofread carefully, double-check the recipient list, and do not include anything in an email that you would not include in other business documents.

Nearly every college student in the 21st century has used email, usually for informal conversation. Email also can be used in formal business correspondence, sometimes in place of a letter or memo.

Although email is less formal than other forms of written communication, it is easy to let an overly familiar style creep into your formal email correspondence. You use different words in speaking to clients and colleagues than you use to speak to friends at a party. Similarly, use more formal language in business email. Following are some rules for business email correspondence:

- Avoid email contractions (e.g., *RU* for *are you* and *°s°* for *smile*).
- Avoid *emoticons*—text characters used to express emotions (such as :-) for a smiley face).
- Proofread carefully before you hit "send." Look for language that may be offensive or inappropriate.
- Double-check the names on the "to" list before you send the email. "Replying to all" with the results of your recent medical check-up (when you intended to forward the results to your roommate) is a serious breach of business protocol.

- Emails are as much a part of the technical and legal record as are other documents. Do not include anything in an email that you would not include in other business documents.

An example of a business email message is shown in Figure 13.4.

Email is not the only kind of electronic written document in engineering today. For a look at the future of written technical communication, see the *Focus on Writing: Whither Paper Reports?*

To: Roger Yee (rty@azaengineers.com)
From: John H. Seal (jhs@azaengineers.com)
Subject: Pump cradle design for Rivertown Project
Cc: Cynthia Cronin (cronin@rgoldpumps.com)
Bcc:
Attached: Draft Rivertown report.doc

Roger -

Rivertown has questioned our use of a 2.5 safety factor for the cradle design in Option 4 of the stormwater project. Attached is the draft report.

Are we sure about this safety factor? If so, please help me to justify it. I am copying Cindy Cronin at Rheingold Pumps on this message. We are specifying Rheingold Pumps and Cindy might be able to help.

Please get back to me by the end of the day on this, Roger.

Thanks,

John H. Seal
Senior Associate Engineer
AZA Engineering

Figure 13.4. Example of a Business Email

FOCUS ON WRITING: WHITHER PAPER REPORTS?

BACKGROUND

Probably since the first pyramid was built, engineers have been summarizing their work by writing reports. This chapter was devoted to helping you write better reports and other engineering documents. There are many cases in which the results of engineering work are better communicated by electronic documents. Many clients now are requesting electronic or on-line reports.

ELECTRONIC MANUALS

As an example of electronic reporting, many industries are replacing entire bookshelves of operation and maintenance (O&M) manuals with *on-line manuals*. The on-line O&M manuals typically are written in HTML, XML, or other programming languages used in Web page development. In addition, manuals and other electronic engineering documents commonly are written in Adobe's proprietary *portable document format* as PDF files.

Electronic manuals have a number of advantages over traditional documentation. First, they reduce the need for operations and maintenance training. Perez et al. (2001) estimated that the effectiveness of O&M training at a drinking-water treatment plant was increased four- to sixfold using on-line materials as compared with paper manuals. More effective training results in fewer errors and cost savings.

Second, electronic manuals are much easier to keep current. Engineers struggle to maintain current sets of plans about engineered systems. Facilities personnel need to know the actual conditions of the structure (as-built conditions), not the system as originally designed (design conditions). Electronic manuals allow engineers to update material very quickly and accurately. The underlying database of equipment and other system attributes can be updated centrally, allowing users to access up-to-date information from any location.

Third, electronic manuals are easier to access. Facility personnel sometimes dread the thought of flipping through literally thousands of pages of manuals in three-ring binders to find the information they need. Electronic manuals are written with *hyperlinks* (as on Web pages). This allows the user to find related information quickly. In fact, electronic manuals look like Web pages. As with the Web itself, e-manuals can be very graphically oriented, with liberal use of drawings, photographs, and videos. In addition, electronic manuals can be linked to manufacturer's Web pages. If, say,

you need a new gasket for a pump, you can click on the manufacturer's link and find the part easily.

Fourth, electronic manuals are portable. Many electronic manuals are mounted on company intranets, allowing for secure access by facility personnel from any location. In other cases, the manuals are burned onto CD-ROMs. One CD-ROM can replace up to 1,540 pounds of paper manuals (Perez et al., 2001).

WILL YOU EVER SEE A PAPERLESS OFFICE?

For the foreseeable future, engineers probably will continue to produce reports on paper. The "paperless office" continues to be frustratingly just out of reach. However, the engineer's life is becoming increasingly "webcentric" (i.e., centered on the World Wide Web). As an engineer of the future (and as a person brought up to think of the Internet as an important resource), you should think creatively about how information needs in engineering can be addressed by electronic sources. Always ask whether electronic documents will add value to the information (by allowing linkage to other data sources or by providing real-time data or by using multimedia formats).

Perhaps in your lifetime, paper reports will become as quaint as slide rules and manual typewriters. Regardless of the delivery medium, engineering reports will still be based on the principles outlined in this text: organization, signposting, and clarity.

13.6 SUMMARY

The key to good written technical documents is *organization*. The typical structure of an engineering report includes several aspects: the abstract (or executive summary), introduction/background/literature review, methods, results, discussion, conclusions/recommendations, and references.

Technical documents also must be organized at the paragraph, sentence, and word levels. Choose words to make your writing concise, simple, and specific. In your technical writing, be aware of the rules of grammar and spelling. Strive to use the active voice and avoid gender-specific language. Always proofread your work before allowing it to leave your hands.

In addition to reports, engineers produce letters, memos, and emails almost daily in their working lives. Letters have a heading, a closing, and structured paragraphs. The first paragraph summarizes previous correspondence and states the purpose of the letter. The next paragraphs present supporting information. The last paragraph summarizes the main points and states the required actions or follow-up communication. Memos have the same paragraph structure, with a different heading and no closing. Business emails are part of the business record and should be created and sent in a professional manner.

SUMMARY OF KEY IDEAS

- Organize technical documents from the largest to smallest scale: outline level, paragraph level, sentence level, and word level.
- Common elements of technical documents include the abstract (or executive summary), introduction/background/literature review, methods, results, discussion, conclusions/recommendations, and references.
- The abstract should contain a summary of each element of the report.
- The introduction should take the reader from the report title to an understanding of why the report was written.
- In the methods section, justify the study approach, present data collection techniques, and discuss data analysis methods.
- In the results section, present the results and note the general trends.
- Interpret data in the discussion section.
- Conclusions and recommendations are often in list form and should be written very carefully.
- Although many reference formats are acceptable, the references must be complete and consistent.
- Use section headings or numbering schemes as signposts in technical documents.
- Each sentence should express a single idea.
- Choose words to make your writing concise, simple, and specific.
- Avoid making up new words or new uses of words in conventional technical writing.
- Make sure that the subject and verb agree in number (i.e., they must be both singular or both plural).
- Use a consistent voice, with preference for the active voice.
- Generally use the present tense, unless describing work done in the past.
- Avoid the use of gender-specific pronouns.
- Use *who* as a pronoun for human subjects, *that* for specific nonhuman subjects, and *which* for nonhuman subjects in clauses set off by commas.
- With adjectives and adverbs, avoid adjective chains and make sure the adverb or adjective modifies only the verb or noun you intend to modify.
- Avoid nonstandard capitalization and abbreviations.
- Never assume a document is free of errors because it passes the spell checker.
- Make sure you give credit (by use of a citation) when presenting someone else's words or ideas.
- Avoid double negatives in formal writing.
- Avoid contractions in formal writing.
- Always proofread your work.

- Reports should include a cover page and a transmittal letter.
- Letters should have a heading (including the date, recipient's name, and title), closing (including your name, title, and signature), and structured paragraphs (the first paragraph should summarize previous correspondence and state the purpose of the letter, the next paragraphs should present supporting information, and the last paragraph should summarize the main points and state the required actions or follow-up communication).
- Memos should have a heading (including to whom the memo is written, who wrote the memo, the memo topic, the date, and the word *Memorandum*) and the same structured paragraphs as a letter.
- When writing business email, avoid contractions and emoticons, proofread carefully, double-check the recipient list, and do not include anything in an email that you would not include in other business documents.

Problems

13.1.   Pick two textbooks other than this one. What kinds of signposting are used in the texts? Describe the scheme used to show hierarchy in the signposting.

13.2.   What are the characteristics of a good sentence?

13.3.   What are the three aspects of good word choice in technical writing? Find good and poor examples of word choice in a newspaper or technical journal.

13.4.   List whether the following nouns should take a singular or plural verb form: engineer, axes, phenomena, axis, datum, criterion, thermodynamics, phenomenon, Microsoft, and criteria. You may need to use a dictionary.

13.5.   Find five examples of the use of passive voice in this text. Rewrite them in the active voice.

13.6.   Repair the following paragraph, if necessary. [**Warning: The following material may contain errors!**]

Plans and specifications who lack careful preparation may be faulty. The engineer must use all his skill to find and correct the problems. The engineer that refines her own design is more likely to find their own errors.

13.7.   For each of the following, identify the problem or problems in the use of adjectives or adverbs, if any, and correct the errors. [**Warning: The following material may contain errors!**]

a.   "The mass-produced germanium junction transistor was a major advance."

b.   "The project manager attempted to slowly accelerate the production rate."

c.   "The contract only required plant construction, not the operation of the plant."

d.   "Alternating current power transmission first occurred at Niagara Falls in 1895."

13.8.   Select any paragraph in this text. Paraphrase the idea without a direct quotation, and include a citation and a reference. Repeat with a paragraph from a technical journal of interest to you.

13.9.   Write a letter to your professor asking for permission to take a make-up exam.

13.10. Write a memo to a classmate to organize a study session for one of your courses.

14

Oral Technical Communications

14.1 INTRODUCTION

Few activities intimidate new engineers more than public speaking. Technical oral presentations need not be painful. They can be tamed by focusing on three kinds of activities:

- What to do *before* the talk,
- What to do *during* the talk, and
- What to do *after* the talk.

Many people think that the *delivery* is the key to technical talks. While the delivery is important, the truth is that oral presentations are made or broken by the work put in *before* the talk is delivered. A good technical talk is well organized, with instructive visual aids. It is delivered with the help of useful but nonintrusive memory aids. The talk will be rehearsed, but not overly practiced. These critical activities—organization, visual aids design, memory aids design, and practice—take place well before the oral presentation is made to the audience. The details of talk organization and preparation will be presented in Sections 14.2 through 14.4.

What is your gut reaction to the thought of standing up before a handful or dozens or hundreds of people and delivering technical material? If your palms are sweating already, then Section 14.5 may help. In Section 14.5, your plan of action immediately before the talk (including how to deal with nervousness) will be reviewed. You will learn what to say and how to say it.

Finally, improvement in your technical speaking skills is made only by what you do after the talk. Section 14.6 will provide hints on obtaining feedback and implementing good speaking habits.

OBJECTIVES

After reading this chapter, you will be able to:

- organize a technical oral presentation;
- design visual aids;
- design memory aids;
- deliver an effective technical oral presentation.

14.2   BEFORE THE TALK: ORGANIZATION

Key idea: Technical presentations can be improved by considering the activities before the talk, during the talk, and after the talk.

Key idea: Identify the presentation goals, target audience, and constraints on the presentation (especially time constraints).

visual aids: media used to accompany oral presentations (e.g., slides and overhead transparencies)

title slide: visual aid containing the presentation title and information about the authors

Recall that before writing a single word of the oral presentation, you must identify the goals of the presentation, the target audience, and the constraints on the presentation. The main constraint on oral presentations is the time allotted for the talk. In your career, almost every oral presentation you give will have time constraints. One key to good oral presentations is to respect your audience's time and use their time wisely.

Only after identifying goals, audience, and constraints can an outline be written. With an outline in place, the individual **visual aids** can be designed. Technical talks usually begin with a **title slide**.* The title slide contains the title of the talk and the names and affiliations of the authors. The title slide is the oral presentation equivalent of the cover page. Any example title slide is shown in Figure 14.1.

In many technical presentations, the second slide is an outline or overview of the talk. While an outline slide is optional, it serves as a good road map for the remainder of the talk. Audiences may feel more comfortable if they know where the presentation is going. The outline slide is the first opportunity for signposting in an oral technical presentation. An example outline slide is shown in Figure 14.2.

Biochemical Engineering of Artifical Skin

A.D. Leising, PhD
Chief Chemical Engineer
DermaTech, Inc.

Presented at the VentureCap Expo, Sept. 8, 2005

Figure 14.1.   An Example of a Title Slide

Outline

- Background
  - History of artificial skin
  - Barriers to commercialization
- Approach
- Results
  - Synthesis of smart plastics
  - Results of animal trials
  - Commercialization potential
- Conclusions

Figure 14.2.   An Example of an Outline Slide

---

*To simplify the language here, visual aids in general will be called "slides." Information of the types of visual aids may be found in Section 14.3.2.

Key idea: Use an outline to organize the talk and an outline slide to show your organization.

The remaining sections of a technical talk vary with the goals and target audience. A generic structure that includes an introduction/background, methods, results, discussion, conclusions, and recommendations is a good place to start. Technical talks rarely include an abstract, formal literature review, or list of references.

14.3 BEFORE THE TALK: DESIGNING VISUAL AIDS

Key idea: The number of visual aids should be about $\frac{3}{4}$ times the number of minutes allotted to the presentation.

Once the outline has been established, you can start to design the visual aids. A major difference between written and oral presentations is the reliance on visual aids in oral communication. You must select the number, type, and content of visual aids.

14.3.1 Number of Visual Aids

The number of visual aids depends most strongly on the length of the presentation. *To estimate the maximum number of visual aids, multiply the number of minutes in the presentation by 0.75.* For example, a 30-minute talk should have no more than 21 to 23 slides.

The natural tendency is to prepare too many visual aids. After all, if the number of slides is $\frac{3}{4}$ of the number of minutes, then the average time per slide is $\frac{4}{3}$ minutes = 80 seconds. Many first-time speakers reason that several of the slides in the presentation (e.g., the title and outline slides) will take much less than 80 seconds to present. They conclude that they can have *many more* slides than the number calculated from $\frac{3}{4}$ (number of minutes). *This logic almost always leads to very rushed and incoherent presentations.* Until you become very experienced in oral technical presentations, use the "$\frac{3}{4}$ times the number of minutes" value as a firm guide.

14.3.2 Types of Visual Aids

Several types of visual aids are available, including slides, overhead transparencies, poster boards and flip charts, blackboards and whiteboards, and computer displays and projectors. Physical models and material to be passed around the audience also are used as visual aids.

Key idea: In selecting the type of visual aid, consider image quality, eye contact, and production cost and time. Then use only one type of visual aid in a talk.

In selecting a type of visual aid, three factors are important: image clarity, maintenance of eye contact, and production cost and time. The advantages and disadvantages of several types of visual aids are summarized in Table 14.1. In many professional presentations, image clarity may be paramount. It may be worth the money to produce the highest quality images available.

Eye contact is important for two reasons. First, it allows you to get feedback from the audience. Are they bored? Engaged? Having trouble hearing you? Second, eye contact allows the audience to be drawn into your words. Try listening to a movie or television program with your eyes shut. The magic is reduced when the eye contact is lost. Still not convinced? The next time you speak before a group of people, notice how much time the audience spends looking at your *eyes* rather than the screen.

Visual aids also can be expensive and time-consuming to produce. Always estimate the cost and time required to produce any visual aid before committing to a type of visual aid. If the turnaround time for producing visual aids is long, you may have to adjust your schedule to meet the presentation deadline.

Regardless of the type of visual aid selected, it is important to use only one type of visual aid. Switching back and forth between two types can be distracting to the audience, especially if the room lights are turned on and off repeatedly. For the vast majority of technical talks, stick to one type of visual aid. Each type of visual aid will be discussed in more detail.

Slides

Color photographic slides provide the sharpest images. Slides come with a major disadvantage: they require the room to be darkened. In a dark room, you risk losing eye contact with the audience. Slides also can be expensive and time-consuming to produce.

Overhead Transparencies

Overhead transparencies, also called *overheads*, provide a good trade-off between image clarity and eye contact. The images may be poorer than slides (although color laser printers are capable of producing very high quality overheads on special transparency film).

In presenting overheads, the room lights generally are on, but dimmed. Thus, eye contact is still possible. Unless you have an assistant, overheads require you to stand

TABLE 14.1   Types of Visual Aids and Their Characteristics

Type	Image Quality	Eye Contact	Cost and Time	Other
Slides	Very high	Moderate (room dark)	Moderate	Image very sharp
Overhead transparencies	High	Good	Small	Good compromise
Poster boards and flip charts	Very high	Excellent	Moderate to large	Good for smaller audiences
Whiteboards and blackboards	Low	Excellent	Very small	For informal work
Computers	Can be very high	Moderate (room dark)	Small	Watch compatibility problems

near the projector. Avoid blocking the audience's view of the projection screen with your body.

Poster Boards and Flip Charts

Poster boards and flip charts are large-format visual aids, displayed on an easel. They are used frequently by consulting engineers because they allow the lights to be on; thus, they maximize eye contact with the audience and increase audience participation. Poster boards and flip charts are not appropriate for large audiences.

Blackboards and Whiteboards

Blackboards and whiteboards are appropriate for informal technical presentations. Their use allows the audience to write notes at the same pace as the speaker/writer. They are a good choice when note taking is important or when audience participation is critical.

Computers

Computer-based presentations quickly are becoming the most common delivery mode for technical presentations. Computer-based presentations have a number of advantages over other media:

- They can be changed at the last moment.
- They can include Internet-based materials, videos, and animations.
- They avoid the expense and lead time required to make photographic slides.

Computer-based presentations have several disadvantages as well. Compatibility problems often arise between notebook computers and projection devices. It is important to make sure that your notebook computer interfaces properly with the intended projector. The ability to change computer-based presentations at the last minute may tempt you to throw together the talk at the last minute. As always, do not let the technology control the message.

Computer-based presentations offer their own challenges regarding the content of the slides. Information on content specific to computer-based presentations is presented in Section 14.3.5.

14.3.3   Content of Visual Aids: Word Slides

There are two types of visual aid content: word slides and data slides. Word slides typically contain only words, symbols, and/or equations. Data slides communicate data and may include tables or figures.

Key idea: Word slides should contain as few words as possible.

Word slides should contain as few words as possible to communicate the required information. *It is undesirable to fill a word slide with text*: the audience will read the words rather than look at you.* *You* want to take control of the material and present it to the audience yourself.

Sometimes, symbols or equations can be used in place of words. The choice of equations or words depends on the audience. For a technical audience, a word slide about Newton's Second Law of Motion might contain the equation $F = ma$. For a less technical audience, the gist of the Second Law may be more clearly made with words:

---

*You can prove this point to yourself with a simple experiment. Gather a group of 20 or so people. Prepare two overheads: one with a wordy message and one with an abbreviated form (e.g., "The rain in Spain falls mainly in the plains" and "Spain: Rains in plains"). Show the first overhead and present the message word for word. Show the second overhead and use the same word-for-word speech as the first overhead. You will notice that the audience's eyes are on the screen when you show the first overhead. Their eyes are more likely to be on you when you show the second overhead.

"Force is proportional to both acceleration and mass." For a nontechnical audience, perhaps a cartoon would best illustrate the point.

For a technical audience:

Newton's Second Law

$$F = ma.$$

For a less technical audience:

Newton's Second Law

Force is proportional to both acceleration and mass.

For a nontechnical audience:

Newton's Second Law

The force doubles when the mass doubles.

Word slides should take into account the shape of the visual aid. For example, overhead transparencies and computer-based presentation slides have a length-to-width ratio of 11:8.5 ≈ 1.3:1. Photographic slides usually have a ratio of about 0.7:1. It is pleasing to the eye to have the word shape match the visual aid shape.

Matching the word shape to the visual aid shape also means that the font size can be as large as possible. It is important in word slides to use a large font size. Typically, slides and overheads should have font sizes from about 28 to 44 point. Use consistent font sizes (i.e., major headings all in one size and minor headings all in another size). Presentation software (such as Microsoft PowerPoint or Corel Presentations) can help in maintaining a consistent presentation format. Two word slide examples may be found in Figures 14.3 and 14.4.

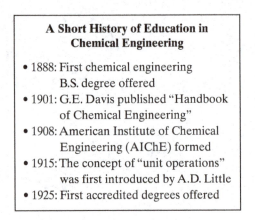

Figure 14.3. Word Slide Example 1 (dates from http://www.cems.umn.edu/~aiche_ug/history/h_toc.html)

A Short History of Education in Chemical Engineering

- 1888: First chemical engineering B.S. degree offered
- 1901: G.E. Davis published "Handbook of Chemical Engineering"
- 1908: American Institute of Chemical Engineering (AIChE) formed
- 1915: The concept of "unit operations" was first introduced by A.D. Little
- 1925: First accredited degrees offered

Figure 14.4. Word Slide Example 2

PONDER THIS

Critique the examples in Figures 14.3 and 14.4. Which would be more appropriate for an oral presentation? How could both examples be improved?

Note the use of abbreviations in Figure 14.3. Abbreviations allow for a larger font size to be used. Small words (*the* and *of*) are eliminated to avoid having the audience read the text rather than listen to the words. In both examples, the slide needs to be *presented*. Figure 14.3 would make a better slide in an oral presentation. Figure 14.4 might be better in a written document, where no additional words are used to explain the text.

14.3.4   Content of Visual Aids: Data Slides

Key idea: Create tables specific to the point you wish to make.

Data slides can be tables or figures. In oral presentations, it is critical that *tables contain only the data required*. Speakers sometimes photocopy large tables onto overhead transparencies and present the tabular material as follows: "I know you can't read all the numbers in this table, but note that the gear ratio of 20-to-1 was optimal." If you wish to speak about a gear ratio of 20:1, design a data or word slide specific to that point.

Properties of Air

Temp. (°C)	Density (kg/m³)	Viscosity (N·s/m²)	Speed of Sound (m/s)
−40	1.514	1.57	306.2
−20	1.395	1.63	319.1
0	1.292	1.71	331.4
20	1.204	1.82	343.3
40	1.127	1.87	349.1
60	1.060	1.97	365.7

"I know you can't read all the tiny numbers, but the speed of sound in air is less than 350 m/s in the temperature range of 0 to 20°C."

"As you can see, the speed of sound in air is less than 350 m/s in the temperature range of 0 to 20°C."

Properties of Air

Temperature (°C)	Speed of Sound (m/s)
0	331.4
20	343.3

Make tables specific to the points you wish to emphasize.

14.3.5   Special Notes about Computer-Based Presentations

Today's software allows you to prepare amazing computer-based presentations, with vibrant colors, inspiring animations, and hundreds of fonts. While all those embellishments are possible, you must ask yourself if they are right for your presentation and your audience.

PONDER THIS

How can you decide if animations and other embellishments are appropriate?

Key idea: With computer-based presentations, watch the colors, number of fonts, and animations.

Use the same criteria that you applied to all other aspects of your presentation: Do the embellishments help you to deliver your message to the target audience?

You should keep a few thoughts in mind as you design computer-based presentations. First, *go easy on the color combinations*. Start with the prepackaged color combinations in the presentation software. Stick with two to four colors, using them consistently for signposting. If you have poor color vision or are unsure of your artistic skills, then you may wish to have a friend review your work prior to presentation.

Second, *use a small number of font families*. You can use font size and font weight (bold, italic, etc.) to create a style, but using many font families is distracting. For example, some textbooks are written with only two font families (Times New Roman and Arial), but over a dozen combinations of font size and weight. If you use nonstandard fonts, then you can run into font availability problems if you use a different computer for the presentation than you used to create the talk.

Third, *be very careful about animations* (e.g., flying text and swirling slide transitions). Some people find animations very annoying. Use them sparingly unless you know your audience well.

14.4   BEFORE THE TALK: PREPARING TO PRESENT

14.4.1   Practicing Oral Presentations

Several tricks can make your practice time more valuable. First, practice your talk for the first time *before* the visual aids are finalized. In this way, you can identify and edit

Key idea: Practice before the visual aids are finalized.

any slides that do not make your points as cleanly as you want. Last-minute changes in visual aids can be expensive and stressful (although computer-based presentations are making last-minute changes easier).

Second, record the duration of *each section* of your talk during the first few practice rounds. This approach allows you to judge the *balance* of the talk. The meat of the talk (e.g., the results and discussion if you are presenting project results) should occupy at least half the time. Timing the talk also helps you to know where cuts or additions should take place if the first run-throughs show that the talk is too long or too short.

Key idea: When practicing, time each section as you practice alone and in front of others.

Third, practice the talk both by yourself and in front of others. When practicing by yourself, always *speak aloud* so you can rehearse any troublesome phrases or transitions. In addition, try to practice in front of others to get feedback about the talk before the main presentation (see also Section 14.6).

How often should you practice the talk before the big day? This is a matter of personal preference. Some people require many practice runs before they feel comfortable with the material, while others become stale after just a few practice sessions. *Experiment with different degrees of practicing to determine what level of preparation suits your personality.*

memory aids: notes used to help remember the main points in the talk

14.4.2   Memory Aids

Memory aids are the notes or devices that help ensure a smooth talk. Memory aids should be designed to help you remember the main points in the talk. Always practice the talk with the same memory aids you intend to use in the final presentation. Common memory aids include

- An outline of the talk
- Note cards containing a list of the key points for each slide
- Speaker's notes in presentation software

An outline lets you see quickly where you are in the presentation. Note cards are useful for making sure that you cover the important points before you go to the next slide. Most computer-based presentation software allows you to put your speaker notes near a miniature version of the slide so you can remind yourself to make the main points you wish.

Speaker's notes in a computer-based presentation

It pays to take a few seconds to glance at your notes or outline before you change slides. Although a few seconds may feel like an eternity when you are in front of an

Key idea: Avoid memorizing or reading oral presentations.

audience, the slight pause will help you gain confidence that you are not forgetting anything important. Audiences generally appreciate the small respites as well.

A final note about memory aids. Do not read the talk or memorize it completely. We all write differently than we talk. A read speech usually sounds "written." A memorized talk almost always sounds mechanical and forced. Read or memorized talks have another pitfall. If you lose your place or become flustered when reading or reciting by memory, general meltdown often occurs. If you use streamlined notes, it is much easier to get back on track.

14.5   DURING THE TALK

14.5.1   Pre-Talk Activities

Key idea: Learn about the facilities and coordinate introductions well before the talk.

Before walking to the podium or to the front of the conference room to give your talk, it is important to know what to expect. Examine the podium or speaking area well before the talk begins. Before you begin speaking, you want to know the answers to several questions:

- Is there a pointer?
- Is the projection equipment in working order?
- Who is responsible for changing computer or slide-projector slides? Are personnel available to help with your overhead transparencies?
- What type of microphone is in use?
- Is there a podium light to allow you to read your notes when the room lights go off? (Memory aids are useless if the room is in complete darkness, a fact you do not want to learn during your first technical presentation!)

It is also helpful to find and introduce yourself to the person who will introduce you. He or she may require some background information from you and may be able to help answer questions about the availability of pointers and the like. You should ask whether he or she will signal you when the allotted time has nearly expired.

14.5.2   Group Presentations

Key idea: Practice transitions between speakers in group presentations.

Group presentations raise their own set of challenges. It is critical to practice the transitions between the speakers. In general, it is better not to have too many speakers in a short period of time. Make sure the responsibilities of each speaker are understood, including whether one speaker will introduce the next speaker.

14.5.3   Nervousness

The main concern of most neophyte speakers is the control of nervousness. Being nervous means you care about the presentation. This is a positive attribute, as long as you can control the outward signs of nervousness.

Key idea: Do not worry about *being* nervous; learn to control or avoid the *signs* of nervousness.

The key to dealing with "the jitters" is to determine how nervousness affects you. If being nervous makes you speak more quickly, then focus on slowing your pace. If nervousness makes your hands shake, then avoid holding anything (such as notes or a pointer) during the talk. *It is natural to be a little apprehensive, but desirable to minimize the manifestations of nervousness.*

Remember also that for many talks you will give, the audience *wants* you to succeed. You are giving the talk for a reason. It is likely that the members of the audience desire the information you will share with them. Engineers face truly hostile audiences only rarely in their career.

14.5.4 What to Say

Key idea: Paraphrase information in word slides and point to each item in a list.

Technical presentations consist of two elements: presentation of word or data slides and making transitions between slides. When presenting word slides, it is often useful to paraphrase the material rather than reading it to the audience (see also Section 14.3.3). Again, you are trying to control the message. With lists, gesture to each item as you present it to remind the audience where you are in the slide.

Keep a mental checklist of the items to be covered during the presentation of figures. You should

Key idea: When presenting figures, tell what the figure is showing, identify the axes, communicate the meaning of each plot, and enumerate the main points to be made.

1. Tell the audience what the figure represents.
2. Identify the axes (with units).
3. Communicate the meaning of each plot (i.e., state the legend information).
4. Enumerate the main points to be made.

Figure 14.5 contains a sample figure and text showing how the figure would be presented orally. Look at the text in Figure 14.5 carefully and note the elements presented: a description of what the figure is showing ("removal of dye over time using the new technology"), identification of the axes with units ("time in minutes" and "dye concentration in milligrams per liter"), the meaning of each plot ("solid squares are the experimental data and the line is the first-order model fit"), and enumeration of the main points ("two points to notice in this figure. First, the technology ...").

Tell 'em Rule: the idea that you present information three times in a talk: you tell 'em what you will tell them, then tell 'em the information, and finally tell 'em what you just told them

Recall that it is necessary to remind the audience of your organization. This is crucial in oral presentations. If audience members feel lost, they will tune out completely. There is an old doctrine in public speaking called the **Tell 'em Rule**. According to the Tell 'em Rule, you present information three times in a talk: you tell 'em what

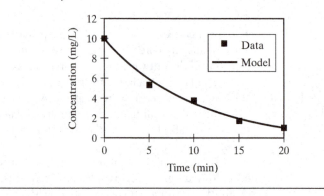

Sample Presentation Text:

Shown here is the removal of dye over time using the new technology. The x-axis is time in minutes and the y-axis is the dye concentration in milligrams per liter. The solid squares are the experimental data and the line is the first-order model fit. There are two points to notice in this figure. First, the technology could reduce the dye concentration to below two milligrams per liter in 20 minutes. Second, the exponential model fits the data reasonably well.

Figure 14.5. Example of the Presentation of Data in a Figure

you will tell them, then tell 'em the information, and finally tell 'em what you just told them. Following this rule allows for smooth transitions (also called *segues*) between parts of the talk. For example, you may say,

> I wish to give you a little background information on rotary engines. Rotary engines were used first in the automotive industry in .... [text omitted for clarity]. Now that I've told you about the history of rotary engines, let's turn to the modern versions of this unique engine type.

Notice the three presentations of the information ("I wish to give you ...," "Rotary engines were used first ...," and "Now that I've told you about ..."). Also note the transition to the modern versions of the rotary engine ("... let's turn to the ...").

Key idea: Use signposting liberally in technical oral presentations.

The process of telling the audience where you are during transitions between major portions of a talk is called *signposting*. The word *signposting* comes from the analogy with road signs: well-spaced markers tell the audience where you are in the talk. Most technical speakers do not signpost enough. Audiences are much more comfortable when they know they are in synch with the speaker. To assist in signposting, it is helpful to present an outline of the presentation near the beginning of the talk. By referring to the outline, you can keep the audience with you through the talk. Intermediate outlines can be placed in the middle of the talk for complicated sections. For example, you may wish to have an outline of the results to guide the audience through the results section.

14.5.5   How to Say It

The audience responds to two features of a speaker: the speaker's voice and body. The voice should vary in pitch and intensity: a monotone voice leads to a sleeping audience. Speak through each sentence to avoid swallowing words at the end of the sentence. Be aware of the *speed* and *volume* of your voice. It is useful to have a colleague in the audience signal you (discreetly, of course) if you are speaking too quickly or too softly. The volume of your speech depends on the room and amplification.

Key idea: Speak loudly and slowly. Use meaningful hand gestures and move your body without pacing.

Your body movements should be purposeful and strong. The main problem for most speakers concerns what to do with the hands. Use them to your advantage! Hand gestures are a great way to emphasize important points. For the most important messages, make your gestures higher. Avoid holding pens, pencils, or other mental crutches, and **never** leave your hands in your pockets.

Your legs can work for you as well. Avoid standing stock-still. Walk toward the audience and engage its members at critical points in the talk. While mechanical pacing should be avoided, small steps can make a speaker seem more human to the audience.

In spite of your best preparation, things sometimes go wrong in oral presentations. A few true stories are shared in the *Focus on Talks: Horror Stories*.

14.6   AFTER THE TALK

Key idea: Seek feedback and incorporate changes into your speaking style.

After a talk, seek out feedback from colleagues in the audience. Listen to their constructive criticism and think about modifications to your speaking style that will make communication more effective. Do not be afraid to identify weaknesses in your speaking style and practice ways to overcome them.

Finally, be an attentive listener. Listen critically to other speakers (such as colleagues, professional speakers, actors, and your professors) and note what you like and dislike about their speaking styles. Ask yourself why you like or dislike their speaking style. Why do good speakers engage you personally? Are they friendly, open, and confident? Incorporate the good aspects and avoid the bad in your next presentation.

FOCUS ON TALKS: HORROR STORIES

INTRODUCTION

Even after reading this chapter, you may still approach your first professional oral presentation with some trepidation. In this section, a few true stories of oral presentations gone awry are shared. Do not panic when you read these stories. They are offered in the spirit of comic relief and to show you that bad things sometimes happen to good presenters. (*Note:* Stories labeled "Lytle" come from the "Stress of Selling" articles compiled by Chris Lytle on the Monster.com Web site. Stories labeled "Hoff" come from Ron Hoff's (1992) very readable book on oral presentations. All other stories come from my experiences or the experiences of my colleagues.)

FROM THE "DRESS FOR SUCCESS" DEPARTMENT

Numerous speakers have walked back to their seat after an oral presentation, only to discover in horror that their pants or skirt zipper was in the down position. Perhaps "check your zipper" is as important as "check your slides." During a graduate course, I noticed my students giggling every time I turned to face them after writing on the blackboard. When I asked them what was going on, they gleefully informed me that I had a sticker of a lamb on my derrière (courtesy of my then-two-year-old daughter). I keep the sticker on my class notebook to this day to remind me to check my attire.

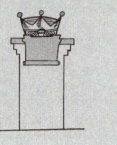

FROM THE "LOCATION, LOCATION, LOCATION" DEPARTMENT

A colleague of mine relates the tale of a presentation he gave for a job interview. He used a long wooden pointer to emphasize his points. But being a good speaker, he kept good eye contact with the audience. Part way through the talk, he realized that he was pointing *behind* the screen with the pointer.

Podiums also can be a source of frustration. Lytle collected the story of a presenter who stood on a stool behind the podium during a speech to 1,500 people. Shortly after the presentation started, the heel of her pump broke. She fell off the stool and crashed onto the concrete floor. Her sympathetic audience gave her the courage to complete the presentation.

Hoff reports that the Queen of England stepped up to a podium during a visit to the United States, only to find that the podium was higher than her head.

FROM THE "NEVER LET THEM SEE YOU SWEAT" DEPARTMENT

Obviously nervous presenters make the audience a little uncomfortable, so never draw attention to your nervousness. I witnessed a student presentation at a state conference where the speaker was using a laser pointer. The pointer danced all over the screen as the speaker's hand shook. Rather than letting it pass, he said, "Well, look at that—I must be really nervous!"

Hoff reports a company treasurer starting a speech with "I'm so nervous this morning. I hope you can't see how badly my knees are shaking." Guess where the audience's eyes were glued for the remainder of the speech. Of course, you should avoid bringing attention to overconfidence as well. Al Gore probably regrets the sighs picked up by microphones during the first presidential debate of 2000.

FROM THE "EQUIPMENT MALFUNCTION" DEPARTMENT

A colleague of mine gave a technical talk in another country, where a more powerful slide projector bulb was in use. She stared in shock as her first slide literally melted before her eyes. Needless to say, she completed the talk without slides.

Lytle reports a presenter panicking when the overhead projector did not turn on. Reaching down to plug in the power cord resulted in a loud ripping noise as the seam of his pants gave out.

FROM THE "WATCH YOUR LANGUAGE" DEPARTMENT

Word choice is important in oral presentations. Lytle relates a story from a salesman giving a presentation before a defense contractor with a product representative (rep). The product rep had a way of choosing the worst possible words to express himself. Quoting from the Web site: "Discussing the ease with which you can use the product, the other rep stated, 'You don't have to be a rocket scientist to use this.' Twenty rocket scientists [in the audience] sat back in their chairs and crossed their arms. After 20 minutes of weasel words to get their interest back ... [the rep said], 'We just have to get your propeller heads to talk to our propeller heads to work it out.' With that, their propeller heads stood up and walked out." Phrases like *propeller head* or *gear head*—both derogatory terms for technical staff—are inappropriate in formal speech.

Written words on slides can bite you, too. A consulting engineer reports that she made up slides with the client's names based on a telephone call she made to the client. Unfortunately, several of the names were misspelled, leading to embarrassment and, not surprisingly, an unsuccessful bid for the project.

The moral of these stories? Be prepared, be relaxed, and go with the flow. While you never may face the discomfort of the speakers in these stories, remember that audiences often are pulling for you. If something unusual happens, finish with grace and hope for the best.

14.7   SUMMARY

Most of the work in an oral presentation occurs before the talk is presented. Before the talk, take the time to organize the material, construct the visual aids (which should number no more than $3/4$ times the number of minutes), and practice. Learn about the facilities in the room before you walk to the podium.

During the talk, do not worry about being nervous, but learn to control or avoid the signs of nervousness. Take your time in presenting data slides (especially figures). During transitions from one part of the talk to another, be sure to "tell 'em" three times: preview the material, present the material, and summarize the material. Modulate the speed and volume of your voice and use your hands effectively.

After the talk, seek feedback to become a better speaker. Remember, the best way to become an effective technical speaker is to take every opportunity to give technical talks.

SUMMARY OF KEY IDEAS

- Technical presentations can be improved by considering the activities before the talk, during the talk, and after the talk.
- Identify the presentation goals, target audience, and constraints on the presentation (especially time constraints).
- Use an outline to organize the talk and an outline slide to show your organization.
- The number of visual aids should be about $3/4$ times the number of minutes allotted to the presentation.
- In selecting the type of visual aid, consider image quality, eye contact, and production cost and time. Then use only one type of visual aid in a talk.
- Word slides should contain as few words as possible.
- Create tables specific to the point you wish to make.

- With computer-based presentations, watch the colors, number of fonts, and animations.
- Practice before the visual aids are finalized.
- When practicing, time each section as you practice alone and in front of others.
- Avoid memorizing or reading oral presentations.
- Learn about the facilities and coordinate introductions well before the talk.
- Practice transitions between speakers in group presentations.
- Do not worry about *being* nervous; learn to control or avoid the *signs* of nervousness.
- Paraphrase information in word slides and point to each item in a list.
- When presenting figures, tell what the figure is showing, identify the axes, communicate the meaning of each plot, and enumerate the main points to be made.
- Use signposting liberally in technical oral presentations.
- Speak loudly and slowly. Use meaningful hand gestures and move your body without pacing.
- Seek feedback and incorporate changes into your speaking style.

Problems

14.1. Pick an engineering topic of interest to you and identify a target audience. How will the target audience influence the visual aids you select and the material you present in your talk?

14.2. Write an outline for a 15-minute talk on the topic and target audience selected in Problem 14.1.

14.3. How would your outline change if you were asked to prepare a five-minute talk? A two-minute talk?

14.4. How many visual aids will you need for the talk?

14.5. Prepare visual aids for the talk using the principles presented in this chapter.

14.6. Practice the 15-minute talk. Prepare a table showing the percentage of time in the talk devoted to each of the major sections of the presentations. Refine the talk to better use the time allotted and summarize your refinements.

14.7. Before presenting the talk, what questions do you anticipate from the audience?

14.8. Present the talk to a group of people pretending to be your target audience. Did you predict (in Problem 14.7) the questions that were asked? What feedback did you receive from the audience?

14.9. Rewrite your talk using double the number of visual aids that you prepared in Problem 14.7. Present the revised talk to a group of people pretending to be your target audience. What feedback did you get about the number of visual aids?

14.10. Attend lectures delivered by three different public speakers. For each speaker, list and explain two aspects of his or her speaking style that appeal to you the most and two aspects that appeal to you the least.

PART V
Engineering Profession

Chapter 15: Introduction to the Engineering Profession and Professional Registration
Chapter 16: Engineering Ethics

pro · fes · sion —a calling requiring specialized knowledge and often long and intensive academic preparation
Merriam-Webster

[The first response of Bob Lund, Vice President for Engineering at Morton Thiokol, designers and builders of solid rocket boosters for the space shuttle] was to repeat his objections [to the launch of the space shuttle *Challenger* due to concerns about O-ring failure at low air temperatures]. But then [Jerald] Mason [a senior manager with Morton Thiokol] said something that made him think again. Mason asked him to THINK LIKE A MANAGER INSTEAD OF AN ENGINEER. (The exact words seemed to have been "take off your engineering hat and put on your management hat.") Lund did and changed his mind. The next morning the shuttle exploded, killing all aboard. An O-ring had failed.
M. Davis

15

Introduction to the Engineering Profession and Professional Registration

15.1 INTRODUCTION

In this part of the text, you will explore engineering as a profession. To do so, you must consider two questions. First, what does *profession* mean? If you are getting paid, is that enough to qualify a job as a profession? This area will be explored in Section 15.2.

Second, does engineering qualify as a profession? To answer this question, the characteristics that make engineering unique and important in today's society will be discussed.

15.2 PROFESSIONAL ISSUES

15.2.1 What Is a Profession?

A profession is more than just a job requiring specialized knowledge and intensive study.

Common elements include (Martin and Schinzinger, 1989)

- Compensation
- Professional practices that result in public good
- Need for formal education
- Requirement for judgment, discretion, and skill in performing your job
- Requirement to be admitted to the profession
- Self-policed conduct

Do these concepts make sense? You can apply them against known professions, such as medicine and law. Both medicine and law require all these elements. The fact that doctors and lawyers satisfy the first four criteria is obvious: they are paid, they do good, they are trained formally, and their jobs require judgment, discretion, and skill.

OBJECTIVES

After reading this chapter, you will be able to:

- explain why engineering is a profession;
- explain the benefits of becoming a licensed professional engineer;
- list the steps in the registration process.

PONDER THIS

What other elements come to mind when you think of the word *profession*?

Key idea: Common elements of a profession include compensation; performing a public good; a need for formal education; a requirement for judgment, discretion, and skill; a requirement to be admitted to the profession; and self-policing.

The fifth criterion also is satisfied, since both jobs require admission to the profession (passing medical board examinations or the bar examination). The admission to both professions is regulated by each state. In other words, doctors are licensed and lawyers are admitted to the bar by the state in which they practice. Finally, practicing professionals sit on the boards regulating professional conduct (for example, medical and legal ethics panels). Thus, the last criterion is satisfied.

15.2.2 Engineering as a Profession

For engineering to be a profession, it must satisfy the elements listed in Section 15.2.1. That engineers are compensated is obvious. Further, it is sincerely hoped that you are convinced by now that engineering practice results in public good. What about formal education? You are at the beginning of your formal education now.

Key idea: Engineers must exercise professional judgment in the practice of their profession.

What about the requirements for judgment and discretion, admission to the profession, and self-policing? In fact, these elements are at the heart of engineering as a profession.

15.2.3 Judgment and Discretion in Engineering

Engineering is all about generating alternative solutions to a problem and choosing from among alternatives. Thus, judgment is a critical part of engineering decision-making. Professional judgment is at the core of "thinking like an engineer."

What about discretion in engineering?

PONDER THIS

Why should engineers be thoughtful and discreet?

There are two reasons why discretion and thoughtfulness are imperative for engineers. First, engineering has a significant impact on public safety and health. A poorly designed automobile or bridge or electric transmission line could result in terrible suffering and loss of human life. Any profession that affects *every* citizen *every* day (as is the case with engineering) should be held to the highest standards of ethical behavior.

Key idea: Discretion in engineering is important because (1) engineering affects public safety and health, and (2) public trust in engineers must be preserved, since engineering work is not easy to understand.

Second, most engineering work is difficult for the average person to comprehend. As a result, the public must trust the word of engineers. If an engineer declares that the probability of the collapse of a high-rise building is extremely small, then the public is reassured. Reassurance comes *without* an understanding of the details of the materials employed or the construction techniques or even elementary statics. To *preserve the public trust*, engineers must be held accountable to a very strict professional code.

15.2.4 Admission to the Profession

Engineering also satisfies the criteria that you must be formally accepted into the profession. To use the title "engineer" to make money, you must be *admitted* to the profession. This process is called *registration*. The steps required for professional registration are discussed in more detail in Section 15.4.

15.2.5 Self-Policing

Engineers police themselves. Self-policing is accomplished by having engineers sit on state review boards that oversee licensing and the removal of licenses.

For professionals to police themselves, the profession needs a list of guidelines. Engineering has such guidelines, called *professional ethics*.

Thus, engineering fulfills all the requirements normally associated with being a profession. This analysis does not capture the awesome potential you have for influencing society or the serious responsibilities you will carry with pride as an engineer. As a source of inspiration, read the *Focus on Professionalism: Standing on the Shoulders of Giants*.

FOCUS ON PROFESSIONALISM: STANDING ON THE SHOULDERS OF GIANTS

The formal analysis of whether engineering is a profession misses the most important point: *you* are joining this profession, this band of brothers and sisters devoted to changing the world in a positive way. You will reap the satisfaction of helping others and you will make part of your being the obligations of excellence.

You are not alone. The contributions you will make build on the advances made by engineers before you. As Isaac Newton (1642–1727) said, "If I have seen further, it is by standing on the shoulders of giants." By standing on such a tall perch, you should feel that your potential to help others is unlimited.

In your everyday world of studying and working, it is easy to lose sight of the traditions of engineering, of the pride and obligations implicit in the profession. Presented here are two statements of engineering pride and obligations. The first statement, the "Obligation of an Engineer," is recited as part of the ceremony of joining the American association called the Order of the Engineer. The "Obligation of an Engineer" was written as an engineering version of the Hippocratic Oath taken by physicians. The Order of the Engineer was formed in 1970.

Obligation of an Engineer

(from the Order of the Engineer)

I am an Engineer, in my profession I take deep pride. To it I owe solemn obligations.

Since the Stone Age, human progress has been spurred by the engineering genius. Engineers have made usable Nature's vast resources of material and energy for Humanity's benefit. Engineers have vitalized and turned to practical use the principles of science and the means of technology. Were it not for this heritage of accumulated experience, my efforts would be feeble.

As an Engineer, I pledge to practice integrity and fair dealing, tolerance and respect, and to uphold devotion to the standards and the dignity of my profession, conscious always that my skill carries with it the obligation to serve humanity by making the best use of Earth's precious wealth.

As an Engineer I shall participate in none but honest enterprises. When needed, my skill and knowledge shall be given without reservation for the public good. In the performance of duty and in fidelity to my profession, I shall give the utmost.

The American ceremony was based on the Canadian "Ritual of the Calling of an Engineer," which dates back to 1926. The Canadian engineer's oath, called "The Obligation," was written (in three weeks!) by Nobel Laureate Rudyard Kipling.

The Obligation

(from the Iron Ring Ceremony in the Canadian Ritual of the Calling of an Engineer)

I (your name), in the presence of these my betters and my equals in my Calling, bind myself upon my Honour and Cold Iron, that, to the best of my knowledge and power, I will not henceforward suffer or pass, or be privy to the passing of, Bad Workmanship or Faulty Material in aught that concerns my works before mankind as an engineer, or in my dealings with my own Soul before my Maker.

My Time I will not refuse; my Thought I will not grudge; my Care I will not deny towards the honour, use, stability and perfection of any works to which I may be called to set my hand.

My Fair Wages for that work I will openly take. My Reputation in my Calling I will honourably guard; but I will in no way go about to compass or wrest judgement or gratification from any one with whom I may deal. And further, I will early and warily strive my uttermost against professional jealousy and the belittling of my working-colleagues in any field of their labour.

For my assured failures and derelictions I ask pardon beforehand of my betters and my equals in my Calling here assembled, praying that in the hour of my temptations, weakness and weariness, the memory of this my Obligation and of the company before whom it was entered into, may return to me to aid, comfort, and restrain.

Upon Honour and Cold Iron, God helping me, these things I purpose to abide.

When times are tough—when you have two tests on Tuesday and two homework assignments due tomorrow and you wonder why you ever wanted to be an engineer—reread the "Obligation of an Engineer" and "The Obligation." Remind yourself of your potential and your obligations to help.

15.3   PROFESSIONAL ENGINEERS

15.3.1   Introduction

registration: the act of being admitted to the engineering profession

Engineering is a profession in part because engineers are admitted to practice. Every state regulates the practice of all professions. Examples include medicine (e.g., physicians, dentists, optometrists, and nurses), law, and engineering. In engineering, admission to the profession is called **registration** (or licensing). A registered engineer is called a **professional engineer** or PE. In this chapter, the registration process will be described.

professional engineer (PE): a licensed (registered) engineer

15.3.2   Why Become a Professional Engineer?

Key idea: People become licensed to show their competence and commitment, to perform duties unavailable by law to unlicensed personnel, and to gain personal benefits.

The registration process (described in Section 15.4) is long, typically requiring at least seven years. Why go to the trouble of becoming licensed? The most important reason to become licensed is that it demonstrates your extra commitment to the profession. In addition, it is a badge of honor that shows your increased level of competence in the field.

Another reason to become licensed is that a PE may conduct professional activities that unlicensed personnel are forbidden to do. For example, only a licensed engineer may prepare, approve, and submit engineering plans and drawings to a public authority. Only a licensed engineer may approve engineering work for public and private clients. Consulting engineers in positions of authority over work performed in their office must, by law, be licensed. In general, you cannot use the term *engineer* in your title unless you are licensed. State engineering boards are successfully using the courts to impose civil penalties on individuals using the term *engineer* without an engineering license. As a example, penalties have been imposed on people without a PE doing business as *software engineers*.

In addition, PEs reap personal benefits from registration. Surveys have shown that professional engineers can expect salaries 15% to 25% higher than unlicensed colleagues. Some jobs are available only to professional engineers. This is particularly true in some government positions, where obtaining a PE license often is required for promotion. PEs also may find themselves more desirable to another firm in the event of downsizing. For advice about registration from professional engineers, see the *Focus on Registration: PE or Not PE?*

In spite of these arguments, not all engineers become licensed. Registration is more common in some disciplines than others. In particular, engineers working primarily in private practice (e.g., civil and environmental engineers) are more likely to become licensed than engineers working primarily in industry.

15.4 THE REGISTRATION PROCESS

15.4.1 Overview

Key idea: The steps in the registration process are as follows: obtain a baccalaureate degree from an accredited engineering department, pass the Fundamentals of Engineering Exam, acquire the requisite experience, and pass the Principles and Practice Exam.

Engineer-in-Training: a prelicensure certificate given to individuals completing the first steps in the registration process

The registration procedure for engineers is a four-step process. First, would-be engineers must obtain a baccalaureate degree from an accredited engineering department. Second, after earning the degree (or in the final stages of the program), potential engineers must take and pass a written test called the Fundamentals of Engineering Examination (or FE Exam). At this point, individuals can earn a prelicensure certificate. This level of licensing generally is called the ***Engineer-in-Training*** (or EIT). Third, EITs must acquire experience under the supervision of a licensed engineer. Finally, the would-be engineer must take and pass the Principles and Practice Examination in a specific engineering discipline. At this point (after the requisite licensing fees are paid), a person may be called an engineer and have the title *professional engineer*. You indicate the title by the initials *PE* after your name. Each of these steps will be discussed in more detail.

15.4.2 The Accredited Degree

The registration process starts at a college or university.* In accordance with the idea that professions require extensive training, potential engineers must obtain an undergraduate degree through an accredited engineering program.

What does *accredited* mean? To ensure the high technical level of the profession, engineering programs must request accreditation through the Accreditation Board for Engineering and Technology (ABET). Engineering programs must be reaccredited periodically, typically every six years. ABET's Engineering Accreditation Commission conducts site visits of each department seeking accreditation.

In 2005, there were over 1,600 accredited engineering programs in the United States. For a list of accredited programs, see the ABET Web site (www.abet.org).

The accreditation process is the beginning of the self-policing feature of the engineering profession. To be eligible for professional registration later, it is important for you to seek a degree from an accredited engineering program. Accreditation is awarded to each program, not the college or university in general.

15.4.3 Fundamentals of Engineering Examination

Fundamentals of Engineering (FE) Exam: an eight-hour exam covering general science, general engineering, and discipline-specific engineering topics

The ***Fundamentals of Engineering Examination*** (FE Exam) generally is taken in the last semester of the undergraduate program. The FE Exam[†] is administered by the National Council of Examiners for Engineering and Surveying (NCEES). The FE Exam consists of two parts. The morning portion deals with general science and engineering concepts and is common to all disciplines. The morning portion consists of 120 one-point questions to be answered in four hours. Topics include (with the percentage of questions in parentheses)

- Chemistry (9%)
- Computers (6%)

---

*In most states, the requirement for a baccalaureate degree can be satisfied by many years of experience. However, the most common pathway for professional registration includes an undergraduate degree from an accredited engineering program.
[†]You may hear the FE Exam referred to by its previous name: the EIT Exam.

- Dynamics (7%)
- Electrical circuits (10%)
- Engineering economics (4%)
- Ethics (4%)
- Fluid mechanics (7%)
- Materials science/structure of matter (7%)
- Mathematics (20%)
- Mechanics of materials (7%)
- Statics (10%)
- Thermodynamics (9%)

The afternoon session is discipline-specific. It consists of 60 two-point questions to be answered in four hours. Afternoon sessions are offered in chemical, civil, electrical, environmental, industrial, and mechanical engineering. A general engineering section also is available.

Following successful completion of the FE Exam, the potential engineer is eligible for certification as an Engineer-in-Training (or EIT; in some states, it is also called an Intern Engineer or Engineer Intern). The EIT receives a registration number from the state.

15.4.4 Experience

Key idea: For registration purposes, your work experience must be under the direction of a PE.

For an EIT to become a PE, you first must acquire engineering experience. Most states require four years of experience under the supervision of a PE. This distinction is important. If a newly minted EIT takes his or her first job at a firm that has no PEs on staff, it would be difficult to acquire experience under the direction of a PE. Thus, the first job after becoming an EIT is very important for future licensure. A portion of the time spent obtaining a graduate degree in engineering typically may be applied towards the experience requirement.

15.4.5 Principles and Practice Examination

Principles and Practice (PP) Exam: an eight-hour exam of a specific engineering discipline

After acquiring the necessary experience, potential engineers can apply to take the **Principles and Practice Examination** (PP Exam).* Like the afternoon session of the FE Exam, the PP Exam is discipline-specific. Exams are offered in 17 areas. Not all areas are offered in every state.

The PP Exam is eight hours long and may be all essay questions, all multiple-choice, or a mixture of essay and multiple-choice questions. The exam, administered by NCEES, is the same nationwide, but each state sets its own passing grade.

15.5 AFTER REGISTRATION

To maintain registration, an engineer must follow the ethical guidelines of his or her field and maintain competence in the field. Some states require PEs to engage in continuing education activities. Even without such requirements, maintaining state-of-the-art knowledge of the field is mandated by the ethical codes. In addition, fees must be paid periodically to the state to maintain the license.

reciprocity: the recognition of a license in one state by another state

There is general agreement between the states on the licensing process. Most states recognize licenses obtained in another state (a process known as **reciprocity**). Thus, it is often possible for a engineer licensed in one state to gain licensure in another state relatively easily.

---

*The PP Exam is sometimes referred to by its previous name: the PE Exam.

In this section, advice will be presented from registered professional engineers. The professional engineers offer sound comments on why to become licensed and suggestions for the licensing process. The quotations are from interviews published on one of the Web sites run by the National Council of Examiners for Engineering and Surveying (www.engineeringlicense.com).

Kathy Caldwell, PE; civil engineer

President, JEA Construction Engineering Services

... (T)ake the FE exam just as soon as you possibly can before you leave the college environment. It's very difficult to go back and take the fundamentals exam once you have been practicing in your major area of practice. A lot of fundamentals slip away from you and it becomes much more difficult to go back and relearn the fundamentals later.

Jim Parrish, PE; electrical engineer

General Manager, Duke Energy, Energy Delivery Services International

Achieving a PE also means that I have more career choices and can benefit from higher financial rewards that are commensurate with industry-recognized professional status. In my [unit of Duke Energy], engineers without a PE license can be promoted to only certain levels of engineering-related management jobs.

Brett Pielstick, PE; civil engineer

Vice President, PTG Construction Services

I think people know that as an engineer with a PE I will provide an honest day's work, an honest answer, and that they can trust in what we're going to do for them.

Deborah Grubbe, PE; chemical engineer

Corporate Director of Safety and Health, The DuPont Company

Ethics is an important aspect of professional licensure. Obtaining a PE (license) registration not only targets technical competence, but also requires a clear understanding of both ethics and professionalism. Ethics is essential to one's personal reputation, and once compromised, can almost never be reclaimed. Personal credibility becomes even more important when working outside the United States.

15.6 SUMMARY

In all aspects of the word, engineering qualifies as a profession. First, engineers are compensated for their work. Second, the general public benefits significantly from engineering practice. Third, engineers require a formal education prior to practicing their craft. Fourth, engineering practice requires judgment, discretion, and skill. Fifth, to legally use the title *engineer*, a person must be admitted to the engineering profession. The process of admission to the profession in engineering is called registration. Sixth, the professional conduct of engineers is self-policed through strong written codes of professional ethics.

Professional registration as a professional engineer (PE) is a demonstration of competence and commitment to the engineering profession. Registration also allows engineers to perform duties unavailable by law to unlicensed personnel and to gain personal benefits. Registration is a four-step process: obtain a baccalaureate degree from an accredited engineering department, pass the FE Exam, acquire the requisite experience (generally four years under the supervision of a PE), and pass the PP Exam. To maintain registration, PEs must follow the ethical guidelines of their field and maintain professional competence in the field.

SUMMARY OF KEY IDEAS

- Common elements of a profession include compensation; performing a public good; a need for formal education; a requirement for judgment, discretion, and skill; a requirement to be admitted to the profession; and self-policing.
- Engineers must exercise professional judgment in the practice of their profession.
- Discretion in engineering is important because (1) engineering affects public safety and health, and (2) public trust in engineers must be preserved, since engineering work is not easy to understand.
- People become licensed to show their competence and commitment, to perform duties unavailable by law to unlicensed personnel, and to gain personal benefits.
- The steps in the registration process are as follows: obtain a baccalaureate degree from an accredited engineering department, pass the Fundamentals of Engineering Exam, acquire the requisite experience, and pass the Principles and Practice Exam.
- For registration purposes, your work experience must be under the direction of a PE.

Problems

15.1. Write a short essay to justify why engineering is a profession.

15.2. List the steps in the professional registration process. Make a timeline to show when you plan to complete each step.

15.3. From the ABET Web site, make a list of accredited engineering programs in your discipline of interest in your state.

15.4. Can a person without a PE license use the title "Web design engineer" in a business? Why or why not?

15.5. Using data from an interview with a practicing engineer or from the Internet, determine the impact of earning a PE on salary in your area.

15.6. Find and report the distribution of questions on the disciple-specific portion of the Fundamentals of Engineering Examination for an engineering discipline of interest to you.

15.7. Find and report the distribution of questions on the disciple-specific portion of the Principles and Practice Examination for an engineering discipline of interest to you.

15.8. Some states have continuing education requirements for professional engineers. Report on the continuing education requirements, if any, for professional engineers in your state.

15.9.   Some states do not offer reciprocity without further examination. Research and report on any additional examination for obtaining a PE license in civil engineering by reciprocity in California.

15.10.   Using the Web site of your state licensing office, find a case of an engineer who lost his or her license (for reasons other than nonpayment of fees). What did the engineer do to lose it?

16

Engineering Ethics

16.1 INTRODUCTION

Engineering is a profession. This means in part that engineers must practice their profession in accordance with high ethical standards. In this chapter, you will explore *why* engineers must act ethically and examine the basic codes of engineering ethics. Examples will be used to illustrate the ethical conflicts that occur on the job.

Ethics can be studied on many levels. Please understand that the discussions in this chapter are on a very practical level. The philosophy of ethics (sometimes called *formal ethics*) is a beautiful and interesting field of study. This is not a chapter on formal ethics. Rather, the purpose of this chapter is to provide you with guidance on how to live your life as an engineer.*

16.2 PROFESSIONAL ISSUES

Ethics refers to a system of moral principles. All professions have standards of ethics to which their members are bound. In fact, a code of ethics allowing practitioners to "police themselves" is really a prerequisite for a profession.

Ethical issues in the medical, legal, and political arenas appear almost daily in the newspaper. Engineers also must follow ethical standards.

OBJECTIVES

After reading this chapter, you will be able to:

- explain why engineers should be ethical;
- list the canons in the NSPE Code of Ethics;
- identify the correct ethical choices in engineering applications.

---

*A student of moral philosophy might say that this chapter is about *normative ethics* (i.e., the principles that guide how we should live our lives), rather than *metaethics* (i.e., the study of what is good).

| **Why should engineering follow ethical standards?**

ethics: a system of moral principles

Recall that engineering ethics is important because (1) engineering affects public safety and health, and (2) public trust in engineers must be preserved, since engineering work is not easy to understand.

16.3 CODES OF ETHICS

16.3.1 Introduction

Many engineering societies have their own codes of ethical behavior. Engineering societies having codes of ethics (and other ethics documents) are listed in Table 16.1. The differences between the codes are fairly minor. Most of the ethical standards are modeled after the code of ethics of the National Society of Professional Engineers (NSPE). To understand the concepts underlying the various ethics doctrines, the NSPE Code of Ethics will be examined in more detail.

16.3.2 NSPE Code of Ethics

NSPE Fundamental Canons of Ethics: the six basic principles outlining the professional responsibilities of engineers

Board of Ethical Review: an NSPE committee that offers commentary on ethics cases

NSPE has established six principles called the ***Fundamental Canons of Ethics***. The principles refer to the responsibilities of engineers when conducting their work and when approving documents. When a PE approves a document, he or she is *personally certifying* that the work meets professional standards. An NSPE committee called the ***Board of Ethical Review*** (BER) offers commentary on ethics cases to show engineers how the fundamental canons have been interpreted in the past. Each of the fundamental canons will be presented with examples on their interpretation. (*Note*: The entire NSPE Code of Ethics is given at the end of the chapter. More examples of the interpretation of the fundamental canons can be found in the Rules of Practice and Professional Obligations sections of the NSPE Code of Ethics.)

Key idea: Engineers shall hold paramount the safety, health, and welfare of the public.

Canon 1: Engineers shall hold paramount the safety, health, and welfare of the public. This is the most important of the fundamental canons. It states that nothing—not profit, not inconvenience, not personal gain—comes before the safety, health, and welfare of the public. Remember, safety, health, and welfare are paramount. If you face a contradiction in your work life between canons, Canon 1 prevails.

Canon 1 has three important interpretations. First, the canon has been interpreted to mean that individual engineers must notify their employer, client, or the proper authorities when life or property is endangered. Notification must occur *even if the client requests that data or any other engineering work be held back*. For example, if you discover a potentially dangerous situation in the course of your professional work, you **must** report it even if the paying client *demands* that you squelch the information.

TABLE 16.1 Engineering Societies with Codes of Ethics

Organization	Last Revision	Other Ethics Documents
American Consulting Engineers Council (ACEC)	1980	none
American Institute of Chemical Engineers (AIChE)	1989	none
American Society of Civil Engineers (ASCE)	1993	ASCE's Guidelines on Practice
American Society of Mechanical Engineers (ASME)	1991	ASME Criteria for Interpretation of the Canons
Institute of Electrical and Electronics Engineers (IEEE)	1990	Employment Guidelines

Safety outweighs all other considerations.

A second implication of the first canon is that engineers must approve only documents that conform to applicable standards. Approving plans that you know are not in compliance with standards is an ethical violation and grounds for loss of your engineering license.

A third implication of the first canon is that engineers have a responsibility to report any violations of the code of ethics. Reporting known violations is a form of *whistle-blowing*. Professionals who come public with mistakes made by themselves, their firm, or their colleagues risk being fired or ostracized. However, because of the overriding importance of the safety, health, and welfare of the public, engineers have a moral duty to report ethical violations committed by themselves or others.

Key idea: Engineers shall perform services only in the areas of their competence.

Canon 2: Engineers shall perform services only in the areas of their competence. At present, engineers usually are licensed in a particular discipline. This canon makes it clear that engineers must stay in their areas of expertise. Thus, a chemical engineer cannot approve structural engineering plans.

One ramification of this canon is that you must approve only documents prepared under your supervision. For example, a mechanical engineer cannot approve heating, ventilation, and air conditioning (HVAC) plans drawn up by people being supervised by another engineer. As a practical matter, the second canon means that each technical portion (e.g., electrical, mechanical, and structural portions) of a set of plans usually must be approved individually.

Key idea: Engineers shall issue public statements only in an objective and truthful manner.

Canon 3: Engineers shall issue public statements only in an objective and truthful manner. As stated in Section 16.2, the engineering profession thrives because of the public trust. This canon states that you must be honest in your dealings with the public. For example, you must acknowledge if you are being paid to issue a public statement about an engineering issue. Suppose you are being paid by a developer to lay out a new neighborhood. If you speak before a city council meeting in favor of the development, then you must identify yourself as an engineer paid by the developer.

Key idea: Engineers shall act for each employer or client as faithful agents or trustees.

Canon 4: Engineers shall act for each employer or client as faithful agents or trustees. The practice of engineering depends on the trust between the engineer and the public. Engineering also is dependent on the trust between the engineer and the client. This canon says in part that the client has the right to expect that the engineer will use his or her best engineering judgment in solving the client's problems.

This canon also is interpreted to mean that engineers must disclose to the client all known or *potential* conflicts of interest. Identification of potential conflicts of interest can be difficult. Clearly, the engineer cannot represent two clients who may come into conflict. For example, you could not represent both a potentially polluting industry and the town owning the wastewater treatment facility or both a developer and a city where the developer does work.

Key idea: Engineers shall avoid deceptive acts.

Canon 5: Engineers shall avoid deceptive acts. This canon has implications in the procurement of work. For example, it is a violation of the code of ethics to falsify your qualifications. In addition, it is considered unethical to offer or give a contribution to influence a public authority's decision about who should be awarded a contract.

The difference between ethical and unethical behavior can be very small. Clearly, it is improper to offer a bribe to the head of a housing authority to get a contract. What about taking the town engineer out to lunch right before a contract is awarded? What about giving the daughter of the mayor extra playing time on the soccer team you coach?

Key idea: Engineers shall conduct themselves honorably, responsibly, ethically, and lawfully so as to enhance the honor, reputation, and usefulness of the profession.

Canon 6: Engineers shall conduct themselves honorably, responsibly, ethically, and lawfully so as to enhance the honor, reputation, and usefulness of the profession. This canon greatly affects the lives of engineers. It means that engineers must advise their clients if the engineers believe a project will *not* be successful. In addition, engineers are prohibited by this canon from accepting free material from suppliers or contractors in return for specifying their products or services.

Language similar to that in the sixth canon is used by the Accreditation Board for Engineering and Technology (ABET) in the fundamental principles section of their code of ethics. (One responsibility of ABET is to certify undergraduate engineering programs.) The ABET fundamental principles state that engineers can "uphold and advance the integrity, honor, and dignity of the engineering profession by"

I. Using their knowledge and skill for the enhancement of human welfare;

II. Being honest and impartial, and serving with fidelity the public, their employers, and clients;

III. Striving to increase the competence and prestige of the engineering profession; and

IV. Supporting the professional and technical societies of their disciplines (Wright, 1994).

Although the NSPE canons remain silent on the issue, both the ABET and ASCE codes of ethics include ethical obligations for professional development. In other words, engineers are ethically obligated to continue their technical training and education throughout their career. Thus, even though some states do not require continuing education for relicensing as a professional engineer, continued technical training is part of your ethical obligation as an engineer.

16.4 EXAMPLES OF ENGINEERING ETHICS

Engineering ethics can become very complex. Sometimes, one ethical canon contradicts another. Two examples will be presented to illustrate the applications of the fundamental canons. These examples are taken from case studies developed by NSPE's Board of Ethical Review. Take a moment to consider your responses before reading the comments that follow the discussion questions. For examples of ethics that crop up in business practice, see the *Focus on Ethics: Workplace Ethics*.

16.4.1 Not Reporting Violations

Case: A civil engineer is hired to assess the structural integrity of a 60-year-old apartment building. The structure of the building is determined to be sound. However, during the inspection, mechanical and electrical problems are noted. The problems are severe enough that they may lead to safety concerns. The engineer tells the client of the problems, but does not include them in the report. The client reminds the engineer of his obligations regarding client confidentiality and then moves to sell the building quickly without repairs. The engineer decides not to report the violations to the authorities.

Discussion: This example represents a conflict between the responsibility to protect public safety (Canon 1) and client confidentiality (Canon 4). Does Canon 2 (not working outside your area of expertise) play in this case?

PONDER THIS

> **Which factor is more important? Could the engineer have avoided the conflict by including the problems in his original report?**

Board of Ethical Review comments: The Board concluded that the engineer had an ethical obligation to report the problems to the authorities because of the paramount importance of public safety (Canon 1). The engineer was correct not to include the problems in his report, since the problems were outside of the engineer's area of expertise (Canon 2).

16.4.2 Whistle-Blowing

Case: The city engineer/director of public works for a medium-sized city is the only licensed professional engineer in a position of responsibility within the city government. The city has several large food processing plants that discharge large amounts of waste into the sewage system during canning season. The engineer is responsible for the wastewater treatment plant. She reports to her supervisor about the inadequate capacity of the treatment plant to handle potential overflow during the rainy season and offers several possible solutions. The engineer also privately notifies other city officials about the plant problem, but her supervisor removes the responsibility for the wastewater treatment plant from her. She also is placed on probation and warned not to discuss the matter further or she will be fired.

Discussion: This example explores what engineers must do to meet their ethical obligations.

PONDER THIS

> **Has the engineer discharged her ethical responsibilities by notifying city officials of the potential problem?**

Board of Ethical Review comments: The Board concluded that removal of responsibility for the treatment plant did not terminate the engineer's ethical obligations, even when threatened with loss of employment. Again, public health and safety are of highest importance (Canon 1). The Board thought that the engineer should have reported the potential problem to higher authorities in the state or federal government.

16.5   SUMMARY

Engineers are professionals and must meet high ethical standards. Engineering ethics is required for two reasons. First, engineering directly influences public safety, health, and welfare. Second, engineering work is not easy to understand, so the trust between the public and engineers must not be diminished by unethical behavior.

Several codes of engineering ethics exist. Most are similar to the six NSPE Fundamental Canons of Ethics. The six canons require that engineers shall (1) hold paramount

FOCUS ON ETHICS: WORKPLACE ETHICS

Engineering ethics often is taught with the high-stakes ethics cases discussed in Problems 16.4 through 16.6. While interesting and instructive, the large cases do not provide you as a new engineer with tools you can use to address everyday ethics questions.

In this section, you will be presented with a series of ethics questions and possible solutions. You may wish to discuss the questions in a group, so that you can exchange ideas. One of the lessons is that ethics questions often do not have one correct answer. Responses to ethical dilemmas in the workplace can be ranked from better to worse. (Of course, some responses are just plain wrong.)

The questions below are from an ethics game called *Gray Matters*. The game was devised by George Sammet, Jr., to teach business ethics to the employees of Martin Marietta (now Lockheed Martin). Sammet was Vice President of International Ethics and Business Conduct for Lockheed Martin and now works for Grainer and Associates in Ottawa. The questions presented here were selected for their relevance to the engineering workplace and interest to entry-level engineers.

For each question below, a Web site is given to allow you to select from potential responses. The questions are quoted from the onlineethics.org Web site.

Enjoy the game and your discussion of the answers.

GrayMatters Case #8
Appropriating Office Supplies for Personal Use

"Two of your subordinates routinely provide their children with school supplies from the office. How do you handle this situation?"

Web site: http://onlineethics.org/corp/graymatters/case8.html

GrayMatters Case #64
Instructed to Distort the Truth?

"You are on a proposal-writing team. In the orientation briefing, the head of the team gives the following guidance: 'We really have to win this one. I want you to be really optimistic in what you write.' How do you interpret her advice?"

Web site: http://onlineethics.org/corp/graymatters/case64.html

GrayMatters Case #87
Supplier Offers You a Discount

"You are in Production Control. Planning on adding a porch onto your house, you visit a lumberyard to get ideas and a price. During the discussion, the sales manager says, 'Oh, you work for XYZ Company. They buy a lot from us, so I'm going to give you a special discount.' What do you do?"

Web site: http://onlineethics.org/corp/graymatters/case87.html

GrayMatters Case #24
HIV Positive Employee

"A female employee tells you, her manager, that a fellow employee is HIV positive. What do you do?"

Web site: http://onlineethics.org/corp/graymatters/case24.html

GrayMatters Case #72
He Calls All the Women "Sweetie"

"When a male supervisor talks to any female employee, he always addresses her as 'Sweetie.' You have overheard him use this term several times. As the supervisor's manager, should you do anything?"

Web site: http://onlineethics.org/corp/graymatters/case72.html

How did you do in your group? The maximum and minimum possible scores for the five scenarios above are +50 and −55, respectively.

the safety, health, and welfare of the public; (2) perform services only in the areas of their competence; (3) issue public statements only in an objective and truthful manner; (4) act for each employer or client as faithful agents; (5) avoid deceptive acts; and (6) conduct themselves honorably, responsibly, ethically, and lawfully to enhance the honor, reputation, and usefulness of the profession.

SUMMARY OF KEY IDEAS

- Engineers shall hold paramount the safety, health, and welfare of the public.
- Engineers shall perform services only in the areas of their competence.
- Engineers shall issue public statements only in an objective and truthful manner.
- Engineers shall act for each employer or client as faithful agents or trustees.
- Engineers shall avoid deceptive acts.
- Engineers shall conduct themselves honorably, responsibly, ethically, and lawfully so as to enhance the honor, reputation, and usefulness of the profession.

Problems

16.1. What are the six canons in the NSPE Code of Ethics?

16.2. Is there a hierarchy among the canons in the NSPE Code of Ethics?

16.3. Review the Professional Obligations section of the NSPE Code of Ethics at the end of the chapter. Discuss the ethical implications of engineering consulting firms making campaign contributions.

16.4. Discuss the ethical behavior of engineers in the Space Shuttle *Challenger* disaster. For background information, see the World Wide Web Ethics Center for Engineering and Science, hosted by Case Western Reserve University at http://www.onlineethics.org. This site contains a fine discussion of engineering ethics, including detailed examples of whistle-blowing in engineering.

16.5. Discuss the ethical behavior of engineers in the walkway collapse at the Kansas City Hyatt Regency Hotel. For background information, see Pfrang and Marshall (1982).

16.6. Discuss the ethical behavior of engineers with regard to the structural problems in New York City's Citicorp Towers. For background information, see the World Wide Web Ethics Center for Engineering and Science, hosted by Case Western Reserve University at http://www.onlineethics.org.

16.7. Pick a field of engineering and discuss how an ethical violation would adversely affect public health, safety, or welfare.

Problems 16.8 through 16.10 are taken from actual ethics cases that came before NSPE's Board of Ethical Review. For each scenario, discuss the applicable parts of the NSPE Code of Ethics and decide whether the behavior was ethical.

16.8. Due to potential dangers during construction, an engineer recommends to a client before a project begins that a full-time person should be hired for on-site monitoring of the project. The client rejects the request for on-site monitoring, stating that the monitor would increase project costs to an unreasonable level. The engineer starts work on the project anyway. Was it ethical for the engineer to begin the project knowing that the client would not agree to hire an on-site monitor?

16.9.   A company was advised by a state agency to get permission to discharge waste into a river. The company hires a consulting engineer to help them respond to the state's request. The engineer finds that the waste will degrade the quality of the river below established standards and that treating the waste will be expensive. The engineer tells the company of these findings orally, before the final report is written. The company then pays the engineer, terminates the engineering service agreement, and tells the engineer not to write a report. The engineer hears later that the company reported to the state agency that the waste discharge will not degrade the quality of the river. Does the engineer have an ethical obligation to report the findings to the state agency?

16.10. A state agency hires an engineer to conduct a feasibility study on a proposed highway spur. The spur will go through an area near where the engineer owns property. The engineer tells the state agency of the potential conflict of interest, but the agency does not object to the engineer working on the project. The engineer completes the study and the highway spur is built. Did the engineer act ethically by performing the study, even though the engineer's property may be affected?

NSPE CODE OF ETHICS FOR ENGINEERS
Preamble

Engineering is an important and learned profession. As members of this profession, engineers are expected to exhibit the highest standards of honesty and integrity. Engineering has a direct and vital impact on the quality of life for all people. Accordingly, the services provided by engineers require honesty, impartiality, fairness, and equity, and must be dedicated to the protection of the public health, safety, and welfare. Engineers must perform under a standard of professional behavior that requires adherence to the highest principles of ethical conduct.

I. Fundamental Canons

Engineers, in the fulfillment of their professional duties, shall:

1. Hold paramount the safety, health and welfare of the public.

2. Perform services only in areas of their competence.

3. Issue public statements only in an objective and truthful manner.

4. Act for each employer or client as faithful agents or trustees.

5. Avoid deceptive acts.

6. Conduct themselves honorably, responsibly, ethically, and lawfully so as to enhance the honor, reputation, and usefulness of the profession.

II. Rules of Practice

1. Engineers shall hold paramount the safety, health and welfare of the public.

    a. If engineers' judgment is overruled under circumstances that endanger life or property, they shall notify their employer or client and such other authority as may be appropriate.

    b. Engineers shall approve only those engineering documents that are in conformity with applicable standards.

c. Engineers shall not reveal facts, data, or information without the prior consent of the client or employer except as authorized or required by law or this Code.

d. Engineers shall not permit the use of their name or associate in business ventures with any person or firm that they believe are engaged in fraudulent or dishonest enterprise.

e. Engineers shall not aid or abet the unlawful practice of engineering by a person or firm.

f. Engineers having knowledge of any alleged violation of this Code shall report thereon to appropriate professional bodies and, when relevant, also to public authorities, and cooperate with the proper authorities in furnishing such information or assistance as may be required.

2. Engineers shall perform services only in the areas of their competence.

   a. Engineers shall undertake assignments only when qualified by education or experience in the specific technical fields involved.

   b. Engineers shall not affix their signatures to any plans or documents dealing with subject matter in which they lack competence, nor to any plan or document not prepared under their direction and control.

   c. Engineers may accept assignments and assume responsibility for coordination of an entire project and sign and seal the engineering documents for the entire project, provided that each technical segment is signed and sealed only by the qualified engineers who prepared the segment.

3. Engineers shall issue public statements only in an objective and truthful manner.

   a. Engineers shall be objective and truthful in professional reports, statements, or testimony. They shall include all relevant and pertinent information in such reports, statements, or testimony, which should bear the date indicating when it was current.

   b. Engineers may express publicly technical opinions that are founded upon knowledge of the facts and competence in the subject matter.

   c. Engineers shall issue no statements, criticisms, or arguments on technical matters that are inspired or paid for by interested parties, unless they have prefaced their comments by explicitly identifying the interested parties on whose behalf they are speaking, and by revealing the existence of any interest the engineers may have in the matters.

4. Engineers shall act for each employer or client as faithful agents or trustees.

   a. Engineers shall disclose all known or potential conflicts of interest that could influence or appear to influence their judgment or the quality of their services.

   b. Engineers shall not accept compensation, financial or otherwise, from more than one party for services on the same project, or for services pertaining to the same project, unless the circumstances are fully disclosed and agreed to by all interested parties.

c. Engineers shall not solicit or accept financial or other valuable consideration, directly or indirectly, from outside agents in connection with the work for which they are responsible.

d. Engineers in public service as members, advisors, or employees of a governmental or quasi-governmental body or department shall not participate in decisions with respect to services solicited or provided by them or their organizations in private or public engineering practice.

e. Engineers shall not solicit or accept a contract from a governmental body on which a principal or officer of their organization serves as a member.

5. Engineers shall avoid deceptive acts.

a. Engineers shall not falsify their qualifications or permit misrepresentation of their or their associates' qualifications. They shall not misrepresent or exaggerate their responsibility in or for the subject matter of prior assignments. Brochures or other presentations incident to the solicitation of employment shall not misrepresent pertinent facts concerning employers, employees, associates, joint venturers, or past accomplishments.

b. Engineers shall not offer, give, solicit or receive, either directly or indirectly, any contribution to influence the award of a contract by public authority, or which may be reasonably construed by the public as having the effect of intent to influencing the awarding of a contract. They shall not offer any gift or other valuable consideration in order to secure work. They shall not pay a commission, percentage, or brokerage fee in order to secure work, except to a bona fide employee or bona fide established commercial or marketing agencies retained by them.

III. Professional Obligations

1. Engineers shall be guided in all their relations by the highest standards of honesty and integrity.

a. Engineers shall acknowledge their errors and shall not distort or alter the facts.

b. Engineers shall advise their clients or employers when they believe a project will not be successful.

c. Engineers shall not accept outside employment to the detriment of their regular work or interest. Before accepting any outside engineering employment they will notify their employers.

d. Engineers shall not attempt to attract an engineer from another employer by false or misleading pretenses.

e. Engineers shall not promote their own interest at the expense of the dignity and integrity of the profession.

2. Engineers shall at all times strive to serve the public interest.

a. Engineers shall seek opportunities to participate in civic affairs; career guidance for youths; and work for the advancement of the safety, health, and well-being of their community.

b. Engineers shall not complete, sign, or seal plans and/or specifications that are not in conformity with applicable engineering standards. If the client or employer insists on such unprofessional conduct, they shall notify the proper authorities and withdraw from further service on the project.

c. Engineers shall endeavor to extend public knowledge and appreciation of engineering and its achievements.

3. Engineers shall avoid all conduct or practice that deceives the public.

a. Engineers shall avoid the use of statements containing a material misrepresentation of fact or omitting a material fact.

b. Consistent with the foregoing, engineers may advertise for recruitment of personnel.

c. Consistent with the foregoing, engineers may prepare articles for the lay or technical press, but such articles shall not imply credit to the author for work performed by others.

4. Engineers shall not disclose, without consent, confidential information concerning the business affairs or technical processes of any present or former client or employer, or public body on which they serve.

a. Engineers shall not, without the consent of all interested parties, promote or arrange for new employment or practice in connection with a specific project for which the engineer has gained particular and specialized knowledge.

b. Engineers shall not, without the consent of all interested parties, participate in or represent an adversary interest in connection with a specific project or proceeding in which the engineer has gained particular specialized knowledge on behalf of a former client or employer.

5. Engineers shall not be influenced in their professional duties by conflicting interests.

a. Engineers shall not accept financial or other considerations, including free engineering designs, from material or equipment suppliers for specifying their product.

b. Engineers shall not accept commissions or allowances, directly or indirectly, from contractors or other parties dealing with clients or employers of the engineer in connection with work for which the engineer is responsible.

6. Engineers shall not attempt to obtain employment or advancement or professional engagements by untruthfully criticizing other engineers, or by other improper or questionable methods.

a. Engineers shall not request, propose, or accept a commission on a contingent basis under circumstances in which their judgment may be compromised.

b. Engineers in salaried positions shall accept part-time engineering work only to the extent consistent with policies of the employer and in accordance with ethical considerations.

c. Engineers shall not, without consent, use equipment, supplies, laboratory, or office facilities of an employer to carry on outside private practice.

7. Engineers shall not attempt to injure, maliciously or falsely, directly or indirectly, the professional reputation, prospects, practice, or employment of other engineers. Engineers who believe others are guilty of unethical or illegal practice shall present such information to the proper authority for action.

   a. Engineers in private practice shall not review the work of another engineer for the same client, except with the knowledge of such engineer, or unless the connection of such engineer with the work has been terminated.

   b. Engineers in governmental, industrial, or educational employ are entitled to review and evaluate the work of other engineers when so required by their employment duties.

   c. Engineers in sales or industrial employ are entitled to make engineering comparisons of represented products with products of other suppliers.

8. Engineers shall accept personal responsibility for their professional activities, provided, however, that engineers may seek indemnification for services arising out of their practice for other than gross negligence, where the engineer's interests cannot otherwise be protected.

   a. Engineers shall conform with state registration laws in the practice of engineering.

   b. Engineers shall not use association with a nonengineer, a corporation, or partnership as a "cloak" for unethical acts.

9. Engineers shall give credit for engineering work to those to whom credit is due, and will recognize the proprietary interests of others.

   a. Engineers shall, whenever possible, name the person or persons who may be individually responsible for designs, inventions, writings, or other accomplishments.

   b. Engineers using designs supplied by a client recognize that the designs remain the property of the client and may not be duplicated by the engineer for others without express permission.

   c. Engineers, before undertaking work for others in connection with which the engineer may make improvements, plans, designs, inventions, or other records that may justify copyrights or patents, should enter into a positive agreement regarding ownership.

   d. Engineers' designs, data, records, and notes referring exclusively to an employer's work are the employer's property. The employer should indemnify the engineer for use of the information for any purpose other than the original purpose.

   e. Engineers shall continue their professional development throughout their careers and should keep current in their specialty fields by engaging in professional practice, participating in continuing education courses, reading in the technical literature, and attending professional meetings and seminars.

   —As revised January 2003

Part VI
Case Studies in Engineering

Case studies are stories.
Clyde Herreid

What should engineering educators do to graduate enthusiastic students who have a good foundation in the fundamental engineering sciences and the ability to think and communicate?
J. Henderson, L. Bellama, and B. Furman

17

Introduction to the Engineering Case Studies

17.1 INTRODUCTION

In Part VI of this text, you will have the opportunity to practice and integrate the material you have learned in this course through what is known as the *case study*. Case studies are stories based on historical facts or present-day realities. In the case studies presented here, young engineers are faced with questions stemming from real-world challenges. The case studies will allow you to bring fundamental engineering principles to bear on interesting and realistic problems.

17.2 CASE STUDIES IN THIS TEXT

17.2.1 Introduction

Seven case studies are presented in this text. Descriptions of the case studies may be found in Table 17.1.

Each case study has the same format. An introduction is provided in the first section of each chapter to give you the general idea behind the case study.

The second section of the chapter is devoted to a story about young engineers involved in the case study. The young engineers are fictitious, but the facts behind the case study are valid.

The third section of each chapter explains the science behind the engineering analysis. In some cases, design parameters are developed to allow the evaluation of design alternatives. Also in the third section, case study activities are discussed. Data collection is an important part of most of the case studies. In addition, some ideas about report format are given in the third section of the chapter. Please follow the report format given to you by your instructor. Examples of report formats are provided at the end of each chapter.

OBJECTIVES

After reading this chapter, you will be able to:

- identify the case studies in this text;
- understand the value of case studies in engineering education.

Key idea: Case studies are stories based on historical facts or present-day realities.

Key idea: Case studies in this text include an introduction, a story, background science, case study activities, a reporting format, references, and acknowledgments.

TABLE 17.1   Case Study Characteristics

Name	Description	Case Study Type
Millennium bridge	Analysis and repair of a swaying footbridge in London	Historical/current problem
Controllability	Comparison of computer and pencil interfaces	Current problem
Dissolution	Factors controlling the rate of dissolution	Current problem
Computer workstation	Analysis and design of a computer desk/chair	Current problem
Power transmission	Evaluation of power transmission alternatives	Historical
Walkway collapse	Analysis of the catastrophic collapse of a hotel walkway	Historical
Trebuchet	Analyze, design, build, test, and optimize a medieval siege weapon	Historical/current problem

The final section of each chapter provides sources of information about the case study. In addition, the input of other engineering professionals is acknowledged.

17.2.2 Using the Case Studies

Each case study focuses on specific elements from the text. The main elements in the case studies are listed in Table 17.2. Elements include analysis, design, data collection, engineering calculations, and technical communications. Note that four of the case studies include hands-on data collection activities. The focus and target disciplines of the case studies are presented in Table 17.3. By judicious selection of the case studies, most of the elements in the text and all of the main engineering disciplines can be covered.

The case studies are designed to be explored in groups of about four students. In a typical use, project teams will turn in small group-generated reports. In some case studies, data are shared between groups to create larger data sets.

The case studies can be used as a way to *integrate* the material from the text or *present* the material from the text. If used to present material, then the readings listed in Table 17.4 should be read prior to beginning each case study. If the material has been assigned in a previous case study but is listed in Table 17.4 with a new case study, please review the reading prior to beginning the new case study. Each case study contains *text links* with specific areas to review as the case study is read.

TABLE 17.2   Case Study Elements

Case Study	Analysis	Design	Data Collection	Engineering Calculations	Technical Communications
Millennium bridge	• •	• •		• •	• • •
Controllability	• •	•	• • •	• •	• • •
Dissolution	• • •		• • •	• • •	• • •
Computer workstation	• • •	• •	• •	• •	• • •
Power transmission	• • •	• • •		• • •	• • •
Walkway collapse	• • •	• •		• •	• • •
Trebuchet	• • •	• • •	• • •	• •	• • •

Note: If a case does not have any bullets, then there is little relationship with the listed element; by contrast, three bullets indicate a significant relationship with the listed element.

TABLE 17.3 Case Study Foci and Disciplines

Case Study	Focus	Discipline(s)
Millennium bridge	Group dynamics	General, civil
Controllability	Use of models	Industrial, general
Dissolution	Use of models	Chemical
Computer workstation	Optimization	Industrial, general
Power transmission	Alternatives	Electrical, mechanical
Walkway collapse	Ethics	Civil
Trebuchet	Models, data collection	Mechanical

TABLE 17.4 Case Study Readings

Case Study	Readings
Millennium bridge	Chapters 1–4, 6, 13, and 14
Controllability	Section 4.6; Chapters 5, 6, 8, 9, 13, and 14
Dissolution	Section 4.3; Chapters 5, 6, 8, 9, 13, and 14
Computer workstation	Section 4.6; Chapters 5–9, 13, and 14
Power transmission	Sections 4.5, 4.7; Chapters 5, 6, 8, 13, and 14
Walkway collapse	Section 4.4; Chapters 13–16
Trebuchet	Section 4.7; Chapters 6–9, 13, and 14

17.3 SUMMARY

Case studies are useful tools for learning about engineering and applying engineering fundamentals to interesting problems. Each case study in this text includes an introduction, a story, background science, case study activities, a reporting format, references, and acknowledgments. The case studies in this text are designed to be explored in groups of about four students. They have been developed to encompass the major engineering disciplines and the themes in the text.

SUMMARY OF KEY IDEAS

- Case studies are stories based on historical facts or present-day realities.
- Case studies in this text include an introduction, a story, background science, case study activities, a reporting format, references, and acknowledgments.

18

Millennium Bridge Case Study

18.1 INTRODUCTION

Engineers frequently work in groups. For group work to be effective, you must master several skills, including sharing information with others; setting up and running meetings; and making group decisions. This case study will allow you to practice each of these skills in an interesting case concerning a famous pedestrian bridge over the Thames River in London.

This case study is based on the true story of the Millennium Bridge. In the actual event, engineers were faced with unexpected problems after the bridge was opened to the public. They had to identify and correct the problem under withering public scrutiny. The story of the Millennium Bridge is a lesson in the values of perseverance, exacting analysis, and creative design.

18.2 THE STORY

Millicent was excited. June 2000—what a month! Just last week, she received her acceptance letter from the university. "The first one in the family to go to university," crooned her mother. "And in engineering, too!" Now here she stood with her best friends from school, awaiting the arrival of the Queen.

"I can't believe the Queen is going to be here today," exclaimed Millie's friend Jeremy.

"And why not?" retorted Angela. "After all, this is the first new bridge over the Thames in London in over 100 years." Not just any bridge, thought Millie. She recalled the words of Ms. Evans, their technology teacher from last term, who had told them the story of the Millennium Bridge. Ms. Evans said that the designers had sought to build a 325-meter "blade of light" from near St. Paul's Cathedral in the City to the Tate Modern Gallery at Bankside.

OBJECTIVES

After reading this chapter, you will be able to:

- work more effectively in groups;
- explain the challenges facing the designers of the Millennium Bridge.

Screen capture from a video shot on the opening day of the Millennium Bridge. Note the high density of pedestrians. (Photo courtesy of Arup.)

Millie sighed, "A blade of light."

"I was thinking the same thing," said Jeremy. "They got it right, didn't they?"

Angela looked around nervously at the throng of 80,000 gathered for the opening of the bridge. "I hope the engineers knew what they were doing," she said suspiciously. Angela's tastes ran to architecture rather than structural engineering. "This 'blade' doesn't look like it would support our school rugby team, let alone this mob."

"Oh, Angie. It's perfectly safe." Millie squeezed her hand.

"I'm suppose you're right," agreed Angela reluctantly. "But it doesn't look like a real bridge. On holiday last summer in San Francisco, we drove over the Golden Gate Bridge. Now that looks like a bridge—it has big fat cables hanging around all over the place . . . "

Millie broke in, "This bridge works the same way. But just a bit differently." Angela sighed and rolled her eyes, knowing that Millie was warming up for one of her long-winded speeches again. Millie ignored her. "The Millennium Bridge is a suspension bridge, just like the Golden Gate. The deck of the Golden Gate Bridge doesn't fall down because the big fat cables you saw are in tension, pulling it back up." Angela's quizzical look made Millie back up. "Um, like a rubber band, Angie. Suspend it loosely on your finger and it can't support much weight. But stretch it across two fingers and it could almost hold up, well, almost all of Jeremy's piercings."

Jeremy lunged at Millie playfully. She was always teasing him about his jewelry and tattoos. "Millie's right, even if she is a pain sometimes," said Jeremy, picking up the thread. "In the Millennium Bridge, the cables are located on the *sides* of the deck. They are still under tension and still hold up the deck. But they only sag a little bit." He consulted a brochure he had picked up earlier in the day. "It says here that they sag only 2.3 meters over the 144 meters of the central span."

"Central Spam?" echoed Angela, incredulously.

"No, silly, central span. The distance between the two piers."

Angela squinted at the bridge. "Oh, I see. Those cables do sag down a bit, like a miniature version of the Golden Gate."

"Exactly," said Jeremy triumphantly.

"But why do it that way?" said Angela. "I thought you engineers liked big heavy things, with lots of gears and pulleys." A mischievous smile crept across Angela's face.

"It *could* have been built like that," said Millie patiently. "But the design team didn't want to spoil the view. Engineers need to think about a lot of things besides gears and pulleys, Angie. This magnificent view, for example."

"And money," chimed in Jeremy.

"Yes," said Millie. "And money. But it looks worth all its £18 million, doesn't it? Look—there's the Queen!" Millie, Angela, and Jeremy joined the crowd in greeting the Queen enthusiastically.

After numerous speeches and much polite applause, the crowd began to move towards the bridge. "We're on!" said Angela. "This is terrific! I thought that this thin little bridge would be bouncy, but it feels solid."

"Yes, strange, isn't it? Even with all these people, there is very little up-and-down motion. I bet if we got everyone to march in step, we could really get this bridge hopping!"

Angela shuddered. "Stop it, Millie—I don't even want to think about that." The three friends had reached about halfway across the span. The bridge was loaded with people.

"Whoa," said Jeremy. "I'm being to feel like a drunken sailor."

"You *look* like a drunken sailor," teased Millie. "But I know what you mean, Jeremy. We're not going up and down, but it feels like we're going side to side." Millie put out a hand to steady herself.

"It's getting worse with every step," blurted Angela. "I really have to brace myself with my feet to keep walking."

Note that the suspension cables of the Millennium Bridge are on the sides of the bridge. The suspension cables sag only slightly. (Photo courtesy of Arup.)

Text link: See the *Focus on Design* following Section 7.8.7 for the names of parts of a suspension bridge.

"Don't stop, everyone. Let's hurry and get off this mad thing!"

That night, Millie read a review of the bridge opening in her favorite on-line newspaper. Pictures of the gorgeous span were accompanied by a description of the swaying that she, Angela, and Jeremy had felt. Two days later, the Millennium Bridge was closed to pedestrian traffic.

Although excited about starting her engineering studies in the next term, Millie couldn't get the bridge out of her thoughts. What had gone wrong? Why did the bridge move from side to side, but not up and down? Why didn't the engineers anticipate that awful swaying? What could they—no, what could *she*—do about it?

18.3   THE CASE STUDY

18.3.1   Introduction

Text link: The failure mode of the Millennium Bridge was different than the Tacoma Narrows Bridge (see the *Focus on Design* following Section 7.8.7).

In this case study, you will share information to perform a simplified analysis of the Millennium Bridge. This section will provide a brief summary of the forces acting on the bridge.

Many objects have *natural frequencies*. The natural frequency is the frequency at which the object naturally vibrates when excited. For example, if you blow across an open bottle, you produce a sound at the natural frequency of the bottle.

When systems are excited at their natural frequency, the response can be amplified and problems can occur. This is called *resonance*. Think back to your carefree days as a child.

PONDER THIS

> **How did you push your friends on a swing?**

To increase the response (i.e., the amplitude of their swing), you would give them *small* pushes at the natural frequency of their swing. If they swung out and back in 2.5 seconds, then their frequency was (1 cycle)/(2.5 seconds) = 0.4 cycles per second = 0.4 hertz = 0.4 Hz.

THOUGHTFUL PAUSE

> **What would happen if you pushed your friend harder, but at a different frequency (say 1 Hz)?**

Key idea: Small excitations at the natural frequency may give larger responses than large excitations at a different frequency.

If you pushed harder, but at a different frequency, your friend would not swing as high. Some of your pushes would occur when they were swinging back towards you, so your pushes would cancel out their swing. As another example, your car may vibrate only at a certain speed—the resonance speed where the natural frequency is excited.

Buildings and bridges behave in the same way. They sway or move in response to applied forces. If the applied force (say, from an earthquake or from footfall) has a frequency near the natural frequency of the structure, then resonance can occur and the amplitude of motion is increased dramatically.

THOUGHTFUL PAUSE

> **What is the frequency of your walking pace?**

A typical walking pace is about one mile in 20 minutes with a six-foot stride. (Here, stride means the time from the right foot touching the ground to the next time the right foot touches the ground.) The frequency is

$$\frac{(5{,}280 \text{ ft})}{(20 \text{ minutes})\left(6\dfrac{\text{ft}}{\text{stride}}\right)\left(60\dfrac{\text{sec}}{\text{min}}\right)} = 0.7 \text{ Hz.}$$

In tests done on a mock-up version of the Millennium Bridge, the natural frequency of the bridge was found to be about 0.5 to 1.1 Hz. Thus, the natural frequency of the bridge was similar to the walking pace of the pedestrians.

PONDER THIS

If you were walking on a swaying bridge (like a rope bridge), how would you adjust your footfall?

Key idea: One way to minimize the response of a structure is to shift its natural frequency to a higher frequency.

As the bridge began to sway, the pedestrians put more pressure on the outsides of their feet. Recall that pressure is force divided by area. Thus, an additional force was imposed in the sideways (lateral) direction. As a result, the bridge was displaced laterally (i.e., moved sideways) about ±75 mm (about ±3 inches).

How do engineers prevent structures from moving in response to a periodic (i.e., cyclic) input? One approach is to shift the natural frequency of the structure to a higher frequency. This can be accomplished by increasing the stiffness of the structure through the addition of braces.

18.3.2 Case Study

This case study consists of two parts. The parts of the case study are summarized in Table 18.1.

Part I. In the first part of the case study, you must meet as a group to accomplish three tasks. The first task is to answer questions 1 and 2 in Section 18.4. You will need to develop and use your analysis and brainstorming skills to answer the questions. In addition, you must share information.

To model the sharing of information, *photocopy the data slides following this chapter before you meet with your group.* Cut the facts along the solid lines, shuffle them, and deal them like playing cards to the members of your group. Then pool your information to answer study questions 1 and 2. You will have to write up your answers (see Section 18.3.3), so take notes on your discussions.

In answering the first two study questions, you will come to realize that the engineers who worked on the Millennium Bridge chose *not* to solve the swaying problem by stiffening the bridge. Instead, they used a number of dampeners to absorb the energy from the pedestrian traffic. The engineers used a combination of *viscous dampeners*

The Capilano Suspension Bridge is 450 feet long and 230 feet above the Capilano River in Vancouver, British Columbia. Imagine how your footfall would change as you crossed this bridge. (Photo courtesy of Capilano Suspension Bridge.)

TABLE 18.1 Scope of Work for the Millennium Bridge Case Study

Part	Task	Description
I	1	Distribute facts, answer questions
	2	Investigate dampeners on the Internet
	3	Assign tasks, set outside meeting time and location
II	1	Discuss work brought to the meeting
	2	Make assignments necessary to complete the report
	3	Complete, proofread, and submit the report

The viscous dampers shown on the left were attached to new steel plates located under the deck. Energy from lateral motion in the deck will be absorbed as the "shock absorber" compresses and extends.

(large shock absorbers) to absorb energy in the lateral direction and *tuned mass dampeners* (very short, heavy pendulums) to absorb energy in the vertical direction.

Your second task is to look up information about viscous and tuned mass dampeners on the Internet. Be prepared to explain the concepts of viscous and tuned mass dampeners to other engineers.

The third and final task of the first part of this case study is to divide the remaining work among the members of your group and set a mutually agreed upon time and location for another meeting. You will have to report your findings (see Section 18.3.3). Assign the writing tasks to members of the group. The writing tasks include documenting the answers to the study questions and the summary of the use of dampeners. Set a time to meet. You may wish to exchange contact information (email addresses, etc.) to allow for contact outside of class.

Part II. The second part of the case study is to meet at the time you set previously. Each group member should bring the completed work assigned to them. Appoint one member of the group as a recorder to take notes on the meeting. As a first task, discuss each person's work in a cooperative atmosphere. Work together to refine the answers to the questions and to sharpen the summary of the use of dampeners.

In the second task, make assignments to complete the report. Decide whether you need to meet again to complete the report or whether the report compilation can be accomplished by email. As a third task, complete, proofread, and submit the report.

18.3.3   Reporting

Use the report format given to you by your instructor. (A default grading scheme is shown at the end of the chapter.) At a minimum, provide answers (including supporting arguments) to the study questions. Also provide a one-half-to one-page summary of the use of dampeners in structures. Append the meeting notes from your outside meeting.

18.4   STUDY QUESTIONS

1. Would you expect the lateral acceleration of the bridge to exceed people's comfort level?
2. Why did the engineers reject the option of solving the swaying problem by increasing the stiffness of the bridge? Justify your answer with appropriate calculations.

18.5   ACKNOWLEDGMENTS AND FURTHER READING

The value of the Millennium Bridge as a potential case study was brought to the author's attention by Prof. Michael E. Ryan of the University at Buffalo. For a summary of the technical issues involved, see Dallard et al. (2001). The interested reader also is directed to the excellent discussion by the engineers in charge (Arup) at their Web site (www.arup.com). An on-line presentation by the design firm for the dampeners (Taylor Devices, Inc.) also is valuable (http://civil.eng.buffalo.edu/webcast/—look for the Douglas P. Taylor presentation dated April 4, 2003).

SUMMARY OF KEY IDEAS

- Small excitations at the natural frequency may give larger responses than large excitations at a different frequency.
- One way to minimize the response of a structure is to shift its natural frequency to a higher frequency.

DATA SLIDES FOR CASE STUDY

Data Slide #1

force = (mass)(acceleration)

or

lateral force from people = (mass of loaded bridge)
$\times$ (lateral acceleration)

Data Slide #2

The lateral velocity was measured in experiments with a mock-up bridge.

Data Slide #3

To avoid resonance, the natural frequency of the bridge must be increased to about 2.0 Hz from about 0.5 Hz.

Data Slide #4

People feel uncomfortable if the lateral acceleration exceeds about 0.02 times the gravitational acceleration.

Data Slide #5

Increasing the stiffness of the bridge by 9–10 times would be very expensive and would destroy the appearance of the bridge.

Data Slide #6

The mass of the loaded bridge was about 840,000 kg.

Data Slide #7

critical damping ratio = 2(mass)(natural frequency)

Data Slide #8

Gravitational acceleration is 9.8 m/s^2.

Data Slide #9

Each of the 2,000 people on the bridge at one time exerted a lateral force of about 1000 N as they walked across the center span of the bridge.

1 N = 1 newton = 1 kg-m/s^2

Data Slide #10

critical damping ratio = $2\sqrt{(\text{mass})(\text{stiffness})}$

Default Grading Scheme: Millennium Bridge Case Study

Topic	Possible Points
Cover Page	
Names, date, title, group number (if appropriate)	5
Background	
Quality of introduction to the problem	10
Quality of review of viscous and tuned mass dampeners	15
Methods	
Brief description of how the group was organized, how the group met, and how the data were shared	15
Results and Discussion	
Lateral acceleration—support of conclusions with calculations	15
Why stiffness was not increased—support of conclusions with calculations	15
Conclusion	
Quality of summary	10
Technical Writing	
Organization (and use of headings), grammar, spelling, tables and figures, and seamlessness of report	15
Total	100

19

Controllability Case Study

19.1 INTRODUCTION

Engineers frequently design interfaces between humans and machines. Examples include computer input devices, automobiles, airplane cockpits, and industrial machinery. A good interface minimizes errors. For example, the design of a personal digital assistant (PDA) keyboard must take into account fingertip size, so that only one button is pressed at a time. Good interfaces also minimize stress on humans. An assembly line requiring heavy boxes to be lifted should be reworked to reduce lifting requirements.

In this case study, you will investigate the ability of a population to control interfaces between humans and two drawing tools: a pencil and a computer mouse. You will find that controllability can be *quantified*, allowing for the selection between the two alternative interfaces.

19.2 THE STORY

Miguel slumped in his seat and groaned. "This is so frustrating!"

"What's wrong, Miguel?" asked Jennifer. The two mechanical engineering students were working at adjacent workstations in the computing lab.

"It's this stupid 104 homework. I'll never get it done."

"104? Oh, yeah—Computer-Aided Design. I took CAD last semester. What seems to be the problem?"

"It's this darn mouse. I can draw these section views freehand no problem." Miguel showed Jennifer his beautiful hand-drawn sketches of the views required in the homework. "But when I try to draw it with the mouse in the CAD program, I can't get the accuracy I need."

"Did you try cleaning the mouse?"

OBJECTIVES

After reading this chapter, you will be able to:

- collect and evaluate data on the use of human–machine interfaces;
- use a model to compare the controllability of two user interfaces;
- quantify the reproducibility of two user interfaces.

Text link: CAD software is discussed in Section 10.4.5.

The Apple iPod has been praised for its user interface.

"Yeah—I took it apart and cleaned the rollers and that little ball inside. It helped a little, but I'm still having problems. When I try to nudge a line just a little bit, I end up sending it too far to the right. See?" Miguel rubbed his eyes. "I've been at this for two hours and it's driving me crazy. Honestly, sometimes I wonder why we need to learn this stuff anyway. What happened to engineers being able to sketch things by hand?"

Jennifer smiled in sympathy. "But Miguel, you know in your heart that CAD is a great tool, right? I mean, with electronic drawings, we can change them easily and import them into other software—even post them to the Web."

"Yeah, I know. But drawing by hand just seems so much more natural to me. I've been sketching since I was 5 years old."

"Not everyone has that talent, Miguel."

"I suppose. But look at this mess on the screen. I know can draw it by hand in a few seconds." Miguel sighed heavily. "I've been working on this one problem on the computer for a half an hour and it still looks like garbage."

"Okay, Miguel. Maybe a little friendly competition will cheer you up. How about we have a drawing contest? Me on CAD and you with a pencil."

"You're on," said Miguel, shaking Jennifer's hand. "Hey Jason, Wei Lin—you guys wanna be judges?" The two juniors walked over to where Miguel and Jennifer were sitting. "Sure," said Wei Lin. "We accept bribes, ya know," chimed in Jason. The four friends laughed.

"Okay, okay. But this is serious." Jennifer adopted a mock somber tone, trying not to smile. "You guys pick what Miguel and I will draw for the Championship Drawing Contest of the World. Then judge the results."

Miguel nodded in agreement. "Yeah—and Jason, don't forget that breakfast I bought for you the other day."

"Hey," cried Jennifer. "No influencing the judges!"

Jason and Wei Lin sat in a corner of the computer lab, heads together, trying to think up a good contest. After a few minutes, they returned to Miguel and Jennifer. "Okay guys," said Wei Lin. "This will be a three-part contest. You can start after I announce the first object to draw. Ready?" The two contestants nodded, intensely gripping their mouse or pencil. "On your mark—get set—draw a rectangular parallelepiped."

"A what?" asked Jennifer and Miguel in unison.

"A box! Now go guys!" Jason started the stopwatch feature on his wristwatch and the two intrepid artists were off. Five seconds later, Jennifer let out a triumphant yell. On her screen was a perfect rectangular box. "No fair," complained Miguel. "You just popped that box in from the preformatted shapes in the software." Jennifer smiled and leaned back in her chair. "Ah, the power of CAD!"

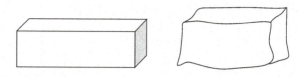

Jennifer's and Miguel's parallelepipeds

Wei Lin cleared his throat. "Time for test number two. On your mark—get set—draw Professor Wiggins in profile." Jennifer and Miguel grinned and started to draw. Professor Wiggins had a very distinctive face. Two minutes went by before Miguel shouted, "Done!" He had captured the professor's profile remarkably well: bulbous nose, high forehead, and flowing head of long hair. Jennifer's screen showed a mess of squashed ovals and nebulous face-like features.

"That was too hard," complained Jennifer. She flung her hand at the computer. "No one can draw a face nicely with this thing." Miguel laced his hands behind his head and grinned. "My point exactly, Jennifer."

Jason interrupted Miguel's gloating. "Wei Lin and I thought that might happen—that CAD would win the box drawing and freehand would win the more artistic challenge."

"Yeah," said Wei Lin. "And now for the tie-breaker. On your mark—get set—draw a bolt with a nut on it." The two mechanical engineering students determinedly turned to their work. Miguel finished first, with Jennifer not too far behind. The two judges examined their work. "These are hard to compare," said Jason. "Miguel finished first, but the perspective on his hand drawing is a little off." Jennifer nudged Miguel in the ribs. "Yes," continued Wei Lin, picking up the thought. "But although Jennifer's drawing is nearly perfect, she was a little slower." Miguel stuck out his tongue in a mock challenge. "So I guess the contest ends in a tie."

Later that night, the four students met in the cafeteria for dinner. Miguel gestured to Wei Lin and Jason. "Thanks for your help with the Championship Drawing Contest of the World, guys."

"No problem," replied Jason. "I'm just sorry that we couldn't break the tie."

"Yeah," said Jennifer. "Now we'll never know which is mightier: the pen or the mouse."

"I suppose," said Miguel thoughtfully. "But we should be able to come up with *some* way to combine the ideas of speed and accuracy to compare the interfaces."

"Watch out, everybody," said Wei Lin playfully. "Miguel's got that look on his face."

19.3　THE CASE STUDY

19.3.1　Introduction

In this case study, you will develop and use a model to compare two drawing interfaces. To compare the interfaces, you need a *test track*. A common test track in comparison studies is a constant-width path, such as a road of constant breadth. Another example of a constant-width path is the roadside sobriety test of "walking a straight line." If you are allowed to move along the path at your own pace along a constant-width path, then the test is called a *self-paced path control task*.

Suppose you asked a friend to ride her bicycle in as straight a line as possible across an unmarked parking lot. It is likely that her path would deviate from a straight line. The deviation from a straight line would increase the farther she rides. If the deviation is called the uncertainty U (in, say, centimeters), then U increases as the distance traveled (D) increases.

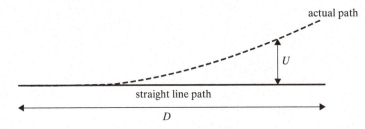

Uncertainty increases as the distance traveled increases.

One model of path control is

$$U = kD \qquad (19.1)$$

Equation (19.1) represents a linear increase in uncertainty with distance.

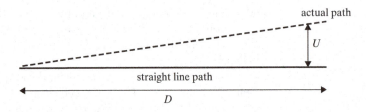

One model of path control: uncertainty increases linearly with the distance traveled.

Note that k is dimensionless if U and D have the same units. From Eq. (19.1),

$$D = U/k \qquad (19.2)$$

Text link: Perform the same operation on each side of the equation (Section 6.5.2).

You know that the average speed S is equal to D divided by the time required, T, or $S = D/T$. Dividing both sides of Eq. (19.2) by T yields

$$\frac{D}{T} = S = \frac{U}{kT}$$

For a path of constant width W, it makes sense to set the maximum uncertainty equal to W. Thus,

$$S_{max} = \frac{W}{kT}$$

Text link: Use units analysis (Section 6.7).

Here, S_{max} is the maximum speed allowable to perform the task with a fixed accuracy rate. The term $\frac{1}{kT}$ is called the controllability C. Note that controllability has units of 1/time (since k is dimensionless). You can write

$$S_{max} = CW \qquad (19.3)$$

The model can be restated in words as follows:

> If the uncertainty increases proportionally with distance, then the maximum speed for a fixed accuracy should be proportional to the allowable width. The proportionality constant is the controllability, with units of 1/time.

Reread the preceding boxed paragraph until it makes sense to you.

19.3.2   Case Study

Equation (19.3) is a model of system performance. To test the model, you will perform experiments on the control of a pencil and computer mouse in tracks of fixed width. Four tracks (i.e., four widths) will be used to test the model. The tracks are the space between two concentric circles, indicated by the shaded space in Figure 19.1.

Each test will consist of five laps around the track (i.e., five successful loops inside the annular space). The model assumes a constant level of accuracy for each test. This means you must specify the number of times that you are allowed to go out of the path. For this case study, allow *no* deviations outside the circles. If you go outside the track, stop the test and start over. *Be sure to record the number of attempts*. In this case study, the diameter of the outer circle is fixed at 80 mm. The path widths are 4, 7, 11, and 15 mm. Thus, the diameters of the inner circles are 72, 66, 58, and 50 mm, respectively.

To test the model in Eq. (19.3), you need to calculate S and W for every trial. You can calculate W easily from the known diameters of the two concentric circles. To determine the speed S, you need the distance traveled and the travel time. Estimate the distance as the circumference of a circle with a diameter equal to the average of the diameters of the concentric circles (as shown by the dotted line in Figure 19.1). *Measure* the time required to make five laps.

Each person in your group should make two successful runs (i.e., staying inside the track for five consecutive laps) for each width with a pencil and two successful runs for each width with a mouse. For each person, calculate the average time required at each width with the pencil and the average time required at each width with the mouse. Tracks for the pencil test are available at the end of this chapter and can be photocopied. Tracks for the mouse test can be made by drawing concentric circles in a software package such as Microsoft PowerPoint.

19.3.3   Modeling

Text link: Review the methods for determining the slope of a line by linear regression (Appendix C). Review the measures of model fits (Section 9.4.5).

To test the model, plot the average speed against the width. Make a separate plot for each member of your group. Each plot should have two sets of data: pencil results and mouse results. Using Eq. (19.3), calculate the controllability for each person for each interface. To calculate controllability, you will have to perform a linear regression. For each regression, calculate a measure of model fit.

Text link: Review measures of variability (Section 11.5).

You also need to determine which interface is more reproducible. To determine reproducibility, select and calculate a measure of variability. Finally, calculate the average (arithmetic mean) controllability for your group for the pencil interface and the mouse interface.

19.3.4   Reporting

Text link: Review the choices of presenting data (Section 12.6).

Use the report format given to you by your instructor. (A default grading scheme is shown in the appendix to this chapter.) You have a choice to make about how to present your data. Be sure to choose the presentation approach appropriate for *trends* (i.e., S versus W). Your instructor may ask you to incorporate some of the study questions in Section 19.4 into your objectives and discussion.

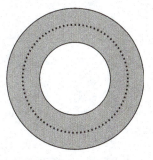

Figure 19.1.   Example Test Track for the Controllability Case Study

19.4   STUDY QUESTIONS

1. Which interface is more controllable? What evidence have you developed to support your conclusion?
2. Which interface had the most reproducible controllability?
3. Did the model in Eq. (19.3) describe the data? If not, hypothesize why the model and data did not match.
4. Would you have reached the same conclusion if you had allowed one error rather than zero errors? What if you allowed 10 errors?
5. How could you incorporate accuracy into the conclusion on controllability? Try using the data you collected on the number of attempts required to obtain two successful runs.
6. Do you think that the widths selected affected the results? In other words, would the controllability be the same over a different range of widths?
7. Your company produces video games. You are making a new maze game for 8- to 10-year-old children. The user must navigate a set of increasingly complex mazes. How could you use controllability data to design the interface?

19.5   ACKNOWLEDGMENTS AND FURTHER READING

This case study is based on material developed by Prof. Colin G. Drury of the University at Buffalo. For an introduction to the analysis of self-paced path control task, see Drury (1971). For a modification of the analysis to take into account rewards for accuracy, see Drury et al. (1987). Applications to vehicle control (DeFazio et al., 1992) and aircraft engine inspections (Drury, 2001) are available.

Default Grading Scheme: Controllability Case Study

Topic	Possible Points
Cover Page	
Names, date, title, group number (if appropriate)	5
Background	
Quality of introduction to the problem	10
Methods	
Brief description of how data were collected	10
Results and Discussion	
Which interface is more controllable? Support of conclusions with calculations	10
Which interface is more reproducible? Support of conclusions with calculations	5
Does the model describe the data?	10
Effects of number of allowable errors, width	5
Incorporation of accuracy	10
Discussion	
Application to the computer maze	10
Conclusion	
Quality of summary	10
Technical Writing	
Organization (and use of headings), grammar, spelling, tables and figures, and seamlessness of report	15
Total	100

PENCIL TRACKS

20

Dissolution Case Study

20.1 INTRODUCTION

You have known for a long time that matter can exist in several phases. Engineers frequently design systems that include or control the transfer of mass from one phase to another. For example, chemical engineers in the agricultural industry must design coatings that control the rate of dissolution of time-release fertilizers. Environmental engineers often must include the transfer of pollutants between the atmosphere and water in their models. Mechanical engineers must account for the vaporization of fuels in combustion processes.

In this case study, you will investigate the factors affecting the dissolution of a proposed coating for a new pharmaceutical.

20.2 THE STORY

"Achoo!"

"Gesundheit, Chantelle." Sarah patted her friend on the back. "Still fighting that cold?"

"Yes," sighed Chantelle miserably. "I can't seem to shake it. And we've got that statics midterm coming up, too."

"I know. I've been trying to study for it a little at a time like they told us, but there is so much material!"

"Oh, Sarah—you'll do fine. You aced the last test, didn't you?"

"Aced it? Are you kidding?" Sarah eyed her friend mischievously. "I need to do well on the midterm or I'm in deep trouble." Chantelle smiled. Sarah always made her feel better. "Have you started the ChemE 101 homework yet, Chantelle?"

"Yeah, but between not being able to sleep at night with this cold and getting drowsy during the day from these"—Chantelle held out a bottle of over-the-counter medication—"I haven't made much progress."

OBJECTIVES

After reading this chapter, you will be able to:

- describe the factors affecting mass transfer between phases;
- collect and evaluate data on mass transfer;
- evaluate a model to describe mass transfer.

An example of phase transfer: gases from volcanoes *dissolved* into rain, forming acid rain which *dissolved* minerals from rock, forming the salty oceans

Sarah took the bottle and looked at the pills inside. "These are all shiny, aren't they?"

"Some sort of new coating to make them pass through the stomach and dissolve in the intestines," said Chantelle casually.

"Eww, Chantelle, I don't want to hear about it. Oh yeah, I keep forgetting that you're doing this engineering thing just as a path to medical school. You're gonna need a long office sign to display all your titles: Chantelle Wilson, MD, PE!" The two friends shared a laugh. "Which homework problem are you stuck on?"

"Number 4." Chantelle read aloud: "'Describe a driving force for mass transfer and give two examples.' I was going to use the concentration gradient as the driving force, but I can't think of any examples other than what Professor Keller talked about in class."

"Explain that driving force stuff again. I think I missed that class and got Michael's notes. I don't quite understand the idea and haven't had a chance to talk with Professor Keller about it during his office hours."

"Well," began Chantelle, "mass can move from one place to another in response to a difference in concentration between the two places. In class, Professor Keller used the example of a drop of dye in a glass of water. The dye dispersed into the water until the concentration was the same everywhere. The dye moved because of the concentration difference—he called it the concentration gradient. When the gradient is zero, the driving force is zero."

"Oh, I see now. Like how the whole pizza tastes fishy when my roommate orders anchovies on half?"

"Maybe we can use that example, Sarah. Now if we could only come up with another one." Chantelle shook the pill bottle gently as she tried to concentrate on finding another example.

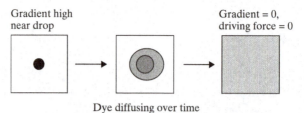

Dye diffusing over time

"Chantelle," said Sarah, eyeing the bottle. "I think I might have found another example."

Chantelle and Sarah finished their homework and went to the Student Union for a study break. Chantelle ordered her usual coffee and Sarah got an iced tea. They went outside to enjoy the sunshine.

"Hey, guys," shouted Michael from across the commons. "Care to join me?" The three budding chemical engineers sat together and chatted about the usual stuff: the impossibility of finding a parking place, the pain of studying while their roommates partied, and their dreams about life after school. Michael asked Chantelle how she was feeling and Chantelle showed him her medication.

"I don't get it—how do those pills of yours survive the stomach?" asked Michael.

"I'm not sure. Their coating is supposed to be resistant to stomach acids." Chantelle reached for the sugar and loaded up her coffee. "I suppose there are other things besides acids that control how fast things dissolve." The sugar was passed around the table as Sarah and Michael sweetened their iced teas.

Michael took a sip of his drink. "Yuck—this still isn't sweet enough. And I added a ton of sugar."

"Look, silly." Sarah pointed to the lump of sugar at the bottom of his glass. "You need to stir, see?" Sarah stirred her iced tea and the sugar crystals danced in the sunlight.

"I don't know what you guys are complaining about," said Chantelle. "This sugar dissolved in my coffee just fine."

"Mine tastes sweet, but I don't think all of the sugar has dissolved. Some of it is just kinda suspended in the glass." Sarah held her glass up to the sunlight to peer inside.

"And my sugar is just sitting there," said Michael glumly.

"A fine bunch of chemical engineers we are," said Chantelle with a smile. "Just by enjoying our drinks, we are learning about what affects the dissolution of solids."

20.3   THE CASE STUDY

20.3.1   Introduction

For this case study, your group will be a research team employed by a large pharmaceutical manufacturer. Your team assignment is to measure and develop a model for the rate of dissolution of a new sugar-based coating for a pill. You decide to use a lollipop and water as physical models of the coating and gastric juices, respectively. For this case study, be sure to use a *solid hard candy lollipop*; do not use a lollipop with a soft, chewy center.

If the concentration gradient controls the mass transfer, then the rate of mass transfer is proportional to the concentration gradient. This statement is expressed as the mathematical equation

$$\frac{dr}{dt} = -\frac{k\Delta C}{\rho} \qquad (20.1)$$

In Eq. (20.1), r is the radius of the lollipop, t is time, k is a mass transfer coefficient, ΔC is the concentration gradient, and ρ is the molar density of the lollipop (i.e., the number of moles of sugar per unit volume of the lollipop). In this case, the concentration gradient is the difference between the concentration of dissolved sugar right at the surface of the lollipop and the concentration of sugar in the water far away from the surface.

PONDER THIS

> **From the experiences of Sarah, Chantelle, and Michael, what factors are expected to influence the mass transfer coefficient (and thus influence the rate of dissolution)?**

Based on the case study dialogue, the mass transfer coefficient is expected to depend upon the temperature and the rate of agitation (i.e., how fast the water is stirred).

The term $\dfrac{dr}{dt}$ in Eq. (20.1) is called the derivative of r with respect to t. It represents the rate of change of the radius with respect to time; that is, the rate at which the radius of the lollipop changes. Equation (20.1) is called a *differential equation* because it involves derivatives. Differential equations are very common in engineering analysis and require special solution techniques to solve. The solution to a differential equation is an *equation*, not a number.

Engineers often use assumptions to simplify models. As a starting point, assume that the temperature is constant during a dissolution experiment. In addition, it may be reasonable to assume that the mass transfer coefficient (k) and molar density of the lollipop (ρ) are constant. If the concentration of sugar dissolved in the water is assumed to be constant, then ΔC is constant. Thus, the right-hand side of Eq. (20.1) may be simply a constant. Under these conditions, Eq. (20.1) can be solved to yield

$$r = r_0 - \alpha t \tag{20.2}$$

In Eq. (20.2), r_0 is the initial radius of the lollipop (i.e., the radius at time $t = 0$) and $\alpha = \dfrac{k\Delta C}{\rho}$ and is assumed to be constant (since k, ΔC, and ρ are assumed to be constant). Equation (20.2) predicts that the radius should vary linearly over time.

PONDER THIS

> **What does "the radius should vary linearly over time" mean?**

Text link: Review the methods for determining the slope of a line (Appendix C).

In other words, if you plot the radius (on the y-axis) versus time (on the x-axis), the resulting graph will be a *straight line* with a slope equal to $-\alpha$ (see Figure 20.1).

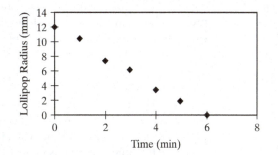

Figure 20.1.   Example Data If Eq. (20.2) Described the Data Reasonably Well

Text link: Review the concept of model fits (Section 9.4).

Text link: Review the concept of an empirical model (Section 9.3.4).

The model embodied in Eq. (20.2) stems from the physical description of the setup and the assumptions. How can you test whether the model describes the data adequately?

The effects of mixing are challenging to model quantitatively. However, there is a simple *empirical model* describing the effects of temperature on chemical rate constants. It is called the *Arrhenius model* (after Svante August Arrhenius, 1859–1927). The Arrhenius model states that the rate constant (α in this case study) should be proportional to e raised to the power of a constant times $1/T$, where T is the temperature in Kelvin:

$$\alpha = ae^{b\frac{1}{T}}, \text{where } a \text{ and } b \text{ are constants}$$

Remember that $e = \ln(1) \approx 2.71828$. According to the Arrhenius model, a plot of $\ln(\alpha)$ versus $1/T$ (T = temperature in degrees Kelvin) should be a straight line.

20.3.2 Case Study

The basic procedure for determining the rate of dissolution is as follows: heat or cool water until it reaches a constant temperature. Record the temperature. Measure and record the diameter of a fresh lollipop. Suspend the lollipop in the water for some period of time *without stirring*. Remove the lollipop from the water and quickly measure and record its diameter. At the same time, measure and record the water temperature. Resuspend the same lollipop in the water for another period of time without stirring. Repeat the measuring and resuspension steps until the diameter is less than one-half its original value.

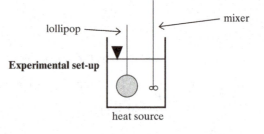

Repeat this procedure several times, using a different temperature or a different agitation rate during the dissolution process. The conditions studied by each lab group should be distributed so that a large database is generated with several (at least four) temperatures at each agitation condition. Do at least two agitation conditions (at least the conditions "unstirred" and "stirred at a constant rate").

Text link: Review the choices of presenting data.

To analyze the data, convert the measurements (if necessary) to have r versus t data for each experiment. Plot the data and decide whether the model in Eq. (20.2) appropriately describes the data. If the model is appropriate, calculate α. Determine the trends in α as temperature and/or agitation are varied. Check if the Arrhenius model fits your data (i.e., fits your values of α).

20.3.3 Reporting

Use the report format given to you by your instructor. (A default grading scheme is shown at the end of this chapter.) The report should include a statement of the context of the project (what you did and why), the approach you followed, a discussion of the results (see the Study Questions in Section 20.4), and your conclusions.

20.4 STUDY QUESTIONS

1. For which data sets is the linear model in Eq. (20.2) **not** appropriate? What measure can you use to decide if the linear model is valid? Can you explain which assumption(s) in the model may be invalid for the data sets?

2. Do the observed qualitative trends in α with respect to mixing and temperature make sense?

3. Does the Arrhenius model explain the data?

4. How would the results have differed if you used a lollipop with a soft center rather than a uniformly hard lollipop as your physical model?

5. How could similar studies be used to assess the effectiveness of coatings on pills?

20.5 ACKNOWLEDGMENTS AND FURTHER READING

This case study is modeled after an experiment suggested by D. M. Fraser (1999). The case study material was modified from a case written by Prof. Carl R. F. Lund of the University at Buffalo.

Default Grading Scheme: Dissolution Case Study

Topic	Possible Points
Cover Page	
Names, date, title, group number (if appropriate)	5
Background	
Quality of introduction to the problem	10
Methods	
Brief description of how the data were collected	10
Results and Discussion	
Fitting a straight line to the data, if appropriate	10
Try to explain why some data are not linear (and hence not a measure of the model fit)	10
Are the trends in the slopes with mixing and temperature as expected?	5
Quality of summary of effects of stirring, temperature	10
Data plots	10
Arrhenius plot	5
Conclusion	
Quality of summary	10
Technical Writing	
Organization (and use of headings), grammar, spelling, tables and figures, and seamlessness of report	15
Total	100

21

Computer Workstation Case Study

21.1 INTRODUCTION

The public thinks of engineers as people who analyze and design machines or equipment. Engineers also analyze and design *workplaces, jobs, and processes.* The goal of this type of engineering is to ensure that people can perform jobs efficiently, effectively, and safely. Although workplace design is part of the area of specialization in industrial engineering called *human factors engineering,* all engineers must be aware of the impact of their work on potential users.

Many jobs in today's world require the use of computers. You may have used computers under poor conditions (e.g., inadequate lighting, poor chair design, and improper wrist support). You know that inflexible and poorly designed computer workstations can lead to tremendous stress on the computer user. (In this case study, the word *workstation* is used to describe the table that supports the computer, monitor, and input devices and the chair that supports the user.)

Engineers can analyze existing workstations to identify problems. In addition, engineers can improve upon workstation design to address potential problems with inefficient or unsafe systems.

21.2 THE STORY

"Ouch! I can't believe I did that again!"

"Are you okay, Robert?" asked Yoshiko. She put her arm around Robert's shoulder as the two friends worked together on a circuits project in the electrical engineering computer lab.

"Yeah, thanks." Robert rubbed his knee. "It's just these stupid workstations. Every time I push my chair in, I bang my knees on the keyboard tray."

OBJECTIVES

After reading this chapter, you will be able to:

- explain the importance of workplace design;
- apply the steps in the engineering analysis method to a real engineering problem;
- apply the steps in the engineering design method to a real engineering problem.

Examples of Computer Workstations

Text link: Review human factors engineering (Section 4.6.2).

"I hate these workstations, too," said Yoshiko. "But I don't have *that* problem." The two friends laughed. Their relationship generated a lot of chiding from their engineering friends—Robert stood 6′2″ in his socks, while Yoshiko barely reached 5′1″. "I feel like a kid in this chair. See? My feet don't even touch the floor. And when I leave the lab, my neck hurts from staring *up* at the monitor."

"You guys have it easy—I've got both your problems." Robert and Yoshiko looked over to see their friend Chris walking in the lab. "My knees are permanently scarred from the desk and my neck hurts, too."

"I can see how you might hit your knees with those long legs of yours, Chris. But how come your neck hurts? You've got to be at least six inches taller than I am."

"I'll show you." Chris walked over and stood next to the seated Yoshiko. "Yoshiko, stand up please. And Robert, measure how tall we are." Yoshiko stood and Robert drew an imaginary level line with his hand from the top of Yoshiko's head to Chris's chin. "Now, sit down, Yoshiko." Chris sat next to Yoshiko.

Yoshiko looked over. "We're the same height sitting down, Chris!"

"Yep. I'm short in the torso, but long in my legs. Look at my knees." Robert and Yoshiko looked down. With Chris's feet flat on the floor, his knees were bent upward at a funny acute angle.

"Reminds me of Danielle in our electromagnetism class," mused Robert. "She and I can look each other in the eye while seated, but I'm a lot taller than she is."

"What are you and Danielle doing, looking each other in the eyes?" teased Yoshiko, with mock seriousness.

"Ummm, well, it's just that she has such a long torso, I guess . . . ."

Chris shook his head slowly. "Better stop while you're not too far behind, Robert."

That afternoon, Robert and Yoshiko were studying in Robert's dorm room. "This is more like it," said Robert, stretching out in his chair. "See? My knees don't bump. The keyboard is at the perfect height for my wrists, too."

"Yes, but you'd be miserable studying in my room," said Yoshiko. "I've got my workplace set up for someone just my size."

"Yeah, we complain about the EE lab all the time, but I guess it's a big challenge to make a workstation that fits everyone."

"You're right. Imagine a workstation that would fit my short legs and your long legs."

"Yes, and it would have to accommodate people who sit low in their chairs like you and Chris and people who sit high in their chairs like me and Danielle."

"Oh, yes—we wouldn't want *Danielle* to be uncomfortable, now would we?" Yoshiko smiled innocently at Robert. He blushed furiously and put his arm around Yoshiko.

"Ummm, do you think we could design a workstation to fit most people? Maybe we could start our own company: Robiko's Universal Workstation, Inc."

"Don't you mean *Yoshibert's* Universal Workstation, Inc., Robert?"

21.3 THE CASE STUDY

21.3.1 Introduction

You have been hired by a small company (the client) to evaluate their computer workstations and design a new computer workstation for them. The company has about 20–25 people using computers. You will use anthropometric measurements of your fellow students to represent the measurements of the workers in the client's office. (Hereafter, the students you measure will be called the *population*.) You will use workstation measurements from two workstations at your university to represent the measurements of the workstations in the client's office.

The existing workstations should be analyzed for their ability to accommodate the client's workers (i.e., your fellow students). You also need to design a new computer workstation for the client's workers (i.e., your fellow students) that minimizes costs.

21.3.2 Case Study

The goals of this case study are to analyze two existing computer workstations for their ability to accommodate a given population and design a new computer workstation for a particular population.

To meet the goals of this case study, you will take anthropometric measurements of your fellow students and use them to analyze and design a computer workstation. You need a population of at least 20–25 people.

First, take four measurements on all members of your population. Have each person sit in a common test chair. Adjust the chair so that the person's feet are flat on the floor and there is a 90° angle between the person's upper body and the thighs. (If necessary, use spacers under the person's feet or seat to make sure the feet are flat and thighs are perpendicular to the torso. Small sections of 2″ × 4″ lumber make good spacers.) Have the person bend elbows at 90° and point his or her arms forward. Take the four measurements shown in Figures 21.1 and 21.2. The four measurements (numbers correspond to the figures) are as follows:

1. Seated eye height (from the eye to the floor when seated)
2. Seated elbow height (from the elbow to the floor when seated)

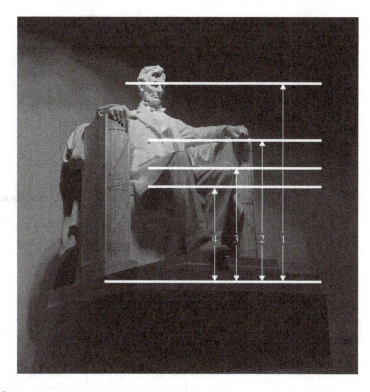

Figure 21.1. Anthropometric Measurements for the Computer Workstation Case Study (numbers correspond to descriptions in the text)

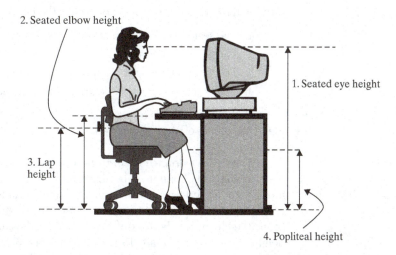

Figure 21.2. Example of the Anthropometric Measurements (in the case study measurements, the elbow should be bent to 90°)

3. Lap height (from the top of thigh to the floor when seated), and
4. Popliteal height (from the underside of the leg near the knee to floor).

Record and distribute the data.

Figure 21.3. Example of the Computer Workstation Measurements

Second, make the following workplace measurements for two computer workstations (see Figure 21.3):

1. Top of the monitor height (distance from the top of monitor to the floor)
2. Keyboard height (distance from the home-row keys—asdf, etc.—to the floor)
3. Desk height (distance from the bottom of the desk to the floor or, if the workstation has a keyboard shelf, from the bottom of the keyboard shelf to the floor)
4. Seat pan height (distance from where you sit to the floor)

To analyze the data, compare the anthropometric data you collected to the measurements you made of the workstation. Apply the following four analysis criteria to determine if the workplace is appropriate for the population.

1. The seat pan height should be less than or equal to the popliteal height.
2. The top of monitor should be at or below the seated eye height.
3. The keyboard height should be approximately at the average elbow height.
4. The desk height should be greater than the lap height.

In the design task, you are asked to design a computer workstation to fit the people in the client's office (i.e., your fellow students). The workstation will consist of a chair and a table. A footrest is optional.

Text link: Review the concept of constrained optimization (Section 2.5).

This is a *constrained optimization* problem. In other words, you must choose between a finite set of options. The characteristics and costs of the different types of chairs, tables, and footrests available to you are listed in Table 21.1.

The design parameters are the keyboard height, seat height of the chair, knee room (height of the bottom of the table or adjustable keyboard shelf), and monitor height. Note that there are 18 possible designs ($3 \times 3 = 9$ designs without a footrest and $3 \times 3 = 9$ designs with a footrest).

The goal of the design task is to design the least expensive workstation for the client. The client is willing to pay more for a workstation that accommodates more people. Therefore, add a \$25 penalty to the cost of each workstation for *each person* that the workstation does not fit.

As an example of the analysis of a workstation, consider the anthropometric data shown in Table 21.2. Suppose that you design a workstation comprising a plain chair

TABLE 21.1 Characteristics and Costs of Workstation Components[a]

Item	Characteristics	Cost
Chair, plain	seat pan height fixed at 17″	$30
Chair, adjustable	seat pan height adjustable from 16″ to 18″	$50
Chair, supremo model	seat pan height adjustable to any height	[b]
Desk, plain (no keyboard shelf)	desk height fixed at 26″; keyboard sits on desk top	$60
Desk, adjustable (no keyboard shelf)	desk height adjustable from 25″ to 27″; keyboard sits on desk top	$90
Desk, supremo model (has an adjustable keyboard shelf)	desk height (with keyboard shelf) adjustable to any height	[c]
Footrest	adjustable to any height	$30

[a]All desks are 2″ thick. Keyboards have home keys $1\frac{1}{2}$″ above the desktop (except units with adjustable keyboard shelves). The tops of all monitors are $17\frac{1}{2}$″ above the desktop.

[b]Cost is $30 + ($10/inch)(adjustment range in inches). Thus, a chair adjustable from 15″ to 19″ (4″ range) would cost $30 + ($10/inch)(4″) = $70.

[c]Cost is $60 + ($15/inch)(adjustment range in inches). Thus, a chair adjustable from 23″ to 28″ (4″ range) would cost $60 + ($15/inch)(4″) = $120.

TABLE 21.2 Example Anthropometric Data

Name	Seated Eye Height (in)	Seated Elbow Height (in)	Lap Height (in)	Popliteal Height (in)
Homer	47.6	23.5	21.1	18.0
Marge	50.5	27.7	20.6	15.0
Ned	49.0	26.9	23.3	18.4
Mo	45.5	29.3	22.2	16.1
Barney	45.5	26.9	20.1	18.0
Sneezy	49.0	30.0	21.1	16.7
Doc	46.9	23.7	25.3	16.5
Grumpy	50.5	29.5	23.7	17.7
Sleepy	43.2	29.3	23.7	17.7
Dopey	49.8	25.3	23.3	16.1

and a plain desk, with no footrest. From the data in Table 21.1 (and the notes at the bottom of Table 21.1), the dimensions of the workstation would be as follows (see Figure 21.4):

$$\text{Top of the monitor height} = \text{desk height} + \text{desktop thickness} + \text{monitor height}$$
$$= 26″ + 2″ + 17.5″ = 45.5″$$

$$\text{Keyboard height} = \text{desk height} + \text{desktop thickness} + \text{keyboard thickness}$$
$$= 26″ + 2″ + 1.5″ = 29.5″$$

$$\text{Desk height} = 26″$$
$$\text{Seat pan height} = 17″$$

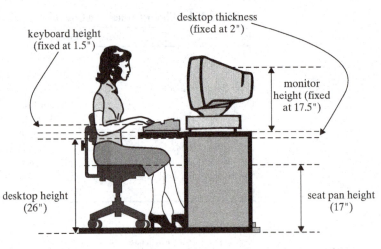

Figure 21.4. Example Computer Workstation Design (not to scale)

In evaluating how the design fits the population, you would see whether it meets the following criteria:

Criterion 1: The seat pan height is greater than or equal to the popliteal height.
Results: This criterion excludes Homer, Ned, Barney, Grumpy, and Sleepy.

Criterion 2: The top of the monitor is at or below the seated eye height.
Results: This criterion excludes Sleepy.

Criterion 3: The keyboard height is approximately at the seated elbow height.
Results: This criterion excludes Homer, Marge, Ned, Barney, Doc, and Dopey (if the keyboard height is allowed to be within ±0.5 inches of the elbow height).

Criterion 4: The desk height is greater than the top of the thigh.
Results: This criterion excludes no one.

Thus, the design fits only Mo and Sneezy. The workstation would cost $30 + $60 = $90 in materials, plus a penalty of $200 for not fitting eight people. Thus, this design would cost $290 per workstation and fit only two people—which is not very impressive. Perhaps a design that fits more people would have lower overall costs.

21.3.3   Reporting

Text link: Review the steps in the analysis process (Chapter 6).

Use the report format given to you by your instructor. (A default grading scheme is shown at the end of the chapter.) The report should include a statement of the context of the project (what you did and why) and the approach you followed. For the analysis goal, discuss your analysis of each of the two computer workstations that you studied. Indicate how you addressed the pertinent steps in the engineering analysis process. Include a table of the workplace measurements you made. This section should end with conclusions about the ability of each workstation to accommodate the target population. Briefly comment on the problems with the existing workstations and how they could be fixed.

For the design goal, discuss how you came up with your design. If you use adjustable equipment, be sure to specify the adjustment range. Summarize your analysis of your design, including its cost and how it accommodated the target population.

21.4   STUDY QUESTIONS

1.   Were your data (i.e., measurements) reasonable and reproducible?
2.   Given your experiences in this case study, discuss the advantages and disadvantages of using Robert and Yoshiko's approach of a *universal workstation*. Would you recommend designing a workstation to fit everyone?
3.   How would your workstation design have to be changed to accommodate users in wheelchairs? The standard seat pan height of a wheelchair is 19 inches.

21.5   ACKNOWLEDGMENTS AND FURTHER READING

This case study was modified from a case written by Profs. Ann M. Bisantz and Victor Paquet of the University at Buffalo.

Default Grading Scheme: Computer Workstation Case Study

Topic	Possible Points
Cover Page	
Names, date, title, group number (if appropriate)	5
Background	
Quality of introduction to the problem	10
Methods	
Brief description of how the data were collected	5
Analysis	
Are the anthropometric/workplace data reasonable and reproducible?	5
Demonstration of engineering analysis steps	10
Use of four analysis criteria	5
Discussion of problems and solutions	10
Discussion of correlations	5
Design	
Discussion of the design approach and evaluation of the design	10
Universal workstation discussion	5
Wheelchair discussion	5
Conclusion	
Quality of summary	10
Technical Writing	
Organization (and use of headings), grammar, spelling, tables and figures, and seamlessness of report	15
Total	100

22

Power Transmission Case Study

22.1 INTRODUCTION

Recall that engineers must decide among alternatives as part of the design process. Challenges arise when *the alternatives must meet many goals*. For example, the choice of buildings to replace the World Trade Center towers destroyed in the attack of September 11, 2001, was very difficult, in part because the new structures had to address so many goals (e.g., respecting the victims, being aesthetically pleasing, and providing commercial space).

Challenges also arise when *the alternatives are very different*. For example, several options are available to you for personal transportation. Since the options are so different (walking, bicycling, driving a car, using a personal transportation device like the Segway), you must find common ground to evaluate the alternatives.

This case study provides an opportunity to compare and select from three very different alternatives for the transmission of power from Niagara Falls to Buffalo, New York. The case study is based on a real choice made by a group of scientists and engineers at the end of the 19th century.

22.2 THE STORY

(*Note*: The characters in this story are fictional. However, the facts behind the story are true. For more details, see the references in Section 22.5. The story opens in a small business office in Buffalo, New York, in 1895.)

Bartholomew Kingston looked up from the papers on his desk as his guests arrived. "Ah, Mr. Smyth. So good of you to come down to my office on such short notice. And whom do I have the honor of meeting?" Kingston gestured to the young woman standing beside Smyth.

OBJECTIVES

After reading this chapter, you will be able to:

- compare alternatives that are very different;
- apply the steps in the engineering analysis method to a real engineering problem;
- apply the steps in the engineering design method to a real engineering problem.

Text link: Review the steps in the design process (Chapter 7).

"My pleasure, Mr. Kingston. May I introduce my daughter, Miss Angela Smyth?"

"Miss Smyth." Kingston kissed her hand. "Welcome to my humble rooms." Turning to his guest, Kingston continued, "I am pleased, of course, to be introduced to your enchanting daughter, but surely, Mr. Smyth, our conversation today would bore her."

Smyth held up his hand in protest. "Mr. Kingston. My daughter has been highly trained in Latin, music, and the mathematical arts. My modest successes in the field of engineering" —he paused to bow humbly—"are due in the main to consultations with my daughter."

"Then I owe Miss Smyth my apologies." Kingston nodded his head towards Angela. "Perhaps we can get to the heart of the matter?" Kingston shuffled some papers on his desk to hide his reddening face. "You know the basic facts from my letter. We wish to harness the great power of Niagara Falls for the purposes of providing power to Buffalo. Lord Kelvin heads the International Niagara Commission, a panel of experts to evaluate proposals regarding the generation and transmission of the power. The winning proposal will garner a prize of $3,000."

Miss Smyth's eyes gleamed with excitement. "And have you had many proposals, Mr. Kingston?"

"Yes, Miss Smyth. We have narrowed the field to 14 proposals. And this is where the trouble begins. We do not know how to compare the proposals."

"But surely the plans are fairly similar," interjected Mr. Smyth.

"Ah, my good sir. Unfortunately, the ideas in the proposals are quite different. Five teams propose to transmit electricity by direct current. Two groups will use this new alternating current technique." Angela nodded enthusiastically. She had read about Nikola Tesla's work on alternating currents and how the technology was being promoted by a certain Mr. George Westinghouse. "Four teams want to use air compressors and tubes to deliver the power by pneumatic means. Of the three remaining proposals, one transmits power hydraulically, one uses belts and pulleys to send power to Buffalo, and one uses a combination of technologies."

"Well," said Mr. Smyth. "Quite a variety of ideas. How can we help you?"

"No decisions have been made by the Commission yet. But I think we can narrow the choices a bit. We would like some help in evaluating three kinds of proposals for power transmission: alternating current, the hydraulic transmission alternative, and the belt-and-pulley option. In addition, I can give you an idea of the power production capacity. We think that we may build three 5,000-horsepower generators at Niagara, some 35 kilometers from here." Kingston gestured in a northward direction. "Here are the particulars." He handed over a thick dossier.

"Thank you, Mr. Kingston. We shall consult with one another and meet with you two weeks hence."

"Thank you, Mr. Smyth, Miss Smyth. On behalf of the Commission, we appreciate your help." As his guests left his office, Mr. Kingston muttered under his breath. "Imagine, a woman engineer!"

That night, Angela sat in the parlor with a thick sheaf of papers in her lap. "Father, this is more difficult that I thought!"

"Indeed, Angela. We have much work to do. How should we start?"

"Well, we need to have some way to compare the proposals that Mr. Kingston gave us." She flipped through the papers. "They are so different."

"Different, yes, but they share the same goal: transmitting power from here"— Smyth pointed to Niagara Falls on the wall map—"to Buffalo. Do you suppose that all of the power from the Niagara will find its way to Buffalo?"

"Oh no, Father. Energy certainly will be lost with every method. For example, the belts and pulleys will heat up a little from friction. The energy used to heat the material will not produce useful work in our city."

"So perhaps we need to find how much power is lost in each case."

"An excellent idea, Father." Angela held up a piece of paper with dense, spidery handwriting. "Here are a few ideas..."

22.3   THE CASE STUDY

22.3.1   Introduction

Text link: Review the Second Law of Thermodynamics in Appendix A.

During transformation of energy from one form to another, some energy is lost as heat. This statement stems from the Second Law of Thermodynamics. Thus, the transmission of power by electric, hydraulic, or mechanical means will not be 100% efficient. In the case study, the power obtained at Buffalo will be less than the power generated at Niagara Falls. Power losses for the electrical, hydraulic, and mechanical transmission systems will be reviewed.

Electrical Transmission Systems

In electrical systems, the power loss is given by

P_{loss} = VI (valid only for purely resistive systems, i.e., no capacitance or inductance)

where V = voltage difference and I = current. Combining power with Ohm's Law (V = IR, where R = resistance) yields

$$P_{loss} = V^2/R \qquad (22.1)$$

A word about units: Ohm's Law holds if V is in volts (V), I in amperes (amps, A), and R in ohms (Ω). With this set of units, P_{loss} in Eq. (22.1) will be in units of watts (W).

The resistance to the flow of electrons, R, depends on the wire. As you might expect, longer wire has a higher resistance. (By analogy, a long hose has more resistance to the flow of water than a short hose.) In addition, a large-diameter wire has less resistance than a small-diameter wire (just as a fire hose has less resistance to flow than a straw for the same water flow rate).

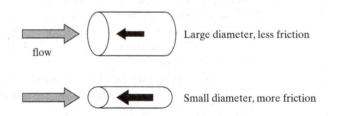

Two wires of the same length and diameter have different resistances if they are made of different materials. Putting these ideas together, we see that the resistance of a wire is given by

$$R = \frac{\rho L}{A} \qquad (22.2)$$

In Eq. (22.2), ρ = resistivity (a measure of the inherent resistance of a material to the flow of electrons), L = length of wire, and A = cross-sectional area of the wire. For wire with a circular cross section,

$$A = {}^1\!/_4 \pi D^2 \qquad (22.3)$$

where D is the diameter of the wire. If ρ is in units of ohm-m (Ω-m), L in m, and A in m^2, then R will be in units of ohms (Ω). Combining Eqs. (22.1) through (22.3)* results in

$$P_{\text{loss}} = \frac{4V^2\rho L}{\pi D^2} \qquad (22.4)$$

Hydraulic Transmission Systems

For hydraulic systems, the energy loss usually is expressed by a decrease in a normalized pressure loss called *head loss*, h_L. The head loss is equal to the pressure divided by the product of the density of the fluid and gravitational acceleration g. Since pressure is force per unit area,

$$\text{head loss} = \frac{\text{pressure}}{g(\text{density})} = \frac{\text{force}}{g(\text{density})(\text{area})} \qquad (22.5)$$

To find the units of head loss, examine the units of each of the terms on the right-hand side of Eq. (22.5). If $\{x\}$ indicates the units of x, then

$$\{\text{head loss}\} = \frac{\{\text{force}\}}{\{g\}\{\text{density}\}\{\text{area}\}}$$

$$= \frac{\{\text{mass}\}\{\text{acceleration}\}}{\{g\}\{\text{density}\}\{\text{area}\}}$$

$$= \frac{(\text{kg})\left(\dfrac{\text{m}}{\text{s}^2}\right)}{\left(\dfrac{\text{m}}{\text{s}^2}\right)\left(\dfrac{\text{kg}}{\text{m}^3}\right)(\text{m}^2)}$$

$$= \text{m}$$

Text link: Recall the value of dimensional analysis to check units and equations (Section 6.7).

Thus, head loss often is expressed in units of length, m or ft. The head loss in a pipe can be calculated from the following empirical equation (a version of a hydraulics equation called the *Hazen–Williams equation*):

$$h_L = 10.70 L D^{-4.87}\left(\frac{Q}{C}\right)^{1.85} \quad \text{(valid for the units below}^\dagger) \qquad (22.6)$$

Text link: Review the concept of an empirical model (Section 9.3.4).

where L is the length of the pipe, D is the pipe diameter, Q is the fluid flow, and C is a friction constant for a given type of pipe (dimensionless). Empirical equations are valid only for a specified set of units. In Eq. (22.6), with L in m, D in m, and Q in m^3/s, h_L will be in m.

You can convert head loss into power loss by

$$P_{\text{loss}} = Q\rho_f h_L g \qquad (22.7)$$

---

*Equation (22.4) assumes that the current is uniformly distributed in the wire. At higher frequencies, currents become concentrated on the outer surface of the wire. This is called the "skin effect." Thus, Eq. (22.4) is not valid at high frequency.
†A number of equations have been developed to relate head loss to flow. The Hazen–Williams equation is appropriate here. However, the Hazen–Williams equation is valid only for water flow with pipe sizes and velocities typical of large-scale water conveyance and distribution systems.

where ρ_f is the density of the fluid. You can confirm that if Q is in m³/s, ρ_f in kg/m³, h_L in m, and g in m/s², then P_{loss} will be in units of watts:[*]

$$\{P_{\text{loss}}\} = \{Q\}\{\rho_f\}\{h_L\}\{g\}$$
$$= (\text{m}^3/\text{s})(\text{kg}/\text{m}^3)(\text{m})(\text{m}/\text{s}^2)$$
$$= \text{kg-m}^2/\text{s}^3$$
$$= (\text{kg-m}/\text{s}^2)(\text{m})/\text{s}$$
$$= \text{N-m/s}$$
$$= \text{J/s} = \text{W} = \text{watts}$$

Mechanical Transmission Systems

As with the electrical and hydraulic means of transmitting power, power losses with a belt-and-pulley approach come from friction. The power loss with a conveyor belt is given by

$$P_{\text{loss}} = fmvLg \tag{22.8}$$

where f is the friction factor (dimensionless), m is the mass of the conveyor belt per unit length (kg/m), v is the belt velocity (m/s), and L is the conveyor belt length (distance between the pulleys, m). With the units given, P_{loss} will be in units of watts.

22.3.2   Case Study

The goal of this case study is to compare electrical, hydraulic, and mechanical alternatives for transmitting power from Niagara Falls. For each of the three alternatives, you must calculate the efficiency of the energy transfer. The efficiency is

$$\text{efficiency }(\%) = \frac{P_{\text{gen}} - P_{\text{loss}}}{P_{\text{gen}}} \times 100\% \tag{22.9}$$

Text link: For a photograph of the hydroelectric facilities on the U.S. side today, see Section 11.6.

In Eq. (22.9), P_{gen} is the power generated at Niagara Falls. As stated in Section 22.2, Niagara Falls originally had three 5,000-horsepower (hp) generators (about 1.1×10^4 kilowatts = kW). In a few short years, the power generation was increased to about 1.6×10^5 kW, all of which was transmitted to Buffalo. Current power generation on both the American and Canadian sides of Niagara Falls is about 4.3×10^6 kW. The original voltage of the power generated was 11,000 VAC (alternating current volts). Niagara Falls now generates power at 345,000 VAC. For the purposes of this case study, use 1.1×10^4 kW generated at 11,000 VAC.

A schematic for the power transmission by electrical transmission lines is shown in Figure 22.1. For the determination of electrical transmission efficiency, assume that the

---

[*]Ironically, James Watt (1736–1819), after whom the *metric* unit of power was named, coined the term for the *English system* unit of power: horsepower. Watt guessed that the average horse could pull 180 pounds (of force). Horses typically were attached to levers called capstans (from the Latin *capere*, to take hold of), with a typical capstan lever length of 12 ft. If a horse completed 144 circuits per hour, then the horse would walk $2\pi(12 \text{ ft})(144 \text{ circuits/hr}) = 10,857$ ft/hr and exert $(180 \text{ lb})(10,857 \text{ ft/hr})/(3,600 \text{ s/hr}) = 542.9$ ft-lb/s of work. Watt rounded this value to 550 ft-lb/s and called it one horsepower. You can convert this to metric units: $1 \text{ ft} = 0.3048 \text{ m}$, 1 lb of force $= (0.45 \text{ kg})(9.8 \text{ m/s}^2) = 4.45 \text{ N}$, so $1 \text{ hp} = 550$ ft-lb/s $= (550 \text{ ft-lb/s})(0.3048 \text{ m/ft})(4.45 \text{ N/lb}) = 746$ N-m/s $= 746$ J/s $= 746$ W.

Figure 22.1. Schematic for Electrical Transmission Option

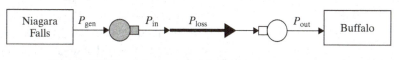

Figure 22.2. Schematic for Hydraulic Transmission Option

transmission lines were 1.8-cm-diameter steel wire. The resistivity of carbon steel is 2.0×10^{-7} Ω-m.

A schematic for the power transmission by hydraulic lines is shown in Figure 22.2. The power generated by the falls must be used to drive a pump (represented by the shaded symbol on the left). Assume that the pump efficiency is 85% (i.e., 85% of the electric power into the pump is transmitted to the water). At the Buffalo end, the water is used to drive a turbine (represented by the open symbol on the right; a turbine is really just a pump operated in reverse). Assume that the turbine efficiency is 90% (i.e., 90% of the hydraulic power into the turbine is converted to electricity).

For the determination of hydraulic transmission efficiency, assume that the C factor is 120 (dimensionless). In the original hydraulic proposal of 1895, 10 2-ft-diameter pipes were proposed. Assume each of the pipes had a flow of 0.04 m^3/s. Use Eq. (22.6) to calculate the head loss in each pipe. Then add the head losses to obtain the overall head loss and convert the head loss to power loss, using Eq. (22.7).

A schematic for the power transmission by a belt-and-pulley is shown in Figure 22.3. The power generated by the falls must be used to drive a motor (represented by the symbol with "M", on the left). Assume that the motor efficiency is 85% (i.e., 85% of the electric power into the motor is transmitted to the pulley). At the Buffalo end, the belt is used to drive a generator (represented by the symbol with "G", on the right). Assume that the generator efficiency is 90% (i.e., 90% of the mechanical power into the generator is converted to electricity).

The original mechanical proposal stated that the power loss in the belt would be only 7 hp per 330 ft. Convert this value to units of W/m and use it to determine the efficiency of the belt-and-pulley system.

22.3.3 Reporting

Use the report format given to you by your instructor. (A default grading scheme is given at the end of the chapter.) The report should include a statement of the context of the project and the approach you followed. Give a summary of your approach to the calculations with the calculation details in an appendix.

Figure 22.3. Schematic for Mechanical Transmission Option

22.4 STUDY QUESTIONS

1. Which transmission system would you recommend and why? Be sure to justify your answer with the results of your calculations.

2. What else would you take in account when recommending a transmission option? (Think about the types of feasibility.)

3. If the belt weighed 90 kg/m and traveled at 1.1 m/s, what would be the power loss per meter if the friction factor were 0.05?

4. How would the efficiency of the electrical power transmission change if copper wire was used instead of steel? (The value of ρ for copper is 1.72×10^{-8} Ω-m. This question is easier to answer if you used a spreadsheet to do the calculations in this case study.)

5. One of the reasons why AC transmission was selected over DC transmission is that AC can be transmitted at high voltages, then "stepped down" to lower voltages to be used at the destination. Why is the transmission of electricity at high voltages desirable? (You may need to use Internet resources to learn more about AC transmission to answer this question.)

22.5 ACKNOWLEDGMENTS AND FURTHER READING

Text link: Review the concept of feasibility (Chapter 11).

The background information for this case study was obtained from Foran (undated). This case study was modified from a case developed by Dr. Jennifer Zirnheld and Mr. Kevin Burke of the University at Buffalo.

Default Grading Scheme: Power Transmission Case Study

Topic	Possible Points
Cover Page	
Names, date, title, group number (if appropriate)	5
Background	
Quality of introduction to the problem	10
Analysis	
Analysis of electrical, hydraulic, and mechanical systems (10 pts each)	30
Quality of selection (including support)	10
Other types of feasibility	15
Other Discussion Questions	
Belt calculation, advantage of AC	10
Conclusion	
Quality of summary	10
Technical Writing	
Organization (and use of headings), grammar, spelling, tables and figures, and seamlessness of report	10
Total	100

23

Walkway Collapse Case Study

23.1   INTRODUCTION

Engineering is an exciting profession in part because engineers have a large impact on the lives of the public. The work you do in your career will help people lead safer and more satisfying lives. However, the potential to make a difference in people's lives means that engineers have a great deal of responsibility. Engineering design typically is conservative (i.e., systems are overdesigned) to protect the safety, health, and welfare of the public.

In spite of the best efforts of engineers, mistakes and accidents do happen. Some errors in design judgment have terrible consequences. The collapse of two suspended walkways in the Kansas City Hyatt Regency Hotel on July 17, 1981, was one of the worst structural failures in United States history. The disaster left 114 people dead and over 200 injured. In this case study, you will evaluate the causes of the disaster.

It should be noted that the purpose of this case study is not to dwell on disasters or to assign blame in the Kansas City incident. The purpose is to expose you to *forensic engineering* (a branch of engineering involving the investigation of failures) and to engineering ethics.

23.2   THE STORY

"What in the world are you doing, Josh?"

Cole looked in amazement at the mess in the dorm room that he and Josh shared. Used pizza boxes littered the floor. Textbooks competed for available floor space with a flat soccer ball, a guitar with only three strings, two weeks of dirty laundry, and Josh's prized 12-speed bicycle. None of this was particularly out of the ordinary for their room. Cole's attention had been captured by the four chains hanging from the ceiling and the dismantled parts of Josh's bed that lay around the room.

OBJECTIVES

After reading this chapter, you will be able to:

- calculate live loads, dead loads, and safety factors in simple structural systems;
- evaluate modes of failure;
- apply ethics to a real engineering scenario.

Text link: Review the lessons of failure in the *Focus on Design* following Section 7.8.

"Got this great idea. Saw it in a magazine. Gonna hang my bed from the ceiling." Josh was a man of few words and even fewer good ideas. Cole sighed. He had played guardian angel with Josh before. Last month, he had to convinced Josh that, no, it *wasn't* a good idea to bleach all your clothes white to cut down on the number of loads of laundry. He also convinced him that you *can't* count beer as a vegetable just because it is flavored with hops.

"Umm, Josh. I'm not sure this looks safe," said Cole. In fact, Cole was convinced that it looked extremely *unsafe*. Josh had looped small chains over the beams near the ceiling of their room. The chains ended in large, messy balls of links looped around the metal frame of the bed.

"What's going on here?" Cole pointed to the twisted mass of links.

"Oh *that*," said Josh with a smile. "That's my idea. Was a Boy Scout for two years, ya know. Never made it past Tenderfoot, but I know my knots. I threw a clove hitch over each corner of the bed frame—tied it up real good. Now give me a hand with this." Josh gestured to the box spring. The two roommates lifted up the box spring and gingerly placed it on the suspended bed frame. As they stepped away, the knot closest to Cole made an ominous creaking sound. Suddenly, the end of the chain slipped through, causing the bed frame to sag precariously. A few seconds later, the frame and box spring crashed to the floor, forcing both students to leap out of the way.

"Holy guacamole!" exclaimed Josh. "What happened?"

Cole sat on the floor next to his friend as they surveyed the wreckage. "Sorry, Josh. What do you remember from statics?"

"Not much, man. Didn't get enough sleep that semester and pulled a D+."

Text link: Review the use of force balances in engineering (Appendix A).

"Maybe you ought to retake it, Josh. Remember the lesson from Sir Isaac? If forces do not balance, then the object will accelerate. I think we just learned that the friction force in the chain knot was not as large as the gravitational force acting on the bed. As Chloe—" Josh's on-again-off-again girlfriend and civil engineering major—"might put it, the capacity of the system was not sufficient to resist the load. And after the first knot slipped, the load was redistributed to the other three knots. That's why they failed so quickly after the first knot pulled through."

"Yeah, I guess you're right." Josh looked around at his broken bed and shook his head slowly. "Gonna have to rethink this, bro," he said sadly.

Two weeks later, Cole returned to school after a quick weekend trip home. He opened the door and casually threw his backpack towards his bed. The backpack landed on the floor with a thud. Cole looked up the unexpected noise. "Josh!" he yelled. "Where did you put my bed?!?"

Josh strolled in the room with a bag of potato chips in his hand. "Hey, you're back. How do you like it?" Cole followed Josh's glance to the makeshift bunk bed on the other side of the room. Josh had resuspended his bed from the ceiling beams, this time bolting the chain to itself after looping it around the bed frame. Not bad, thought Cole. His opinion changed quickly when he noticed that his own bed hung off the frame of the top bed.

"I took your advice and thought about forces this time," said Josh. "The chain has a capacity of 80 pounds. There are four chains supporting each bed. You and I are skinny—each of us plus one bed can't weigh more than 250 pounds. So the chains from the ceiling support *my* 250 pounds and the chains to your bed support *your* 250 pounds."

Cole stared at the monstrosity. He had no intention of sleeping in the free-swinging death trap. "Josh, Josh, Josh. What am I going to do with you?" He walked over to the bunk bed, eyeing it suspiciously. "I can see two problems here, my friend. First, the *safety factor* seems low." Josh had a blank expression on his face.

safety factor: the ratio of the capacity of a system to its load

"Remember the safety factor?" continued Cole. "It's the ratio of the capacity to the load. If you *didn't* hook up my bed, your safety factor would be the capacity of four times eighty or 320 pounds, divided by the load of 250 pounds, or … umm …"

"One point two eight?" For all of Josh's faults, he had a mind for math.

"Right, one point two eight. That means you have only 28% more capacity than load—seems low."

"I suppose you're right." Josh's shoulders slumped. "And the other problem?"

"Well, I think your safety factor is really much, much smaller than one point two eight. Here, let me show you. What is the load on these lower chains, Josh?" Cole tugged on the chains connecting the lower bed frame to the upper bed frame.

"About 250 pounds for you and the bed, I think."

"Okay. Now how about the load on *these* chains?" Cole twanged one of the chains connecting the upper bed to the ceiling beam.

"Two hundred and fifty?" Josh said hopefully.

"Nope, *five hundred pounds*. The upper set of chains supports both your bed *and* mine. See?"

Josh stared at the bed. "Oh yeah—I get it. Thanks, Cole."

"No problem, roommate. Now let's take this thing apart and get our beds back on the floor where they belong."

23.3 THE CASE STUDY

23.3.1 Introduction

You have been hired to evaluate the Kansas City disaster and the walkway design. The suspended walkways are shown in Figure 23.1. The fourth-floor walkway (upper right in Figure 23.1) collapsed and fell on the second-floor walkway (bottom right in Figure 23.1).

Figure 23.1. General Layout of Walkways (hanger rods for third-floor walkway not shown; Pfrang and Marshall, 1982)

The third-floor walkway (on the left in Figure 23.1) was not involved in the collapse. Each walkway was about 120 ft long and 18 ft wide.

Each walkway consisted of four main parts: stringers, box beams, a deck, and hanger rods.

Stringers

The stringers were I-beams running along the length of and under each walkway. They are shown in dark gray in Figure 23.2.

Box Beams

Box beams are hollow beams. Each walkway was supported by three box beams. Each box beam was made by welding together two flanged channel beams as follows: []. The box beams were transverse to and under the stringers. One box beam is shown in Figure 23.2.

Deck

The concrete deck provided the walkway surface. It is shown as a transparent box in Figure 23.2.

Hanger Rods

Six threaded hanger rods, each $1^{1}/_{4}$ inches in diameter, passed through the welds in the box beams and connected the box beams to the ceiling. (Two ceiling hanger rods for one box beam are shown in black in Figure 23.2.) Six additional hanger rods passed from the fourth-floor box beams to the second-floor box beams. The box beams rested on washers and nuts attached to the hanger rods. The nuts are shown as white rectangles in Figure 23.2. The fourth-floor walkway collapsed when the washers and nuts pulled through the box beams.

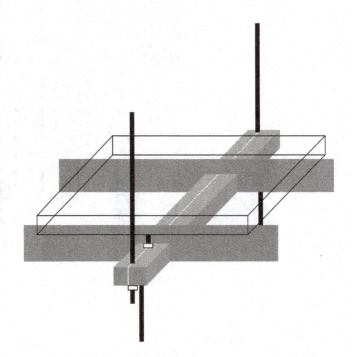

Figure 23.2.   Details of the Walkway Support System

Text link: Think about the connection between the case study story (Section 23.2) and the change in the hanger rod design.

In the original design, the hanger rods were continuous and ran from the ceiling to the bottom of the second-floor box beams (Figure 23.3, left). Thus, the design called for two hanger rods per box beam. During construction, changes were suggested and approved to change the hanger rod configuration. Each continuous rod was replaced with two rods: one running from the ceiling to the bottom of the fourth-floor box beams and one running from the top of the fourth-floor box beam to the bottom of the second-floor box beam. The "as-built" hanger rods are shown on the right side of Figure 23.3.

Modifications during construction are called *change orders* and are very common during large construction projects. The construction team may recommend changes to correct minor problems or facilitate construction. The change orders **must** be approved by the engineering firm. In the Hyatt Regency case, the change order was recommended because the original long hanger rods were difficult to maneuver on the work site.

23.3.2 Case Study

In this case study, you are charged with investigating the Hyatt Regency disaster. Specifically, your consulting firm has been hired to analyze three aspects of the disaster. First, you must determine whether the walkways were designed according to the building code. In addition, you are asked to calculate the number of welds required *if the walkways had been built as designed*.

Second, you must determine whether the walkways *as built* satisfied the building code. In addition, calculate the number of welds required for the walkways *as built*.

Third, comment on the engineering ethics aspects of the collapse.

To answer the case study questions, you must calculate the capacity (strength) of the walkway, the load on the walkways, and code requirements. Each of these factors will be discussed separately.

Capacity

Engineers sometimes express capacities and loads in units of *kips*. One kip is equal to 1,000 lb of force (or 4.45 kN of force). An object weighing one ton (2,000 lb) exerts 2 kips (or 8.9 kN) of force.

The calculation of capacity is difficult. In this case study, assume that the capacity of walkways comes only from the welds. (This is an oversimplification, since some capacity was provided by connections between the walkway and the walls at which they terminated.) From testing performed after the accident, it appears that the weld capacity is related to the area of the welds.

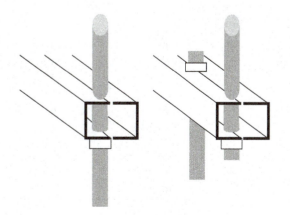

Figure 23.3. Detail of a Fourth-Floor Box Beam/Hanger Rod Connection ("as designed" on left, "as built" on right)

The weld testing generated the data in Table 23.1. There were six connections supporting the fourth floor (see Figure 23.1). The connections had weld areas of 129.6, 157.5, 128.4, 130.2, 150.5, and 132.0 mm². Use these data to estimate the capacity of each connection (i.e., each weld).

Load

Text link: Plot the data and develop a model to relate weld capacity to weld area (see Chapter 9). Review the measures of central tendency (Section 8.4).

The total load is the sum of the *dead load* and *live load*. The dead load comes from the construction materials. For the fourth-floor walkway, the construction materials were the stringers, box beams, and deck. To calculate the weight of the stringers and box beams, you need to know how steel is designated. Structural steel is given a designation that specifies its shape, depth, and weight. (The depth of a beam is its width.) The designation W means a wide flange beam (shaped like an I) and MC means a flanged channel beam (i.e., shaped like a square bracket, [ ). Thus, for example, W30×132 refers to a wide flange beam, about 30 inches deep, with a weight of 132 lb per foot of length. For the walkways, the stringers were W16×26 steel. The box beams were made by welding together two MC8 × 8.5 beams. The deck was $3\frac{1}{4}$-inch-thick concrete. Assume the density of concrete is 150 lb/ft³.

The live load comes from the people on the walkway. From videotape taken at the time of the disaster, it was estimated that about 60 people were on the walkways. For the purposes of this case study, assume that 30 people were distributed evenly on the fourth-floor walkway and 30 people were distributed evenly on the second-floor walkway.

TABLE 23.1 Connection Strength Data from Post-Accident Testing
(Pfrang and Marshall, 1982)

Weld Area (mm²)	Strength (kips)	Weld Area (mm²)	Strength (kips)
1.3	12.7	114.5	19.1
1.3	13.4	122.1	19.8
57.1	13.7	126.0	19.2
54.0	16.0	129.1	18.9
62.4	15.7	126.5	16.2
62.4	16.4	128.7	17.0
65.0	15.2	131.8	17.8
66.3	17.2	136.2	18.3
69.9	15.4	139.3	18.9
73.0	15.7	145.1	20.0
74.3	14.0	146.4	18.8
77.0	15.3	152.6	17.4
88.9	16.7	152.6	18.7
92.4	17.1	153.5	19.1
93.3	16.9	154.3	18.9
100.8	16.3	160.5	19.8
100.4	17.2	178.2	19.0
107.9	18.1		

Building Codes

Two parts of the 1979 Kansas City, Missouri, building code are important. First, the code required that engineers design for a *live load* of 100 lb per square foot of walkway. This means that the capacity must be designed to support a load equal to the dead load plus 100 lb per square foot of walkway.

Second, the safety factor for walkways was 1.67. The *safety factor* is the capacity divided by the load. This means you need a capacity at least 1.67 times the total load. You may wish to calculate the safety factor for the walkway as designed and walkway as built.

Ethics

The contractor submitted 42 drawings with proposed changes (including the change order for the hanger rods) to the engineering firm on February 16, 1979. The contractor did not emphasize the changes in the hanger rods. The lead engineer affixed his seal of approval to the revised engineering design drawings and sent them back to the contractor on February 26, 1979.

23.3.3   Reporting

Use the report format given to you by your instructor. (A default grading scheme is given at the end of the chapter.) The report should include a concise description of the walkways, disaster, and purpose of the report. In addition, you should summarize your calculations of the capacity of the system, the design load, and the actual load.

23.4   STUDY QUESTIONS

Remember, *all your work on this case study was performed with the conservative assumption that the walkways were held up only by the welds.*

1. Was the code satisfied by the design? Be sure to comment on whether the safety factor was up to code and whether the design met the code requirement of 100 lb/ft^2 live load.
2. Is the weld capacity strongly related to the weld area?
3. How many welds would be required to support the code-specified design load? Use the code-specified safety factor and calculate the total load as the design dead load plus the code-specified live load.
4. Was the safety factor satisfied as built?
5. How many welds would be required to support the actual load? Use the code-specified safety factor and calculate the total load as the actual dead load plus the code-specified live load.
6. Comment on the engineering ethics aspects of the lead engineer's actions.

23.5   ACKNOWLEDGMENTS AND FURTHER READING

The background information for this case study was obtained from Pfrang and Marshall (1982). For further discussion of the ethical aspects of this case, see http://ethics.tamu.edu/ethics/hyatt/hyatt1.htm. Photographs can be found at http://ethics.tamu.edu/ethics/ (click on "Hyatt Regency Walkway Photographs" or "Miscellaneous Hyatt Regency Walkway Photographs").

Default Grading Scheme: Walkway Collapse Case Study

Topic	Possible Points
Cover Page Names, date, title, group number (if appropriate)	5
Background Concise description of the walkways, disaster, and purpose of the report	10
Calculations Load: dead load, live load (15) Capacity: approach (relating weld area to capacity) and calculation (10)	25
"As-designed" Conditions Were the safety factors and capacity from the code satisfied as designed? (10) Discussion/calculation of the number of welds required as designed (10)	20
"As-Built" Conditions Were the safety factors satisfied as built? (10) Discussion/calculation of the number of welds required as built (10)	20
Ethics Quality of summary of ethical considerations	10
Technical Writing Organization (and use of headings), grammar, spelling, tables and figures, and seamlessness of report	10
Total	**100**

24

Trebuchet Case Study

24.1 INTRODUCTION

The trebuchet° was a medieval siege weapon. It was used to hurl large objects great distances. To throw an object, any launcher must convert energy from one form into the kinetic energy of the projectile. For example, a cannon converts chemical energy from the explosive charge into the kinetic energy of the projectile and heat.

In medieval times, launchers were used to lay siege against fortified towns. Medieval siege weapons based on launching projectiles included the catapult (and related devices[†]) and the trebuchet. Each of the medieval launchers uses a different source of energy. The catapult (from the Greek *kata*, through, + *pallein*, to hurl) stores energy in an elastic material called a *skein*. The material, usually rope, often is twisted to store additional energy.

The trebuchet works on a different principle. In the trebuchet, a falling counterweight provides the energy to launch the projectile. The potential energy of the counterweight is converted into the kinetic energy of the thrown object.

A typical trebuchet is shown in Figure 24.1. When the counterweight is released (black oval in Figure 24.1), the arm rotates around the pivot point (black dot), and the projectile (gray circle), in a pouch or sling (white oval), swings around the top of the trebuchet. The projectile is flung over the top of the trebuchet and to the right in Figure 24.1.

The trebuchet could be a devastating weapon. A famous medieval trebuchet, called the Warwolf, had a 40-foot arm. It could throw a 250-pound projectile a distance of about 600 feet. In other words, it could easily fling you the length of two football fields!

OBJECTIVES

After reading this chapter, you will be able to:

- design and build a trebuchet;
- apply the steps in the engineering design method to a real engineering problem;
- use models to evaluate the design of a trebuchet.

°The word trebuchet (pronounced "treb-uh-shet") comes from the Latin *trabuchare* to overturn. As you will see, the trebuchet operates by rotating (or overturning) around a pivot point.

[†]Related devices include the *ballista* and *arbalest*. The ballista is similar to the catapult. According to the Oxford English Dictionary, the ballista and catapult originally were different weapons. The ballista (from the Greek *ballein*, to throw) was used to throw stones and the catapult was used to throw arrows. The arbalest (from the Latin *arcus* bow, + *ballista*) resembled a large crossbow.

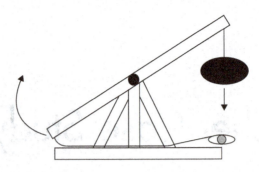

Figure 24.1.   An Example of a Trebuchet

In this case study, you will analyze, design, build, test, and optimize a trebuchet. If you wish, you can organize a contest between your classmates regarding the performance of your trebuchet (longest distance, maximum height, most attractive design, etc.).

24.2   THE STORY

Luke leaned against the giant oak tree and once again counted his blessings. He had left his family on his seventh birthday to become a page for Sir William, Knight of the Marches. Luke had become a squire at age 14, and now, two months before his 20th birthday, he was nearing knighthood. If only he could make a significant contribution to Sir William's land, then perhaps his path to knighthood would be faster.

The Marches, the borderland between England and Wales, were in constant dispute and Sir William suited for battle frequently. Perhaps when I am a knight, mused Luke, I will be as brave as Sir William. Luke's daydreaming was interrupted by the booming voice of his liege.

"Luke, where the devil are you, lad? I have some people here I want you to meet."

Luke dusted himself off and ran to his knight's side. "Yes, Sir William?" The tall, proud knight stood next to an older man with gleaming blue eyes and a woman about Luke's age.

"Luke, this is Giles, a weapons maker from the land of Sir Walter. He and his daughter, Johanna, will be our guests until the winter winds come. Please settle our guests in their rooms and arrange for a feast in the banquet hall."

Luke escorted Giles and Johanna into the castle. An hour later, Sir William, his wife Lady Martha, Luke, and their guests were seated at the enormous banquet table.

"Giles has agreed to build a new siege weapon for us," began Sir William. "It is called a ... "

"A trebuchet, Sir William. It can be used to throw heavy objects at or over the castle walls of your enemies." Giles smiled mischievously. "Some use it to toss heavy stones. Others use it to toss dead cows." Lady Martha frowned at the image of a dead cow soaring over the walls of *her* castle.

"And how far can you throw large stones?" asked Luke.

"Ah, well, the range is not as good as your best archers. We are working to improve the device, of course."

Improvements, thought Luke. Perhaps this is the way I can impress Sir William and finally earn my knighthood. The meal progressed swiftly and soon the group dispersed to their bedchambers. On the walk out of the dining hall, Johanna motioned Luke into the shadows of a dark corridor.

"I noticed your eyes light up when my father said that the trebuchet needed improvements."

Luke was startled at Johanna's perceptiveness. "Why, yes. I was hoping that I could make the device better."

Johanna smiled. "As have I. Meet me in the courtyard tonight when the moon is at its highest. I have something to show you."

Luke arrived at the appointed time and found Johanna with a wooden contraption in her hands. The device looked like a smaller version of the trebuchet that Giles had sketched earlier in the evening.

"Is *that* the weapon that will aid Sir William?" said Luke incredulously.

Johanna laughed. "Not exactly, Squire Luke. It is but a small trebuchet I hold here. I built this myself to see how to improve the larger machines that my father builds." Luke looked at the miniature siege weapon with interest. He could see the value in such a model. Individual parts of the device could be tested independently.

"Pray, what have you found so far, Johanna?"

"Many of the lessons I have learned from my friend here fit a pattern." She patted the small trebuchet. "I use it to throw small pebbles. Of course, lighter pebbles fly farther." She tugged at the counterweight. "And using a larger weight here makes my little pebbles soar. A longer pivot arm also helps, since it makes the large weight fall farther."

"You have discovered much," said Luke, with a respectful voice. "Is there anything left to uncover about the operation of this device?"

"I believe the trebuchet has many secrets still shrouded in darkness. For example, if the pebble leaves the pouch too soon or too late, it does not travel far. My father's machine suffers from the same problem."

Luke smiled at Johanna. "Perhaps we can work together to solve the mysteries of the trebuchet."

A modern model trebuchet with a one-foot arm (photo courtesy of trebuchet.com).

24.3   THE CASE STUDY

24.3.1   Introduction

The operation of the trebuchet may have been a mystery in the time of Luke and Johanna, but you can use engineering analysis tools to optimize the performance of the device.

At the simplest level of analysis, consider the effects of the *release angle* and the *initial velocity* on the trajectory of the projectile. Assume that the projectile is released at an angle θ with the ground and an initial velocity of v_0. (See Figure 24.2.) The initial velocities are $v_0 \cos \theta$ and $v_0 \sin \theta$ in the x and y directions, respectively.

Now perform a force balance in both the x and y directions:

$$\text{sum of forces in the } x \text{ direction} = ma_x$$
$$\text{sum of forces in the } y \text{ direction} = ma_y$$

Text link: Review the use of force balances in engineering (Appendix A).

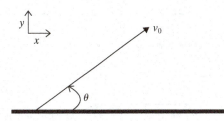

Figure 24.2. Projectile Trajectory Analysis

where m = projectile mass, a_x = acceleration in the x direction, and a_y = acceleration in the y direction. No horizontal forces are acting on the projectile (if air resistance is ignored) and the only vertical force is gravity (force = $-mg$). Acceleration is the second derivative of x or y with respect to time:

$$ma_x = m\frac{d^2x}{dt^2} = 0 \tag{24.1}$$

$$ma_y = m\frac{d^2y}{dt^2} = -mg \tag{24.2}$$

Equations (24.1) and (24.2) can be solved with the initial velocities ($v_0 \cos\theta$ and $v_0 \sin\theta$ in the x and y directions, respectively) and initial positions ($x = 0$ and $y = 0$) to yield[*]

$$x = (v_0 \cos\theta)t \tag{24.3}$$

and

$$y = -\tfrac{1}{2}gt^2 + (v_0 \sin\theta)t \tag{24.4}$$

Text link: See a similar problem with a champagne cork in Section 6.6.3.

Equations (24.3) and (24.4) can be manipulated to predict the performance of a trebuchet. The results are summarized in Table 24.1.

You can use the results in Table 24.1 to select an angle to maximize the range or height.

THOUGHTFUL
PAUSE

What release angles give the maximum range and maximum height?

TABLE 24.1 Trajectory of a Projectile (initial velocity = v_0, initial angle = θ, initial height = 0)

Characteristic	Approach	Result
Maximum distance (= range = R)	Find the landing time by setting Eq. (24.4) equal to 0 (since $y = 0$ when the projectile hits the ground); then plug the landing time into Eq. (24.3) and solve for x	range = $R = \dfrac{v_0^2}{g}\sin(2\theta)$, occurring at time $t = 2\dfrac{v_0}{g}\sin\theta$[a]
Maximum height	Set $dy/dt = 0$ (since the velocity in y direction = 0 at the maximum height)	maximum height = $\dfrac{(v_0 \sin\theta)^2}{2g}$, occurring at time $t = \dfrac{v_0}{g}\sin\theta$
Trajectory	Eliminate t in Eqs. (24.3) and (24.4)	$y = -\dfrac{g}{2v_0^2 \cos^2\theta}x^2 + x\tan\theta$[b]

[a] Range result uses the trigonometric identity: $2\sin\theta\cos\theta = \sin(2\theta)$

[b] The notation $\cos^2\theta$ means $(\cos\theta)^2$.

[*]Confirm that $x = 0$ and $y = 0$ at $t = 0$. If you know differential calculus, differentiate Eqs. (24.3) and (24.4) with respect to time and confirm that $dx/dt = v_0 \cos\theta$ at $t = 0$ and $dy/dt = v_0 \sin\theta$ at $t = 0$. Differentiate again with respect to time to regenerate Eqs. (24.1) and (24.2).

Text link: Review the use of energy balances in engineering (Appendix A).

You can show that the range is maximized at a release angle of $45°$ to the horizontal and the height is maximized at a $90°$ angle to the horizontal.

What about the effects of v_0 on the range and maximum height? From Table 24.1, v_0 should be made as large as possible to maximize the range or height. This can be accomplished by efficiently converting the potential energy from the falling counterweight into the kinetic energy of the projectile. From an energy balance,

initial kinetic energy of the projectile = potential energy of the counterweight

or

$$\tfrac{1}{2}mv_0^2 = Mgh$$

where M = counterweight mass and h = distance that the counterweight falls. From Table 24.1,

$$R = \frac{v_0^2 \sin(2\theta)}{g},$$

so

$$v_0^2 = \frac{Rg}{\sin(2\theta)}$$

Inserting into the energy balance,

$$\frac{1}{2}\frac{mRg}{\sin(2\theta)} = Mgh$$

or

$$R = \frac{2hM \sin(2\theta)}{m} = 2h\left(\frac{M}{m}\right)\sin(2\theta). \tag{24.5}$$

Text link: Review the use of logic to check engineering calculations (Section 6.6.3).

What are the implications for design? To maximize the range, you must maximize the ratio of M to m (M/m) and maximize the distance that the counterweight falls (h). Think through Eq. (24.5) to make sure that these results make sense. The results of this simple analysis are summarized in Table 24.2.

Much more sophisticated analysis of the trebuchet is possible. For more ideas, see the analysis reference in Section 24.5 and in the Appendix.

TABLE 24.2 Summary of the Results of the Trebuchet Analysis

Parameter	Optimum θ	Optimum v_0	Maximum Parameter Value
Maximum range	$45°$	as large as possible	$\dfrac{v_0^2}{g} = 2h\left(\dfrac{M}{m}\right)$
Maximum height	$90°$	as large as possible	$\dfrac{v_0^2}{2g} = h\left(\dfrac{M}{m}\right)$
Maximum time aloft	$90°$	as large as possible	$\dfrac{2v_0}{g} = 2\sqrt{\dfrac{2h\left(\dfrac{M}{m}\right)}{g}}$

Note: The maximum time aloft is the time elapsed in reaching the maximum range (or twice the time required to reach the maximum height).

24.3.2   Case Study

This case study is much more open ended than the other case studies in this text. Your job is to analyze, design, build, test, and optimize a trebuchet to meet an objective given to you by your instructor. A common goal is to build a trebuchet to achieve the longest range with a given projectile, counterweight mass, and arm length. Even with M/m fixed and the arm length fixed, you still have a number of design choices, including the following:

- **Counterweight type**
  You will get different results if the counterweight is fixed or allowed to swing.
- **Pivot position**
  You can adjust the pivot point. The pivot point will determine the proportion of the arm on the counterweight side of the pivot and the proportion of the arm on the projectile side of the pivot.
- **Height of pivot**
  The height of the pivot point also influences the performance of the trebuchet.
- **Sling and pouch construction**
  You will find that proper sling and pouch construction is critical to optimal release of the projectile.

To compare designs, it is important to limit the materials of construction. Common trebuchet construction materials are twigs and twine; cardboard and tape; or dowels and twine. Common projectiles are small candies. Sources of design ideas are listed in Section 24.5. To test and optimize your ideas before construction, use the design equations in the chapter and the simulators listed in Section 24.5. *Always use eye protection and take other necessary safety precautions when testing trebuchets.*

24.3.3   Reporting

Use the report format given to you by your instructor. (A default grading scheme is given at the end of the chapter.) The report should include a concise description of the project objectives, analysis methods, design methods, and results.

24.4   STUDY QUESTIONS

1. What is the effect of gravity on the optimum range? Would your trebuchet perform differently on the Moon?
2. Can you explain the differences, if any, between the design calculations (and/or simulations) and actual performance?
3. How can your trebuchet be improved?

24.5   ACKNOWLEDGMENTS AND FURTHER READING

Thanks to Prof. Christina Bloebaum of the University at Buffalo for suggesting and organizing a trebuchet-building contest. Many Internet resources are available for the analysis, design, and history of the trebuchet. Following are the URLs of a few Web sites:

General introduction and history
Nova Web site (http://www.pbs.org/wgbh/nova/lostempires/trebuchet/)

Trebuchet model ideas
Cutouts for paper models (http://www.fryerskits.demon.co.uk/treb)
Small wooden model (http://www.io.com/~beckerdo/other/trebuchet.html)

Kits to purchase (http://www.trebuchet.com/kit/tabletop);
(http://www.redstoneprojects.com/trebuchetstore/siege_engine_plans.html)

Trebuchet calculators and simulators

Windows and Mac simulators (http://www.algobeautytreb.com);
Java simulator (http://www.algobeautytreb.com/javatreb.html);
Shockwave simulator (http://www.pbs.org/wgbh/nova/lostempires/trebuchet/destroy.html)

Trebuchet analysis

http://www.algobeautytreb.com; see the section on trebuchet mechanics

Default Grading Scheme: Trebuchet Case Study

Topic	Possible Points
Cover Page	
Names, date, title, group number (if appropriate)	5
Background	
Concise description of the problem, context, and objectives	10
Methods	
Analysis method documentation (5)	
Design method documentation (10)	15
Results	
Presentation of at least three simulation run results (10)	
Presentation of at least three projectile throwing results (10)	20
Discussion	
Comparison of simulations and actual results (10)	
Explanation of differences between simulations and actual results (10)	
Potential improvements (5)	25
Conclusions	
Quality of summary and recommendations	15
Technical Writing	
Organization (and use of headings), grammar, spelling, tables and figures, and seamlessness of report	10
Total	100

Appendix A
Review of Physical Relationships

A.1   INTRODUCTION

Text link: The laws of conservation, Newton's Laws of Motion, and some constitutive laws are reviewed in Section 6.3.2.

The purpose of this appendix is to review the physical relationships referred to in the body of this text. This appendix is not a replacement for a class on statics or dynamics or circuits. It is intended to supplement and extend the material in the text. In addition, examples of the physical relationships discussed in the text will be summarized in this appendix. For a review of the laws of conservation, Newton's Laws of Motion, and some constitutive laws, see Section 6.3.2.

A.2   DEFINITIONS

A.2.1   Kinematic Parameters

A number of physical parameters were referred to in this text, often without formal definitions. The definitions of some kinematic measures are given in Table A.1.

A.2.2   Fundamental Forces*

Definitions of terms related to forces are given in Table A.2. Force can be defined from kinematics as the change in momentum with respect to time:

$$F = d(mv)/dt$$

At constant mass, this becomes Newton's Second Law of Motion:

$$F = ma$$

If the main source of acceleration is Earth's gravitational field, then $a = g$ and $F = mg$.

Although Newton's Second Law of Motion is extremely valuable for engineers, it hides where forces come from. We know of only four sources of forces (called the *fundamental forces*): gravity, electromagnetic, strong, and weak forces. The strong force holds the nucleus of an atom together and does not play a role in traditional engineering (but may influence nanoengineering). Similarly, the weak force is responsible for radioactive decay and does not play a role in the professional lives of most engineers.

---

*The discussion of fundamental forces and other forces was influenced in part by Holtzapple and Reece (2000).

TABLE A.1 Definitions of Kinematic Parameters

Parameter	Definition	Text Example (section in parentheses)
Velocity (v)	v = distance/time	controllability (19.3.1)
		settling velocity (7.7.2)
	instantaneous $v = dx/dt$	trebuchet (24.3.1)
Momentum (p)	$p = mv$	See Table A.5
Acceleration (a)	instantaneous $a = dv/dt = d^2x/dt^2$	trebuchet (24.3.1)

x = distance (m) m = mass (kg)
t = time (s) p = momentum (kg-m)
v = velocity (m/s) a = acceleration (m/s^2)

TABLE A.2 Types of Forces

Parameter	Definition	Text Example (section in parentheses)
Newton's Second Law	$F = ma$	novel propulsion system (6.3.2°)
		waterbed (6.3.2,° 6.3.3)
		parachuting (6.4.2)
		moving van ramp (6.6.3)
		forces on a car (7.6)
		in technical communications (14.3.3°)
		bridge lateral acceleration (18.5)
		projectiles (24.3.1)
Gravitational	$F = G\dfrac{m_1 m_2}{r^2}$ (for Earth, $F = mg$)	
Electrostatic	$F = \dfrac{1}{4\pi\varepsilon_0}\dfrac{q_1 q_2}{r^2}$	
Magnetic	$F = \dfrac{3\mu_0}{2\pi}\dfrac{\mu_1\mu_2}{r^4}$	
Friction	$F = \mu$(perpendicular force)	moving van ramp (6.6.3)
		forces on a car (7.6)
Drag	$F = \frac{1}{2}C_d\rho v^2 A$	parachuting (6.4.2)
		forces on a car (7.6)
		moving belt (22.3.1)
Spring	$F = kx$ (Hooke's Law)	Hooke's Law statement (6.3.2°)
		bungee jumping (6.6.2)
	$F/A = E\delta/L$	compression of bed legs (6.8)

° Description only (all other examples are worked examples)

F = force (kg-m/s^2)
a = acceleration (m/s^2)
m, m_1, and m_2 = mass (kg)
G = gravitational constant = 6.6720×10^{-11} N-m^{2}/kg^2
r = separation distance (m)
q_1, q_2 = charges on objects (C = coulombs)
ε_0 = permittivity of a vacuum = 8.854×10^{-12} C^2/(N-m^2)
μ_0 = permeability of a vacuum = $4\pi \times 10^7$ N/A^2
μ_1, μ_2 = magnetic dipole moments (A-m^2)

μ = coefficient of friction (dimensionless)
C_d = drag coefficient (dimensionless)
ρ = fluid density (kg/m^3)
A = cross-sectional area (m^2)
k = spring constant (kg/s^2)
x, δ = displacement (m)
stress = F/A
E = Young's modulus (Pa)
L = length

The main fundamental force used in this text is the force from gravity.[*] Gravitational force is responsible for Newton's Second Law of Motion. The gravitational force is proportional to the product of the masses of two objects and inversely proportional to their separation distance squared:

$$\text{gravitational force} = G\frac{m_1 m_2}{r^2}$$

where G = gravitational constant (6.6720×10^{-11} N-m^2/kg^2); m_1 and m_2 = masses of the objects; and r = separation distance. Most engineers deal with objects near the surface of the Earth, where m_2 = mass of the Earth = 5.98×10^{24} kg and r = radius of the Earth = 6.37×10^6 m. Thus,

$$Gm_2/r^2 = (6.6720 \times 10^{-11} \text{ N-m}^2\text{kg}^2)(5.98 \times 10^{24} \text{ kg})/(6.37 \times 10^6 \text{ m})^2$$

$$= 9.8 \text{ m/s}^2$$

$$= g = \text{Earth's gravitational acceleration}$$

Thus, for objects near the surface of the Earth,

$$\text{gravitational force} = Gm_1 m_2/r^2 = mg$$

A.2.3 Other Forces

Three other forces of interest to engineers are the friction, drag, and spring forces. Your common experience is that you must exert a force to start an object in motion and to keep the object in motion at constant velocity. This suggests you are overcoming another, "unknown," force, the *friction force*. The friction force is proportional to the perpendicular force. The proportionality constant is called the *coefficient of friction*.[†]

Friction also occurs if you push an object through a fluid. The friction force in opposition to the direction of movement is called the *drag force*. In some engineering applications, the drag force is calculated using a dimensionless parameter called the *drag coefficient* (C_d):

$$\text{drag force} = \tfrac{1}{2}C_d \rho v^2 A$$

where ρ = fluid density and A = cross-sectional area of the object.

Finally, engineers often deal with devices and materials that experience the following behavior: the force exerted by the material is proportional to the displacement (i.e., proportional to the distance moved or stretched). A force that is proportional to the displacement is called a *spring force*. An example, not surprisingly, is a spring. The force exerted by a spring is usually given by Hooke's Law:

$$\text{spring force} = kx$$

where k is the spring constant and x is the displacement. Another example is the stress–strain constitutive relationship:

$$\text{stress} = E(\text{strain})$$

where stress = force per unit area, E = Young's modulus, and strain = displacement per unit length.

A.2.4 Energy, Work, and Power

The concepts of energy and work are interrelated. Energy is the ability to do work. Work is the transfer of energy by a force acting over a distance. How are we to break this circle of definitions?

---

[*]The electromagnetic force is usually modeled as electrostatic and magnetic forces. For completion, the electrostatic and magnetic forces are listed in Table A.2.

[†]The proportionality constant is called the *coefficient of static friction* if the object is not moving (but on the threshold of movement) and is called the *coefficient of kinetic friction* if the object is moving. The value of the coefficient of static friction typically is greater than the value of the coefficient of kinetic friction.

TABLE A.3 Energy and Power Examples

Parameter	Text Example (section in parentheses)
Energy	champagne cork (6.6.3)
	rocket (6.3.2*, 6.3.3)
	bungee jumping (6.6.2)
	kinetic energy (8.4.6*)
	trebuchet projectile (24.3.1)
Power	electrical power loss (5.4.3, 6.5.4, 22.3.1)
	automobile power losses (7.6)
	hydraulic power loss (22.3.1)
	friction power loss (22.3.1)

* Description only (all other examples are worked examples)

Perhaps it is better to start with the definition of work as

$$\text{work} = \text{force} \times \text{distance}$$

With this definition, it is clear that work should have units of N-m, where 1 N-m = 1 joule = 1 J. Work is the transfer of energy, so energy also must have units of joules.

Energy often is divided into three types: internal, kinetic, and potential energy. Internal energy is the energy inside the system. Kinetic energy is the energy associated with motion. Potential energy is the energy that comes from the position of an object in a potential field. An example of a potential field is a gravitational field. Common expressions for kinetic energy and the potential energy in the Earth's gravitational field are

$$\text{kinetic energy} = \tfrac{1}{2}mv^2$$
$$\text{potential energy} = mgh$$

Power is the work produced per unit time. The units of power are J/s, where 1 J/s = 1 watt = 1 W. Power loss in electrical systems is given by

$$\text{power loss} = VI$$

Combining power loss with Ohm's Law ($V = IR$, see Section A.5) yields

$$\text{power loss} = (IR)I = I^2R$$

Examples where energy and power are used in this text are shown in Table A.3.

A.3 DECOMPOSITION BY VECTORS

A.3.1 Position Vectors

We typically describe the relationship between two objects in space by a vector. In other words, the difference in position has both magnitude and direction. To get from Point A to Point B below, you would move five units in a direction of about 53.1° from the horizontal. Another way to move from Point A to Point B is to move horizontally three units to the right, then move vertically for four units (see Fig. A.1). In this approach, you have decomposed the position into an x component (+3 units, if positive x is to the right) and a y component (+4 units, if positive y is up).

A.3.2 Other Vectors

If position is a vector, then the other parameters derived from position also are vectors. From Tables A.1 and A.2, it follows that velocity, acceleration, momentum, and force also are vector quantities. Velocity and force were decomposed into x and y components several times in this text, as indicated in Table A.4.

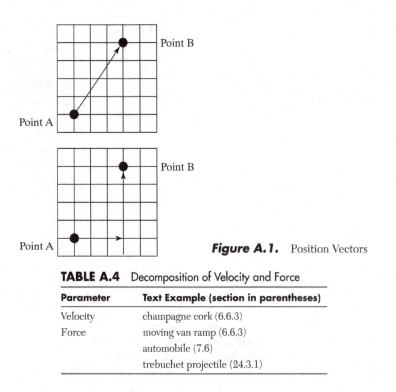

Figure A.1. Position Vectors

TABLE A.4 Decomposition of Velocity and Force

Parameter	Text Example (section in parentheses)
Velocity	champagne cork (6.6.3)
Force	moving van ramp (6.6.3)
	automobile (7.6)
	trebuchet projectile (24.3.1)

A.4 CONSERVATION LAWS

Text link: See Section 6.3.2 for the laws of the conservation of mass, momentum, angular momentum, energy, and charge.

Examples of the laws of the conservation of mass, momentum, angular momentum, energy, and charge are listed in Table A.5.

TABLE A.5 Examples of Conservation Laws

Conservation Law	Text Example (section in parentheses)
Mass	moles as combining proportions (2.3.2)
	disinfection device (number balance) (5.4.3)
	parts inventory (6.3.2°)
	chemical doses (6.3.2°)
	flow in = flow out (6.3.2°)
Momentum	ink flow in inkjet printer (6.3.2°)
	baseball (6.3.2°)
	crash dummy's head (6.3.2°)
	example laboratory report (starts 13.2.3)
Angular momentum	Kepler's Second Law (6.3.2°)
Energy	global warming (5.8)
	refrigeration system (6.3.2°)
	airplane wing lift (6.3.2°)
	Kirchhoff's Voltage Law (6.3.2°)
	rocket escape velocity (6.3.2°, 6.3.3)
	bungee jumping (6.6.2)
	kinetic energy (8.4.6°)
	trebuchet (24.3.1)
Charge	pH of acid rain (6.3.2°)
	Kirchhoff's Current Law (6.3.2°)
	ionization smoke detector (6.3.2)
	current in subcircuits (6.3.3)

° Description only (all other examples are worked examples)

A.5 GRADIENT-DRIVEN PROCESSES

A number of processes of interest to engineers are driven by gradients. In other words, flux of some property is proportional to the difference in the value of a parameter across space. (Flux is the change in the property over time.) The proportionality constant is a kind of conductivity, while the inverse of the proportionality constant is a kind of resistance:

$$\text{flux} = (\text{conductivity})(\text{gradient}) = (1/\text{resistance})(\text{gradient})$$

As an example, consider the flux of electrons (also called the *current density* $= I/A$, where I = current and A = area). Why do electrons flow? Electrons flow because of a difference in potential (voltage). Formally,

$$I/A = \sigma(\Delta V/\Delta x)$$

where V is the voltage, σ is the electric conductivity, and Δ represents a difference in a property. We often use the symbol V to represent the voltage difference, so

$$I = A\sigma V$$

As discussed, the term $A\sigma$ represents the inverse of the *resistance* to the flow of electrons. Thus,

$$I = (1/R)V$$

or

$$V = IR$$

where R = resistance. This is Ohm's Law.

Examples of gradient-driven processes in the text are given in Table A.6. Only one-dimensional spatial gradients are shown in Table A.6. Another common gradient-driven process of interest to engineers is Fourier's Law of Heat Conduction:

$$\text{heat flux} = k(\Delta T/\Delta x)$$

where k = thermal conductivity and $\Delta T/\Delta x$ is the temperature gradient.

TABLE A.6 Gradient-Driven Processes

Name	Equation	Text Example (section in parentheses)
Fick's Law	diffusive flux $= -DA(\Delta C/\Delta x)$	dissolution (20.3.1)
Newton's Law of Viscosity	shear stress $= \mu(\Delta v/\Delta x)$	viscosity units (6.7.2)
Ohm's Law	$I/A = \sigma(\Delta V/\Delta x)$ or $V = IR$	electrical power loss (5.4.3, 6.5.4, 22.3.1)
		definition (6.3.2°)
		resistances in series (6.5.4)

° Description only (all other examples are worked examples)
D = diffusion coefficient (m^2/s) v = velocity (m/s)
A = cross-sectional area (m^2) V = voltage (V = volts)
C = mass concentration (kg/m^3) I = current (A = amperes)
x = distance (m) R = resistance (Ω = ohms)
μ = viscosity (kg/m-s) σ = electric conductivity (1/Ω-m or S/m, S = siemens = 1/Ω)

Appendix B
Greek Alphabet in Engineering, Science, and Mathematics

Letter	Symbols	Common Use in Engineering, Science, and Mathematics
alpha	A, α	α: generalized angle, thermal diffusivity
beta	B, β	β: coefficient of bulk expansion
gamma	Γ, γ	γ: specific gravity or specific weight
delta	Δ, δ	Δ: change, δ: small change
epsilon	E, ε	ε: dielectric constant
zeta	Z, ζ	
eta	H, η	η: Stefan–Boltzmann coefficient
theta	Θ, θ	θ: generalized angle or dimensionless time
iota	I, ι	
kappa	K, κ	
lambda	Λ, λ	λ: wavelength
mu	M, μ	μ: prefix meaning 10^{-6} or viscosity or population mean
nu	N, ν	ν: frequency
xi	Ξ, ξ	ξ: permeability
omicron	O, o	
pi	Π, π	Π: product of a series of numbers
		π: circumference of circle/diameter of circle $= 3.14159\ldots$
rho	P, ρ	ρ: density or resistivity
sigma	Σ, σ	Σ: sum of a series of numbers
		σ: population standard deviation or surface tension or conductivity
tau	T, τ	τ: shear stress
upsilon	Y, υ	
phi	Φ, ϕ	Φ: potential
chi	X, χ	
psi	Ψ, ψ	
omega	Ω, ω	Ω: ohm (unit of electric resistance)
		ω: radial frequency or angular velocity

Appendix C
Linear Regression

C.1   INTRODUCTION

Text link: Review the use of models in engineering (Chapter 9).

A very common model used in engineering is the linear model. In the linear model, the dependent variable is related to the independent variable raised to the first power. Thus, linear models include $y = 2x + 5$ and $y = xe^{3.2} - \pi$. Models such as $y = 4x^2$ or $y = \sin(x)$ are not linear, because the dependent variable (y) is **not** related to the independent variable (x) raised to the first (or zero) power.

Linear models are common for two reasons. First, many natural phenomena are linear. For example, force is linearly related to acceleration in a system of constant mass through $F = ma$.

Second, many nonlinear systems can be linearized through a transformation of variables. For example, a common kinematic equation is

$$y = \tfrac{1}{2}at^2$$

PONDER THIS

Is the equation $y = \tfrac{1}{2}at^2$ linear?

This equation is not linear: y depends on t^2, not on t raised to the first or zero power. It can be made linear by defining a new independent variable x equal to t^2. Thus, a linearized model is $y = \tfrac{1}{2}ax$.

The general form of the linear model is

$$y = mx + b$$

Note that in a linear model, a plot of the dependent variable (on the y-axis) against the dependent variable (on the x-axis) is a straight line, as shown in Figure C.1. The line has a slope of m and an intercept of b. The parameter m has units of (units of y)/(units of x). The parameter b has the same units as y.

C.2   LINEAR REGRESSION ANALYSIS

In Section C.1, the equation for a straight line was written as

$$y = mx + b$$

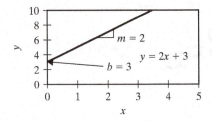

Figure C.1: An Example of a Linear Model

Using the symbolization associated with models, we can write this equation as

$$\hat{y}_i = mx_i + b \tag{C.1}$$

Text link: Review symbols used in models (Section 9.4.3).

$\hat{y}_i$ is the model-predicted value of y when $x = x_i$. The subscript i goes from 1 to n (n data pairs). The square of the error (i.e., the square of the difference between the model prediction and measured data) is equal to $(y - \hat{y}_i)^2$. Thus, the sum of the squares of the errors, SSE, is

$$\sum_{i=1}^{n} (y - \hat{y}_i)^2 \tag{C.2}$$

In Eq. (C.2), $\sum_{i=1}^{n}$ means "the sum from $i = 1$ to n." Combining Eqs. (C.1) and (C.2) yields

$$\text{SSE} = \sum_{i=1}^{n} (y_i - \hat{y}_i)^2 = \sum_{i=1}^{n} [y_i - (mx_i + b)]^2 \tag{C.3}$$

Expanding Eq. (C.3) results in

$$\text{SSE} = \sum_{i=1}^{n} (y_i^2) - 2m \sum_{i=1}^{n} (x_i y_i) - 2b \sum_{i=1}^{n} (y_i) + m^2 \sum_{i=1}^{n} (x_i^2)$$
$$+ 2mb \sum_{i=1}^{n} (x_i) + nb^2 \tag{C.4}$$

Note that terms without subscripts can be brought outside the summation sign. Thus, $\sum_{i=1}^{n} b^2$ means "add b^2 to itself n times" $= nb^2$.

The goal of linear regression is to find the values of m and b that minimize SSE and thus minimize the right-hand side of Eq. (C.4).

The easiest way to minimize SSE with respect to m and b is to use calculus. From differential calculus, the function SSE is minimized when the derivative of SSE with respect to m and the derivative of SSE with respect to b are both zero. If you have knowledge of calculus, you can verify the derivatives below:

$$\text{derivative of SSE with respect to } m = -2 \sum_{i=1}^{n} (x_i y_i) + 2m \sum_{i=1}^{n} (x_i^2) + 2b \sum_{i=1}^{n} (x_i)$$

$$\text{derivative of SSE with respect to } b = -2 \sum_{i=1}^{n} (y_i) + 2m \sum_{i=1}^{n} (x_i) + 2nb$$

Setting the derivatives equal to zero creates two equations in the two unknowns m and b:

$$2m \sum_{i=1}^{n} (x_i^2) + 2b \sum_{i=1}^{n} (x_i) = 2 \sum_{i=1}^{n} (x_i y_i)$$

$$2m \sum_{i=1}^{n} (x_i) + 2nb = 2 \sum_{i=1}^{n} (y_i)$$

Solving this system of two equations in two unknowns, we have

Text link: Review arithmetic means (Section 8.4.2).

$$m = \frac{\sum_{i=1}^{n} x_i y_i - \frac{1}{n}\left(\sum_{i=1}^{n} x_i\right)\left(\sum_{i=1}^{n} y_i\right)}{\sum_{i=1}^{n} x_i^2 - \frac{1}{n}\left(\sum_{i=1}^{n} x_i\right)^2} = \frac{\sum_{i=1}^{n}(x_i - \bar{x})(y_i - \bar{y})}{\sum_{i=1}^{n}(x_i - \bar{x})^2} \qquad (C.5)$$

$$b = \frac{1}{n}\left(\sum_{i=1}^{n} y_i - m\sum_{i=1}^{n} x_i\right) = \bar{y} - m\bar{x} \qquad (C.6)$$

where $\bar{x} = \frac{1}{n}\sum_{i=1}^{n} x_i$ = arithmetic mean of x values and $\bar{y} = \frac{1}{n}\sum_{i=1}^{n} y_i$ = arithmetic mean of y values.

C.3 CALCULATING LINEAR REGRESSION COEFFICIENTS

C.3.1 Spreadsheet Techniques

Text link: Review relative and absolute referencing in spreadsheets (Section 10.3.4).

The expressions on the far right-hand side of Eqs. (C.5) and (C.6) provide useful ways to calculate m and b. As an example of how to use Eqs. (C.5) and (C.6), consider the spreadsheet in Figure C.2. In this spreadsheet, the data are in columns B and C. Column E is calculated by subtracting the mean x (in cell B9) from each x value (in cells B2 through B5). For example, the formula in E3 is **=B3-$B$9**. Column F is calculated by subtracting the mean y (in cell C9) from each y value (in cells C2 through C5). For example, the formula in F3 is **=C3-$C$9**.

Column G is the product of columns E and F. For example, the formula in G4 is **=E4*F4**. Column H is column E squared (e.g., the formula in H4 is **=E4*E4** (or **=E4^2**). For columns E through H, the sum of the cells in rows 2 through 5 is calculated in row 7. The formulas for calculating m and b, Eqs. (C.5) and (C.6), are typed in cells B11 and B12, respectively, and shown in cells C11 and C12, respectively.

Thus, $m = 2.07$ and $b = -0.0453$. A spreadsheet like that shown in Figure C.2 is reusable.

C.3.2 Built-In Spreadsheet Functions

Spreadsheet software offers a variety of ways to find the best-fit slope and intercept. In Microsoft Excel, the functions **SLOPE** and **INTERCEPT** can be used to find the slope and intercept, respectively. In the example in Figure C.2, the slope can be calculated by typing any into cell:

$$= \textbf{SLOPE(C2:C5, B2:B5)}$$

Text link: Review the concept of the correlation coefficient in Section 9.4.5.

	A	B	C	D	E	F	G	H	I
1		x_i	y_i		x_i - mean x	y_i - mean y	E x F	E x E	F x F
2		0	0.1		-1.375	-2.7	3.7125	1.890625	7.29
3		1	2		-0.375	-0.8	0.3	0.140625	0.64
4		1.5	2.8		0.125	0	0	0.015625	0
5		3	6.3		1.625	3.5	5.6875	2.640625	12.25
6									
7	sum	5.5	11.2		0	0	9.7	4.6875	20.18
8	n	4	4						
9	mean = sum/n	1.375	2.8						
10									
11	m	2.069333	=G7/H7						
12	b	-0.04533	=C9-B11*B9						
13	r^2	0.994675	=H7*B11/I7^2						

Figure C.2: Example Spreadsheet for the Calculation of the Slope and Intercept

This formula is of the form: =**SLOPE**(range containing y values, range containing x values). The intercept can be calculated by typing into any cell:

$$=\textbf{INTERCEPT}(\textbf{C2:C5, B2:B5})$$

This formula is of the form: =**INTERCEPT** (range containing y values, range containing x values).

Microsoft Excel combines a number of useful linear regression functions in the Data Analysis tool called *Regression*. To access these tools, select the *Regression* option under *Tools•Data Analysis* on the main menu. Use *Help* for more information.

Finally, regression coefficients can be calculated in Microsoft Excel by plotting the data and using the *Chart•Add Trendline* option. Check *Help* for more details.

C.3.3 Correlation Coefficient for Linear Regression

The correlation coefficient r^2 always can be calculated as

$$r^2 = 1 - \frac{\sum_{i=1}^{n}(y_i - \hat{y}_i)^2}{\sum_{i=1}^{n}(y_i - \bar{y})^2} \tag{C.7}$$

Note that r^2 is dimensionless and always greater than or equal to zero. For linear regression, where $\hat{y}_i = mx_i + b$, you can show that r^2 can be calculated by

$$r^2 = \frac{\sum_{i=1}^{n}(x_i - \bar{x})^2}{\sum_{i=1}^{n}(y_i - \bar{y})^2}m^2 \tag{C.8}$$

The form of r^2 in Eq. (C.8) points out an interesting fact: if the data are described well by a model with a near-zero slope (i.e., the y values are nearly equal), then r^2 also will be small. The linear model may describe the data with no error (suggesting that r^2 should be 1), but the calculated r^2 will be 0. Be careful interpreting r^2 values when the calculated slope is very small.

You can calculate r^2 using similar techniques as you used to calculating m and b. To calculate r^2 using spreadsheet calculations, simply use Eq. (C.7) or Eq. (C.8). The use of Eq. (C.8) is illustrated in the spreadsheet in Figure C.1. The built-in spreadsheet function, =**CORREL**(range containing y values, range containing x values) also can be used, but note that **CORREL** is equal to r, not r^2. To calculate r^2 for the data in Figure C.2, type into any cell: The data from Figure C.1 and the linear regression model are plotted in Figure C.3.

$$=\textbf{CORREL(C2:C5, B2:B5)}\char`^2$$

You also can use the *Regression* option under *Tools•Data Analysis* on the main menu or the *Chart•Add Trendline* option to determine r^2.

There is one a final note on regression: It is a good habit to plot your data to see trends. As with all models, *always plot the data and model on the same graph to see how the model fits the data*. The data from Figure C.1 and the linear regression model are plotted in Figure C.3.

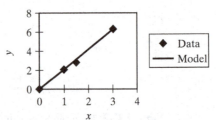

Figure C.3: Plot of Example Data and Linear Regression Model (model: $y = 2.07x - 0.045$)

Appendix D
Using *Solver*

D.1   INTRODUCTION

Text link: Review the use of models in engineering (Chapter 9).

Text link: Review the concept of constrained optimization in Section 2.5.

Engineers often are asked to come up with the optimum solution—the best value of the objective function. You almost never have the luxury of finding the *best* solution. Instead, you seek the optimal value of the objective function, subject to some constraints.

Most spreadsheet packages have a built-in nonlinear solver to analyze constrained optimization problems. In Microsoft Excel, the built-in subprogram is called *Solver*.

Before using *Solver*, it is necessary to set up your spreadsheet properly. Your spreadsheet must have guesses in separate cells for each parameter to be estimated. In addition, your spreadsheet must have the objective function formula in another cell. In other words, your spreadsheet must have a cell (called the *target cell*) that evaluates to the objective function.

Solver can be accessed by selecting the *Solver* option under *Tools* on the main menu. Use *Help* for more information. The *Solver* window, shown in Figure D.1, will appear.

Solver requires four pieces of information. First, *Solver* needs to know the location of the target cell (i.e., the cell evaluating to the objective function). Enter the target cell in the text box labeled: **S̲et Target Cell:**.

Second, *Solver* needs to know if you wish to maximize the objective function, minimize the objective function, or set the objective function equal to a fixed value. Make your selection by clicking on the appropriate radio button after **Equal To:**. If you are setting the objective function equal to a value, then enter the value in the text box after **V̲alue of:**.

Third, *Solver* needs to know the cells where the adjustable parameters are located. Enter the cell locations in the text box following **B̲y Changing Cells:**.

Fourth, any constraints are entered by clicking the **A̲dd** button under **Subject to the Constraints:**. The addition and use of constraints will be discussed in Section D.3.

Engineers use *Solver* in two ways. First, *Solver* can be used to find model parameters that best fit a model. The use of *Solver* in model fitting is discussed in Section D.2. Second, *Solver* can be used to solve constrained optimization problems. This use is illustrated in Section D.3.

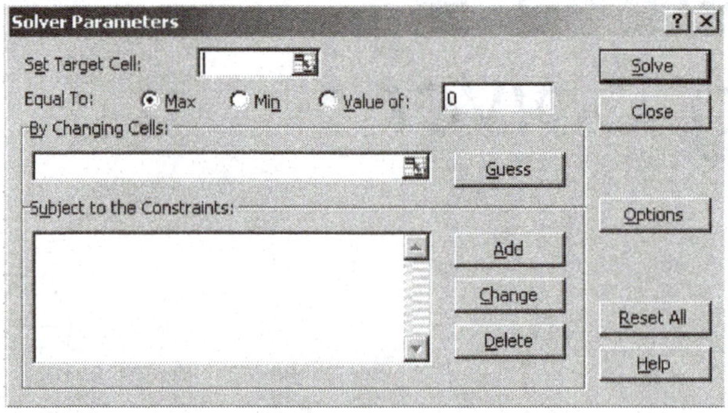

Figure D.1: *Solver* Window

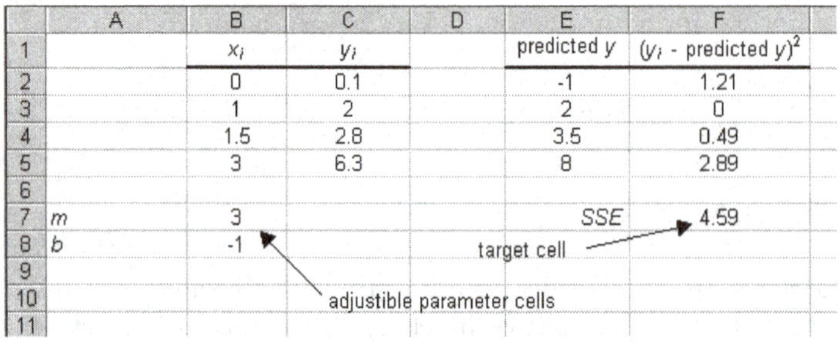

	A	B	C	D	E	F
1		x_i	y_i		predicted y	(y_i - predicted y)^2
2		0	0.1		-1	1.21
3		1	2		2	0
4		1.5	2.8		3.5	0.49
5		3	6.3		8	2.89
6						
7	*m*	3			*SSE*	4.59
8	*b*	-1			target cell	
9						
10			adjustible parameter cells			
11						

Figure D.2: Spreadsheet Set-Up for Model Fitting with *Solver*

D.2 USING *SOLVER* FOR MODEL FITTING

D.2.1 Introduction

Text link: Review the linear model example in Appendix C.

Solver is a very powerful tool for finding the "best-fit" adjustable parameters in a model. The model we will use is the linear model ($y = mx + b$), where m and b are adjustable parameters. (To use *Solver*, it is not necessary that the model be linear.)

D.2.2 Setting Up the Spreadsheet

To use *Solver*, first set up your spreadsheet with four areas. First, enter the data. The data are shown in cells B2:C5 in Figure D.2.

Second, establish cells for the adjustable parameters. It is a little easier to use *Solver* if the cells containing the adjustable parameters are contiguous. Enter a guess for each parameter in its cell. The guesses can be pretty crude. In the example, the parameters m and b are in cells B7 and B8, respectively. Note that the cells are labeled (in cells A7 and A8) and contain initial guesses.

Third, use your model to calculate the predicted value of y ($\hat{y}_i$) for each measured value of y (y_i). In the example, the model is $\hat{y}_i = mx_i + b$. The predicted values of y are in column E. For example, cell E2 contains the formula **=B2*$B$7+$B$8**.

Fourth, create a cell containing the objective function. For most models, an appropriate objective function is the sum of the squares of the errors (SSE). In the example, the squares of the errors [$=(y_i - \hat{y}_i)^2$] are calculated in cells F2 to F5. For example, the formula in cell F2 is **=(C2-E2)^2**. The objective function is simply the sum of the values of $(y_i - \hat{y}_i)^2$. In Figure D.2, SSE is in cell F7, which contains the formula **=SUM(F2:F5)**. Cell F7 is the target cell, since it evaluates to the objective function. Figure D.2 shows the completed spreadsheet, with the value of the objective function equal to 4.59 for the initial guesses of m and b.

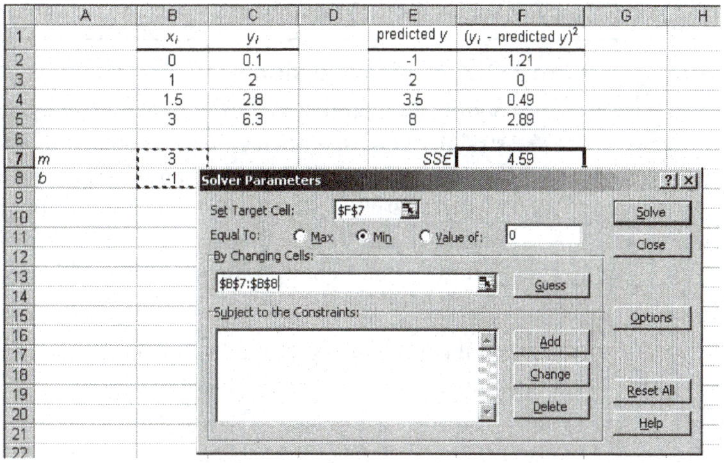

Figure D.3: Sample Spreadsheet with *Solver* Window

D.2.3 Finding Optimal Parameter Values

With the spreadsheet set up, access *Solver*. For most models, you seek to minimize SSE. Therefore, set up *Solver* as follows: First, enter the cell containing SSE in the **Set Target Cell:** text box. This can be accomplished by typing the cell location in the text box (e.g., typing **F7**) or by clicking inside the **Set Target Cell:** text box and then clicking the target cell in the spreadsheet. Second, click the **Min** radio button (to minimize SSE). Third, enter the cells containing the adjustable parameters in the **By Changing Cells:** text box. This can be accomplished by typing the cell or range location in the text box (e.g., typing **B7:B8**) or by clicking inside the **By Changing Cells:** text box and then clicking and dragging across the cells containing the adjustable parameters in the spreadsheet. The final version of the *Solver* window will look like Figure D.3.

To find the values of the adjustable parameters that minimize SSE, click the **Solve** button on the *Solver* window and accept the answer. The final spreadsheet will look like Figure D.4.

Text link: Review the concept of the correlation coefficient (Section 9.4.5).

Note that the minimum value of SSE in this case is about 0.107 (with units of the square of the units of y). You can calculate the correlation coefficient r^2 easily from the formula

$$r^2 = 1 - \frac{\sum_{i=1}^{n}(y_i - \hat{y}_i)^2}{\sum_{i=1}^{n}(y_i - \bar{y})^2} \tag{D.1}$$

The numerator in the second term is SSE (already in your spreadsheet). You can calculate the denominator in the second term by adding a column to your spreadsheet.

	A	B	C	D	E	F
1		x_i	y_i		predicted y	$(y_i$ - predicted $y)^2$
2		0	0.1		-0.04533337	0.021121789
3		1	2		2.02399978	0.000575989
4		1.5	2.8		3.05866636	0.066908284
5		3	6.3		6.16266609	0.018860604
6						
7	m	2.069333			SSE	0.107466667
8	b	-0.04533				
9						

Figure D.4: Final Spreadsheet with Optimal Parameter Values

D.3   USING *SOLVER* WITH CONSTRAINTS

D.3.1   Introduction

Solver also can be used to optimize an objective function subject to one or more constraints. The use of *Solver* with constrained optimization will be illustrated with the following problem: Suppose you have designed a catalyst to minimize byproducts from the manufacture of a solvent. The catalyst is to be made into spherical particles. You wish to maximize the total surface area of the particles. However, the equipment you have at your disposal cannot produce particles with a volume less than 5×10^{-3} mm^3. An analysis reveals that

total particle surface area = A = (number of particles) $\times$ (surface area per particle)

so

$$A = 4\pi r^2 N$$

where N = number of particles and r = particle radius. The number of particles is given by

$$N = \text{(total volume of the particles)}/\text{(volume of one particle)}$$

The total volume of the particles is the total mass of the particles, m, divided by the density of the particles, ρ. If m is in grams and ρ is in g/cm^3, then m/ρ is in cm^3 and the total volume in mm^3 is $1{,}000m/\rho$. Thus,

$$N = \frac{1{,}000\dfrac{m}{\rho}}{\dfrac{4}{3}\pi r^3} \qquad \text{(for } m \text{ in g, } \rho \text{ in g/cm}^3, \text{ and } r \text{ in mm)}$$

The mathematical statement of the problem is as follows: Find the value of r that maximizes $A = 4\pi r^2 N$ subject to $V = (4/3)\pi r^3 < 5 \times 10^{-3}$ mm^3. For your catalyst, use $m = 10$ g and $\rho = 2.3$ g/cm^3.

A spreadsheet for this problem is shown in Figure D.5. The formulas for V, N, and A are shown in cells D6 through D8, respectively. A guess for r of 1 mm has been entered in cell B4.

D.3.2   Finding Optimal Parameter Values with Constraints

In *Solver*, you can find the value of r that maximizes A by selecting cell B8 as the target cell, choosing to maximize it, and selecting cell B4 as the cell to change. You now need to enter the constraint that $V < 5 \times 10^{-3}$ mm^3. To enter the constraint, click the **Add**

	A	B	C	D
1	m	10	g	
2	ρ	2.3	g/cm^3	
3				
4	r	1	mm	
5				
6	V	4.188790205	mm^3	=(4/3)*PI()*B4^3
7	N	1037.96702		=(1000*B1/B2)/B6
8	A	13043.47826	mm^2	=4*PI()*B4^2*B7
9				

Figure D.5: Example Spreadsheet for the Constrained Optimization Problem

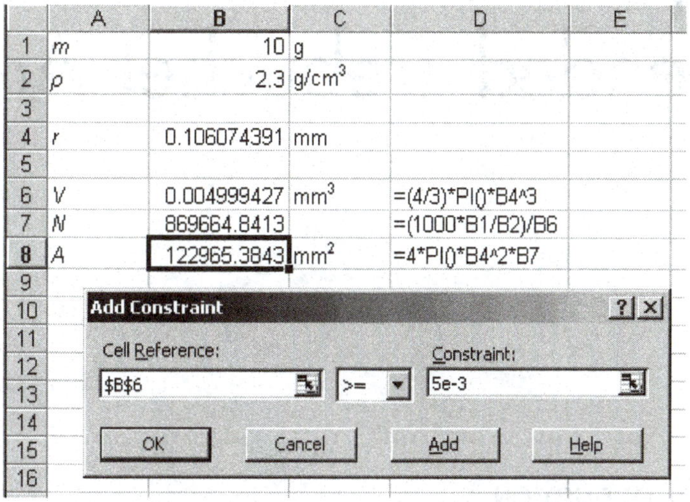

	A	B	C	D	E
1	m	10	g		
2	ρ	2.3	g/cm^3		
3					
4	r	0.106074391	mm		
5					
6	V	0.004999427	mm^3	=(4/3)*PI()*B4^3	
7	N	869664.8413		=(1000*B1/B2)/B6	
8	A	122965.3843	mm^2	=4*PI()*B4^2*B7	

Add Constraint ? ✕

Cell Reference: Constraint:

$B$6 >= 5e-3

OK   Cancel   Add   Help

Figure D.6: Example of the Add Constraint Dialog Box

button under **Subject to the Constraints:**. In the **Cell Reference:** text box, enter (or click on) the cell you want to constrain (B6 in the example). In the drop-down box in the middle of the window, select **>=** (since the constraint is a "greater than or equal to" constraint). In the **Constraint:** text box, type a cell reference or value. In the example, type the value **5e-3**. (See Figure D.6.)

To add the constraint, click the **Add** button. Then click the **Cancel** button to return to the main *Solver* window. The spreadsheet should look like Figure D.7.

Click **Solve** to solve. If you try this on your own, you will find that r is about 0.106 mm, V is just about at its limit of 5×10^{-3} mm^3, and the total surface area is about 1.23×10^5 mm^2.

D.4   FINAL THOUGHTS ON OPTIMIZATION

For linear regression, the exact expressions for the regression coefficients can be written. For nonlinear regression (i.e., regression analysis applied to nonlinear equations), the values of the regression coefficients are approached numerically. Unless special care is taken, the regression algorithm will find a *local* minimum in the SSE rather than the absolute (or *global*) minimum in the SSE. This is analogous to deciding that the hill in your neighbor's yard is the highest point in the country. Here is the bottom line: the nonlinear regression algorithms in spreadsheet programs may not find the global minimum in SSE and therefore may not necessarily find the optimum values of the regression coefficients.

Appendix E
Extended Trebuchet Analysis

E.1   INTRODUCTION

Text link: Review the trebuchet case study (Chapter 24).

The analysis of trebuchet mechanics in Section 24.3.1 was fairly crude. While a complete mechanical description of the trebuchet is beyond the scope of this text, a more detailed analysis may help you with the design of your launcher.

The analysis in Section 24.3.1 revealed that to maximize the range of the projectile, you should (1) maximize the ratio of the counterweight mass (M) to the projectile mass (m), and (2) maximize the distance that the counterweight falls (h).

In theory, the first result suggests that you should make M as large as possible for a fixed projectile. In practice, you will find that the trebuchet is unstable for a very large M.

How should you maximize the distance that the counterweight falls? Simple geometry tells you that the distance that the counterweight falls depends on the height of the pivot (d) and the position of the arm along the pivot (i.e., the fraction of the total arm length L that is on the counterweight side of the pivot). The pivot height and arm position are shown in Figure E.1.

The effects of d and $L - l$ on the range are not obvious. It would appear at first that d should be made large. However, if d is too large, then the counterweight will swing around like a pendulum and not strike the ground. It also may seem logical to make $L - l$ large. This would increase the distance that the counterweight falls (h), in turn increasing the energy available for transfer to the projectile. However, a large amount of the arm on the counterweight side will reduce the release angle (θ) and also reduce the height at which the projectile is released (H). Thus, further analysis is necessary.

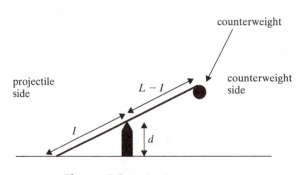

Figure E.1: Trebuchet Geometry

E.2 ANALYSIS

E.2.1 Introduction

In our analysis of trebuchet mechanics, the kinematic equations will include a nonzero release height for the projectile. Further, the release angle (θ), release height (H), and distance that the counterweight falls (h) will be related to the pivot height (d) and the portion of the arm on the projectile side (l/L). This will allow for calculation of the range (R) as a function of d and l/L.

E.2.2 Kinematic Equations

We begin with the following kinematic equations:

$$x = (v_0 \cos \theta)t \tag{E.1}$$

and

$$y = -\tfrac{1}{2}gt^2 + (v_0 \sin \theta)t \tag{E.2}$$

To account for a nonzero release height, Eq. (E.2) must be modified as follows:

$$y = -\tfrac{1}{2}gt^2 + (v_0 \sin \theta)t + H \tag{E.3}$$

To calculate the range, first find the landing time by setting Eq. (E.3) equal to 0 (since $y = 0$ when the projectile hits the ground):

$$\text{landing time} = \frac{v_0 \sin \theta + \sqrt{v_0^2 \sin^2 \theta + 2gH}}{g}$$

Second, plug the landing time into Eq. (E.1) and solve for x to obtain the range R:

$$R = v_0 \cos \theta \left(\frac{v_0 \sin \theta + \sqrt{v_0^2 \sin^2 \theta + 2gH}}{g} \right) \tag{E.4}$$

From energy balance considerations, you know that

$$v_0 = \sqrt{2\frac{M}{m}gh} \tag{E.5}$$

Thus, you can calculate the range by using Eqs. (E.4) and (E.5) once you know how θ, H, and h depend on d and l/L.

E.2.3 Dependency on d and l/L

The dependency of θ, H, and h on d and l/L is a little complicated. To conduct the analysis, it is useful to divide trebuchet designs into four types. The four types are illustrated in Table E.1.

In the first type, the pivot height d is large compared with the arm length L. Thus, the projectile end of the arm starts off the ground. As a result, the counterweight starts in the 12 o'clock position and swings freely clockwise to the 6 o'clock position. (In all designs, it is assumed that the trebuchet has a stopper block, so the counterweight is not allowed to swing past the 6 o'clock position. If the counterweight swings past this position, the projectile will be thrown immediately to the ground.)

TABLE E.1 Trebuchet Types for the Extended Analysis

Type	Description	Starting Position	Release Position
1	CW starts in the 12 o'clock position and ends in the 6 o'clock position		
2	CW starts in the 12 o'clock position and hits the ground		
3	CW does not start in the 12 o'clock position, but ends in the 6 o'clock position		
4	CW does not start in the 12 o'clock position, but hits the ground		

Notes: CW = counterweight.

Counterweight shown by a small bar in the figures.

In the second type, the projectile end of the arm also starts off the ground and so the counterweight starts in the 12 o'clock position. In this case, however, $L - l$ is long enough so that the counterweight hits the ground.

In the third type, the pivot height d is not large compared with the arm length L. The projectile end of the arm starts off resting on the ground, so the counterweight does not start in the 12 o'clock position. However, $L - l$ is short enough that the counterweight swings freely clockwise to the 6 o'clock position.

In the fourth type, the projectile end of the arm again starts off resting on the ground and thus the counterweight does not start in the 12 o'clock position. In this case, $L - l$ is long enough so that the counterweight hits the ground.

An analysis of the geometry of each type reveals how θ, H, and h depend on d, l, and L. The results of the analysis are shown in Table E.2.

You can use this approach to calculate the range for your trebuchet design. First, pick d, l, L, and M/m. Second, use Table E.2 to select the type of trebuchet. Third, calculate the values of θ, H, and h from Table E.2. Fourth, calculate v_0 from Eq. (E.5) for the calculated value of h and your M/m ratio. Fifth, substitute θ, H, and v_0 into Eq. (E.4) to find the range.

Spreadsheets make the calculations much easier, especially if you are simulating a variety of designs. Nested **IF** functions can be used to select the proper type. For example, the following formula will calculate the proper value of h:

$$=\textbf{IF(E6<=E\$4,IF(\$B\$1-E6<=E\$4,2*(\$B\$1-E6),E\$4+\$B\$1-E6),}$$

$$\textbf{IF(\$B\$1-E6<=E\$4,(\$B\$1-E6)*(E6+E\$4)/E6,E\$4*\$B\$1/E6))}$$

(The formula assumes that L is in cell B1, l is in cell E6, and d is in cell E4.)

TABLE E.2 Values of h, H, and θ for Each Trebuchet Type in Table E.1

Type	l	$L - l$	h	H	θ
1	$\leq d$	$\leq d$	$2(L - l)$	$l + d$	0
2	$\leq d$	$\geq d$	$d + L - l$	$\dfrac{dL}{L - l}$	$\dfrac{\pi}{2} - \sin^{-1}\left(\dfrac{d}{L - l}\right)$
3	$\geq d$	$\leq d$	$\dfrac{(L - l)(l + d)}{l}$	$l + d$	0
4	$\geq d$	$\geq d$	$\dfrac{dL}{l}$	$\dfrac{dL}{L - l}$	$\dfrac{\pi}{2} - \sin^{-1}\left(\dfrac{d}{L - l}\right)$

L = total arm length l = arm length on projectile side
$L - l$ = arm length on CW side h = distance CW falls
H = release height θ = release angle in radians

E.2.4   Results

An example of the range results is shown in Figure E.2. The results were calculated for $L = 0.3$ m (about one foot) and $M/m = 20$. Thus, if d is 0.18 m (about 7 inches, $d/L = 0.6$) and if the pivot is halfway down the arm ($l/L = 0.5$), then the theoretical range is about 7.2 m (about 24 feet, $R/L = 23.86$), as expected.

From this analysis, the range increases with increasing d, as expected (see Section E.1). However, you may find that the trebuchet becomes "shaky" at large d, resulting in incomplete energy transfer to the projectile. Also, the range increases as more of the arm is shifted to the counterweight side (i.e., l/L decreases).

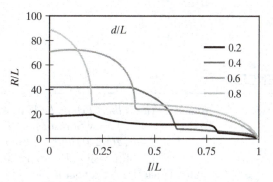
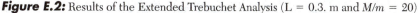

Figure E.2: Results of the Extended Trebuchet Analysis ($L = 0.3$. m and $M/m = 20$)

Appendix F
References and Bibliographies

F.1   REFERENCES

Adams, S. *The Dilbert Principle*. HarperCollins Publ., Inc., New York, NY, 1996.

BLS (Bureau of Labor Statistics, U.S. Department of Labor). *Occupational Outlook Handbook, 2002–03 edition, Engineers*. On the Internet at http://www.bls.gov/oco/ocos027.htm (visited February 20, 2004).

Conan Doyle, A. A Scandal in Bohemia. In *Adventures of Sherlock Holmes*. Harper & Brothers, London, 1892.

Dallard, P., A.J. Fitzpatrick, and A. Flint. The London Millennium Footbridge. *Structural Engineer*, 79(22), 17–35, 2001.

Davis, M. Thinking Like an Engineer: The Place of a Code of Ethics in the Practice of a Profession. *Phil. Public Affairs*, 20(2), Spring, 1991.

DeFazio, K., D. Wittman, and C.G. Drury. Effective Vehicle Width in Self-Paced Tracking. *Appl. Ergonomics*, 23(6), 382–386, 1992.

Drury, C.G. Movements with Lateral Constraint. *Ergonomics*, 14(2), 293–305, 1971.

Drury, C.G., M.A. Montazer, and M.H. Karwan. Self-Paced Path Control as an Optimization Task. *IEEE Trans.*, SMC-17.3, 455–464, 1987.

Drury, C.G. Human Factors Good Practices in Borescope Inspection. *Proceedings of the Fifth Joint NASA/FAA/DoD Conference on Aging Aircraft*, September 10–13, 2001.

Foran, J. (undated) The Day They Turned the Falls On: The Invention of the Universal Electrical Power System. On the Internet at http://ublib.buffalo.edu/libraries/projects/cases/niagara.htm (visited May 28, 2003).

Fraser, D.M. Introducing Students to Basic ChE Concepts: Four Simple Experiments. *Chem. Eng. Ed.*, 33(3), 190–195, 1999.

Griggs, F.E. Amos Eaton Was Right! *J. Prof. Issues in Eng. Educ. Practice*, 123(1), 30–34, 1997.

Harte, J. *Consider a Spherical Cow: A Course in Environmental Problem Solving*. University Science Books, Herndon, VA, 1988.

Harte, J. *Consider a Cylindrical Cow: More Adventures in Environmental Problem Solving*. University Science Books, Herndon, VA, 2001.

Henderson, J.M., L.E. Bellman, and B.J. Furman. A Case for Teaching Engineering with Cases. *J. Eng. Educ.*, January 1983.

Herreid, C.F. What Is a Case? *J. College Science Teaching*, 27(2), 92–94, 1997.

Hoff, R. *I Can See You Naked*. Andrews and McMeel, Kansas City, MO, 1992.

Holtzapple, M.T., and W.D. Reece. *Foundations of Engineering*. McGraw-Hill, Boston, MA, 2000.

Hyman, B. *Fundamentals of Engineering Design*. 2d ed., Prentice Hall, Upper Saddle River, NJ, 2003.

Karlin, S. "The purpose of models is not to fit the data but to sharpen the questions." *Eleventh R.A. Fisher Memorial Lecture at the Royal Society of London, April 20, 1983* (Furman University Mathematical Quotations Server). On the Internet at http://math.furman.edu/%7Emwoodard/ascquotk.html (visited May 28, 2003).

Mackay, A.L. *Dictionary of Scientific Quotations*. A. Hilger, Bristol, UK, 1991.

Martin, M.W., and R. Schinzinger. *Ethics in Engineering*, 2d ed., McGraw-Hill, New York, NY, 1989.

Merriam-Webster OnLine: *Word for the Wise*. On the Internet at http://www.m-w.com/cgi-bin/wftw.pl (visited May 28, 2003).

Moore, T.C. *Ultralight Hybrid Vehicles: Principles and Design*. Presented at the 13th International Electric Vehicle Symposium (EVS-13), Osaka, Japan, October 14, 1996.

NSF (National Science Foundation, Division of Science Resources Statistics). *Characteristics of Recent Science and Engineering Graduates: 2001*, NSF 04-302, Project Officer John Tsapogas, Arlington, VA, 2003.

Paradis, J.G., and M.L. Zimmerman. *The MIT Guide to Science and Engineering Communication*. The MIT Press, Cambridge, MA, 1997.

Perez, A.L., C.D.A. Earle, and S. Ramcharansingh. Development of an On-Line O&M Manual. *Florida Water Resources J.*, 27–28, May, 2001.

Petroski, H. *To Engineer Is Human: The Role of Failure in Successful Design*. Vintage Books, New York, NY, 1992.

Pfrang, E.O., and R. Marshall. Collapse of the Kansas City Hyatt Regency Walkways. *Civil Engineering*, 65–68, 1982.

Revelle, C.S., E.E. Whitlatch, and J.R. Wright. *Civil and Environmental Systems Engineering*. 2d ed., Prentice Hall, Upper Saddle River, NJ, 2003.

Rosenbaum, A. *Measuring Output Prices for Engineering Services in the United States*. Presented at the 17th Voorburg Group Meeting, Nantes, France, September 23–27, 2002.

Saarinen, E. "Always design a thing by considering it in its next larger context—a chair in a room, a room in a house, a house in an environment, an environment in a city plan." In Simpson, J.B. *Simpson's Contemporary Quotations*. Houghton Mifflin Company, Boston, MA, 1988.

Schumacher, E.F. *Small Is Beautiful: Economics as if People Mattered*. Harper and Row, New York, NY, 1973.

Sibly, P., and A. Walker. Structural Accidents and Their Causes. *Proc. Inst. Civil Engineers*, 62(Part 1), 191–208, May, 1977.

Smith, J.G., and P.A. Vesiland. *Report Writing for Environmental Engineers and Scientists*. Lakeshore Press, Woodsville, NH, 1996.

Steiber, J., and B. Surampudi. *Design and Implementation of a Reconfigurable Virtual Electric Vehicle Simulator, 03-9196*. Southwest Research Institute, IR&D Research Summary 03-9196 for 07/01/00-09/30/00. On the Internet at http://www.swri.edu/3pubs/IRD2000/03-9196.htm (visited May 28, 2003).

Strunk, Jr., W., and E.B. White. *The Elements of Style*, 3d ed. Allyn and Bacon, Needham Heights, MA, 1979.

Tufte, E.R., *The Visual Display of Quantitative Information*. Graphics Press, Cheshire, CT, 1983.

Tufte, E.R. *Envisioning Information*. Graphics Press, Cheshire, CT, 1990.

Wright, P.H. *Introduction to Engineering*, 2d ed. John Wiley and Sons, New York, NY, 1994.

F.2 ANNOTATED BIBLIOGRAPHY: TECHNICAL COMMUNICATIONS

There are a number of excellent books on technical communications. A few good sources of information are listed next. This list is by no means exhaustive.

Alley, M., L. Crowley, J. Donnell, and C. Moore (eds.). *Writing Guidelines for Engineering and Science Students*.

> This excellent on-line guide is available on the Internet at http://filebox.vt.edu/eng/mech/writing/index.html (visited June 6, 2003).

Dodd, J.S. *The ACS Style Guide*. American Chemical Society, Washington, DC, 1986.

> Although the emphasis is on technical writing in chemistry, this official guide of the American Chemical Society is a good general reference on technical writing.

McMurrey, D.A. *Online Technical Writing: Online Textbook—Contexts*.

> This is a fine guide to technical writing. It can be found on the Internet at http://www.io.com/~hcexres/tcm1603/acchtml/acctoc.html (visited May 28, 2003).

Paradis, J.G., and M.L. Zimmerman. (See "References" for the full citation.)

> This is a good reference with many examples of technical writing.

Sageev, P. *Helping Researchers Write ... So Managers Can Understand*. Batelle Press, 1968.

> The emphasis here is on technical writing in the corporate setting.

Strunk and White. (See "References" for the full citation.)

> This is a classic reference and very inexpensive. It should be on your reference shelf, along with a good dictionary.

White and Vesiland. (See "References" for the full citation.)

> White and Vesiland cover most aspects of technical writing, with many examples from environmental engineering.

F.3 BIBLIOGRAPHIES FOR *FOCUS ONS*

The Real McCoy?

Gibbs, C.R. *Black Inventors: From Africa to America*. Three Dimensional Press, Silver Spring, MD, 1995.

James, P.P. *The Real McCoy: African-American Invention and Innovation, 1619–1930*. Smithsonian Institution, Washington, DC, 1989.

For a more complete description of the operation of the McCoy lubricator, see http://www.usi.edu/science/engineering/MISC/emccoy/mccoyop1.htm (visited May 28, 2003).

A Square Peg in a Round Hole

Lovell, J., and J. Kluger. *Apollo 13* (previously published as *Lost Moon*). Houghton Mifflin Co., Boston, MA, 1994.

NASA (National Aeronautics and Space Administration). *Apollo 13 Mission Report*, September, 1970. Available on the Internet at http://www.hq.nasa.gov/office/pao/History/alsj/a13/a13mr.html (visited May 28, 2003).

The Multimillion Dollar Units Mistake

Mars Climate Orbital Mishap Investigation Board. *Phase I Report,* November 10, 1999.

Of Plots and Space Shuttles

Robison, W., R. Boisjoly, D. Hoeker, and S. Young. Representation and Misrepresentation: Tufte and the Morton Thiokol Engineers on the Challenger. *Science Eng. Ethics,* 8(1), 59–81, 2002.

Tufte, E.R. *Visual Explanations: Images and Quantities, Evidence and Narrative*. Graphics Press, Cheshire, CT, 1993.

Index